T5-BBB-496

Multiple Criteria Decision Making

A. Goicoechea L. Duckstein S. Zionts
Editors

Multiple Criteria Decision Making

Proceedings of the Ninth International Conference: Theory and Applications in Business, Industry, and Government

With 65 Figures

Springer-Verlag
New York Berlin Heidelberg London Paris
Tokyo Hong Kong Barcelona Budapest

Ambrose Goicoechea
Statcom, Inc.
7921 Jones Branch Drive
McLean, VA 22102
USA

L. Duckstein
Department of Systems and
Industrial Engineering
University of Arizona
Tucson, AZ 85721
USA

Stanley Zionts
School of Management/Jacobs
Management Center
State University of New York at Buffalo
Buffalo, NY 14260
USA

Cover illustration by Nip Rogers Illustrations.

Library of Congress Cataloging-in-Publication Data
Proceedings of the Ninth international conference: theory and applications in business, industry, and government/Ambrose Goicoechea, L. Duckstein, Stanley Zionts [eds.].
p. cm.
Proceedings of the 9th International Conference on Multiple Criteria Decision Making, held Aug. 5–8, 1990, in Fairfax, VA.
Includes bibliographical references.
ISBN 0-387-97805-4 (Springer-Verlag New York Berlin Heidelberg: acid-free paper). — ISBN 3-540-97805-4 (Springer-Verlag Berlin Heidelberg New York: acid-free paper)
1. Decision support systems—Congresses. 2. Management science—Congresses. I. Goicoechea, Ambrose. II. Duckstein, Lucien. III. Zionts, Stanley, 1937– . IV. Title: Theory and applications in business, industry, and government.
TS8.62.P76 1992
658.4′03—dc20 92-2703

Printed on acid-free paper.

Production managed by Francine Sikorski; manufacturing supervised by Robert Paella.
Camera-ready copy supplied by the contributors.
Printed and bound by Edwards Brothers, Inc., Ann Arbor, MI.
Printed in the United States of America.

9 8 7 6 5 4 3 2 1

ISBN 0-387-97805-4 Springer-Verlag New York Berlin Heidelberg
ISBN 3-540-97805-4 Springer-Verlag Berlin Heidelberg New York

ORGANIZATION:

National Organizing Committee:

Conference Chairman:	Ambrose Goicoechea	George Mason University
Program Chairman:	Chelsea C. White, III	University of Virginia
Program Co-Chairman:	Yacov Y. Haimes	University of Virginia
	Miguel Carrio	Teledyne Brown Corporation
	Vira Chankong	Case Western Reserve University
	Jared Cohon	Johns Hopkins University
	Lucien Duckstein	Case Western Reserve University
	Saul I. Gass	University of Maryland
	Eugene Stakhiv	U.S. Army Corps of Engineers
	Ralph E. Steuer	University of Georgia
	Richard M. Soland	George Washington University
	Stanley Zionts	State University of New York at Buffalo

International Executive Committee:

Chairman:	Stanley Zionts	State University of New York at Buffalo
	Valerie Belton	University of Kent (UK)
	Vira Chankong	Case Western Reserve University
	James S. Dyer	University of Texas at Austin
	Gunther Fandel	Fernuniversitat Hagen (FRG)
	Thomas Gal	Fernuniversitat Hagen (FRG)
	Ambrose Goicoechea	George Mason University
	Yacov Y. Haimes	University of Virginia
	Pekka J. Korhonen	Helsinki School of Economics (Finland)
	Jonathan Kornbluth	Hebrew University (Israel)
	E. Jacquet-Lagreze	Universite de Paris-Dauphine (France)
	Oleg I. Larichev	Institute for Systems Studies (USSR)
	A. Geoff Lockett	Manchester Business School (UK)
	Hirotaka Nakayama	Konan University (Japan)
	Yoshikazu Swaragi	Japan Institute for Systems Research (Japan)
	Jaap Spronk	Erasmus University Rotterdam (Netherlands)
	Ralph E. Steuer	University of Georgia
	Mario Tabucanon	Asian Institute of Technology (Thailand)
	A.P. Wierzbicki	Technical University of Warsaw (Poland)
	Po-Lung Yu	University of Kansas

PREFACE

Ambitious as the enterprise was, and as demanding as the planning effort promised to be, we set out to organize the IX International Conference on Multiple Criteria Decision Making (MCDM) which took place in Fairfax, Virginia, on August 5-8, 1990. We knew it was ambitious because the number of participants expected to attend the conference would approach 160, larger than that of earlier conferences, and because it would include a sizable contingent of 39 participants from the Soviet Union and Eastern Europe, possibly the largest yet.

In many ways, it may be appropriate to say, this international conference presented a microcosm of peoples and new ideas that reflected the extraordinary events that were to take place in Europe and other parts of the world during that summer of 1990.

With the theme: **"Multiple Criteria Decision Making and Support at the Interface of Industry, Business and Government"** we wanted to focus on new analytical methodologies and management tools, quantitative and qualitative evaluation of decision techniques, the design of experiments to test existing and proposed methods, and the experience gained in the application of these MCDM methods and tools to real-world problems during the last 10-15 years. Many analytical, behavioral, and technological advances are to be made, we feel, at the interface of MCDM Theory, the Behavioral Sciences, Operations Research, Systems Engineering, Decision Theory, Mathematical Sciences, and Information Technology.

The current proliferation of computer-based decision tools offers new challenges and opportunities. Important topics within the above spheres of research could not have been addressed in depth just five years ago due, in part, to the unavailability of micro-computer technology in office, laboratory and industrial environments. And whereas new analytical developments will continue to be at the core of MCDM research, it is perhaps safe to say that MCDM will benefit from the incorporation of concepts in user interface design, database development, and software system architecture in the design of the next generation of MCDM tools.

Accordingly, the set of specific objectives that we aimed to accomplish included:

(1) Providing a forum for distinguished researchers in the U.S. and abroad to present new and recent scientific developments in decision theory and models for decision making.

(2) Bringing together distinguished researchers, new Ph.D.s, and prominent practitioners from industry, business, government, and academia for purposes of exchanging new ideas, approaches and tools in MCDM.

(3) Identifying new, promising areas of research in analytical methods, cognitive analysis in support of experimental design and evaluation of methods and models, as well as the representation of risk and uncertainty.

(4) Reporting on the experience gained over the last 10-15 years in the application of MCDM methods and tools in industry, business, and government.

Among the many individuals that contributed generously of their time, energy and talents, we like to make special mention of Chelsea C. White and Yacov Y. Haimes, Program Chairman and Co-Chairman, respectively, and William T. Sherer. Their intense effort in the preparation of the Program over a period of several months, organizing the plenary sessions, tutorials, and workshops, was key to the success of the conference. At the center of this effort, and working side by side with Chelsea, Yacov, and Bill we recognize Annelise Tew whose untiring dedication, enthusiasm, humor and wit transformed mountains of mail, fax messages and phone calls into a sucessful information-processing operation.

To the indefatigable team of Ralph E. Steuer and Elliot Lieberman goes the richly deserved credit for bringing the largest contingent yet from the USSR and Eastern European to the conference series. Ralph and Elliot spent countless hours communicating over the phone and by mail with Department of State officials and their European counterparts to secure visas and travel permits for over 30 participants. Their concern extended to personal appearances at airports to receive our colleagues, as well as overseeing their safe departure after the conference.

Any enterprise of this magnitude is often successful because the much needed financial planning went into it from beginning to end. This was certainly the case with our conference, where the challenge of raising all the money resources was accepted by Richard Soland and Jared Cohon. Their effort called for multiple strategies so that proposals would be submitted to institutions and individuals. Again, many proposals were submitted in 1990, a year of substantial budget cuts across the government and private sectors. Yet, their dedication produced a long list of sponsors and contributors, including the National Science Foundation, the U.S. Army Corps of Engineers, The George Washington University, University of Virginia, George Mason University, The Johns Hopkins University, University of Maryland, University of Georgia, State University of New York at Buffalo, and many others (see List of Sponsors and Contributors).

A special vote of thanks to Evan Anderson and Tom Gulledge, Department of Decision Sciences, George Mason University, for their sponsorship and assistance with many of the logistic and organizing activities.

Plenary sessions and speakers provided important focal points of the conference. Our special thanks to our plenary speakers Milan Zeleny, Po-Lung Yu, Yacov Haimes, Jared Cohon, and Ward Edwards for challenging every one of us with their questions, theories, and perspectives about the future of MCDM. As champions of new ideas, and because of their active involvement in other related research fields, they help us all stay at the forefront. To be inventive, relevant, inquisitive, and useful we all strive.

Among the champions of MCDM in the Federal Agencies, we recognize Eugene Z. Stakhiv of the Institute for Water Resources, U.S. Army corps of Engineers, whose untiring effort and vision over the years are responsible in great measure for numerous workshops, seminars, studies, and conferences on MCDM and related subjects.

Also, and most important, the organizing committee is grateful to the small army of students from the various universities cited above, including Amar Hamed and Jingxian Chen who helped with the mail, airport transportation, registration desk, and so many other details. We recognize very specially graduate student Ms. Fu Li whose dedication, talent and boundless energy over a two-year planning period helped make this international conference possible.

Ambrose Goicoechea
STATCOM, Inc.
McLean, Virginia, U.S.A

Lucien Duckstein
Department of Systems and
Industrial Engineering
University of Arizona
Tucson, Arizona, U.S.A

Stanley Zionts
Department of Management
Science and Systems
State University of New
York at Buffalo
Buffalo, New York, U.S.A.

Editors

Fairfax, August 1990

CONTENTS

IX-TH INTERNATIONAL CONFERENCE ON MULTIPLE CRITERIA DECISION MAKING (MCDM)

August 5-8, 1990
Fairfax, Virginia, USA

Conference Sponsors and Contributors

National Science Foundation

U.S. Army Corps of Engineers

The Operations Research Society of America (ORSA)

The Institute of Management Sciences (TIMS)

Cartermill, Inc.

STATCOM, Inc.

Center for International Management Education

Cray Research, Inc.

Expert Choice, Inc.

The MITRE Corporation

Carleton University, Canada

State University of New York at Buffalo

George Mason University

The Johns Hopkins University

University of Georgia

University of Maryland

The RAND Corporation

US Air

Integrated Technologies and Research, Inc.

A FRAMEWORK FOR QUALITATIVE EXPERIMENTAL EVALUATION OF MULTIPLE CRITERIA DECISION SUPPORT SYSTEMS (MCDSS)

Ambrose Goicoechea
STATCOM, Inc.
7921 Jones Branch Drive,
McLean, Virginia, USA

Eugene Z. Stakhiv
Institute for Water Resources
U.S. Army Corps of Engineers
Ft. Belvoir, Virginia, USA

Fu Li
Information Systems and Systems Engineering Dept.
George Mason University
Fairfax, Virginia, USA

ABSTRACT

This paper presents a methodology for the experimental evaluation of four Multiple-Criteria Decision Making (MCDM) methods and their software implementation: MATS-PC, EXPERT CHOICE, ARIADNE, and ELECTRE. These methods are evaluated at four interfaces: (1) the MCDM method and the micro-computer environment interface, (2) the DSS-User interface, (3) the DSS-Problem Domain interface, and (4) the DSS-Organization interface. Eleven planners from U.S. Army Corps of Engineers and nine graduate students with varying levels of skills and experience participated in an experiment, including the application of the methods to a real-world water supply study.

1. INTRODUCTION

There has been a limited number of experimental evaluations of decision support systems (DSS) that make use of analytical multiple-criteria decision making (MCDM) methods thus far, and certainly very few that address engineering settings and applications. The experiment reported in this paper is among the first of its kind because: (1) it involves four computer-based DSS that feature full-scale, market-ready systems and elaborate user-friendly interfaces, with color and graphics capability, (2) it applies both qualitative and non-parametric statistical evaluation methodologies, (3) it involves both professionals and students with varying skills, and (4) it addresses an earlier real-world problem in project planning in the Public Sector, i.e., a water supply study (U.S. Corps of Engineers, 1975). It is probably safe to say that with the current proliferation of microcomputer technologies many more experiments and analyses of this type will be forthcoming in the literature.

The experiment consisted of performing the following tasks:

Task I: (Quantitative analysis) Determine the amount of agreement among decision makers with regards to the rankings of alternative plans with and without the use of MCDM methods. These results have been reported in Goicoechea et al. (1989).

Task II: (Quantitative analysis) Determine the amount of agreement among decision makers when ranking alternative plans with and without the use of computer-based decision support systems (DSS). Non-parametric statistical tests were applied, and the results are reported in Goicoechea (1990) and Goicoechea et al. (1991).

Task III: (Qualitative analysis) Design and apply an analytical framework for the qualitaive evaluation of the four MCDM methods and their respective computer implementations (i.e., DSS):

. MATS-PC	(Brown et al., 1986)
. EXPERT CHOICE	(Forman, 1983, Saaty, 1980)
. ARIADNE	(White et al., 1984)
. ELECTRE	(Roy, 1968)

Contents of this Paper

This paper reports only on Task III above. See the references cited above for a discussion of results in Tasks I and II.

The following sections discuss the design of the experiment, the four interfaces, the methodology, and the experiomental results. The questionnaire for one of the MCDM methods evaluated is presented in the Appendix. For a complete and detailed presentation of these sections the reader is referred to Goicoechea (1990).

2. DESIGN OF THE EXPERIMENT

A total of 12 experiment sessions were designed and conducted. Typically, each session would have two facilitators, half of the participants would evaluate one DSS while the other half would evaluate another DSS, and the session would last four hours, approximately. The experiment sessions took place over a period of six months in 1990.

Experiment Session 2, for example, was facilitated by Stakhiv (S) and Goicoechea (G), such that five of the Corps participants applied MATS-PC to the decision scenario involving the water supply study, while the other six Corps participants applied EXPERT CHOICE. A PC with color-and-graphics monitor was available to each participant, and the two facilitators were available to assist with questions during the sessions. At the end of each session, participants would take 20-30 minutes to fill the evaluation questionnaire associated with the DSS that had just been applied. Appendix presents the questionnaire distributed to evaluate MATS-PC.

3. DSS EVALUATION METHODOLOGY

For the qualitative component of the evaluation it was decided to adopt portions of a model proposed earlier by Adelman and Donnell (1986), as depicted in Figure 1. A fourth interface, interface 1, was added to our model to reflect the fact that both the experiment organizers and the experiment participants cooperated in the design of the computer graphics for two of the four DSS evaluated (ARIADNE and ELECTRE). Interface 1, then, captures the evaluator's preferences for one set of computer graphics over another.

Interface 2, the DSS-User interface, pertains to the extent to which features of the DSS facilitate or hinder its usefulness. Interface 3, the DSS-problem Domain, addresses the potential usefulness of the DSS in handling the nature and complexity of the problem at hand. Finally, interface 4, the DSS-Organization interface addresses the ability of the DSS to assist and support the decision making process within the real-world organization.

A hierarchy of criteria was identified and applied to evaluate the four decision support systems, as shown on Figure 2. The criteria applied to the evaluation of Interface 1 of each system, for example, included choice of computer language, graphics capability, architectural complexity, data file management, monitor screen volume, and information content of color.

Design of DSS User Instructions

About four weeks prior to the experiment sessions, a set of instructions for the use of each DSS, and condensed version of the theories (e.g., utility theory, hierarchical trees and processes, concordance and discordance matrices) were prepared and distributed to the experiment participants.

Each DSS required different instructions to reflect their unique DSS characteristics:

(a) computer language,
(b) graphics and color capability,
(c) method of elicitation of user preferences and weights, and
(d) input-output database requirements.

Design of DSS Questionnaires

For each of the four DSS interfaces a set of relevant questions were articulated. How to proceed with this design task was not immediately apparent. Eventually, it was felt that a number of questions somewhere between 10 and 40 might be appropriate for each interface, and a value function with scale of 1 (e.g., "difficult", "not helpful", etc.) to 5 (e.g., "easy", "very helpful", etc.) was proposed to elicit responses.

Weighting of Responses

In addition to stating his/her responses on a scale of 1 to 5, as noted above, each evaluator was asked to weigh each category of questions.

An example of evaluator responses to the questionnaire for EXPERT CHOICE is provided in Figure 3. These responses address questions and issues at the third interface (the DSS-Problem Domain interface) and have been appended to the hierarchical tree of Figure 4, to yield a "worth" of 0.44 at that interface. Similarly, responses to questions and issues at interfaces 1, 2, and 4 yielded worth values of 0.749, 0.786, and 0.54, respectively. We note that responses to questions for interface 1, 2, 3 and 4 were assigned the weights of 0.2, 0.3, 0.25 and 0.25, respectively. As such, this particular evaluator feels that the categories of questions in interface 2 are more important (e.g., relevant, consequential, etc) than those in interface 1, whereas those in interfaces 3 and 4 are equally important. Responses of 1, 2, 3, 4 and 5 have worth values of 0.2, 0.4, 0.6, 0.8 and 1.0, respectively. Similarly, for other questions with only four discrete responses.

The aggregation of responses to questions addressing issues at the four interfaces was carried out using a linear, weighted worth function. Accordingly, let us say that the responses to the questions at Interface 3 are as shown in Figure 4. It follows that:

$$\begin{aligned} w(u) &= 0.30(1/3)(0.6+0.6+0.6) \\ &\quad +0.4(1/4)(0.6+0.2+0.2+0.4) + 0.3(0.4) \\ &= 0.44 \end{aligned}$$

and for the four interfaces:

$$\begin{aligned} w(u) &= 0.20(0.749) + 0.30(0.786) + 0.25(0.44) \\ &\quad + 0.25(0.54) \\ &= 0.63 \end{aligned}$$

which then represents the total worth attributed by the evaluator to EXPERT CHOICE (highest possible worth is 1.0).

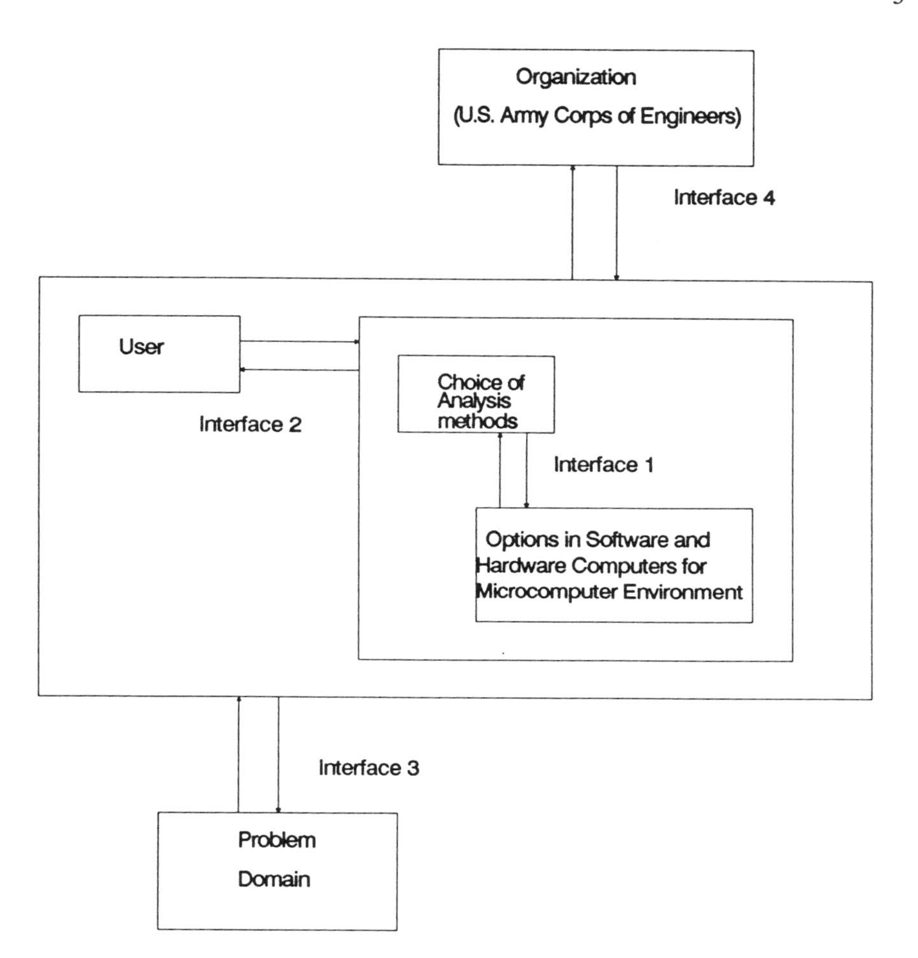

Figure 1. The Four Interfaces to be Evaluated

(After Adelman and Donnell, 1986)

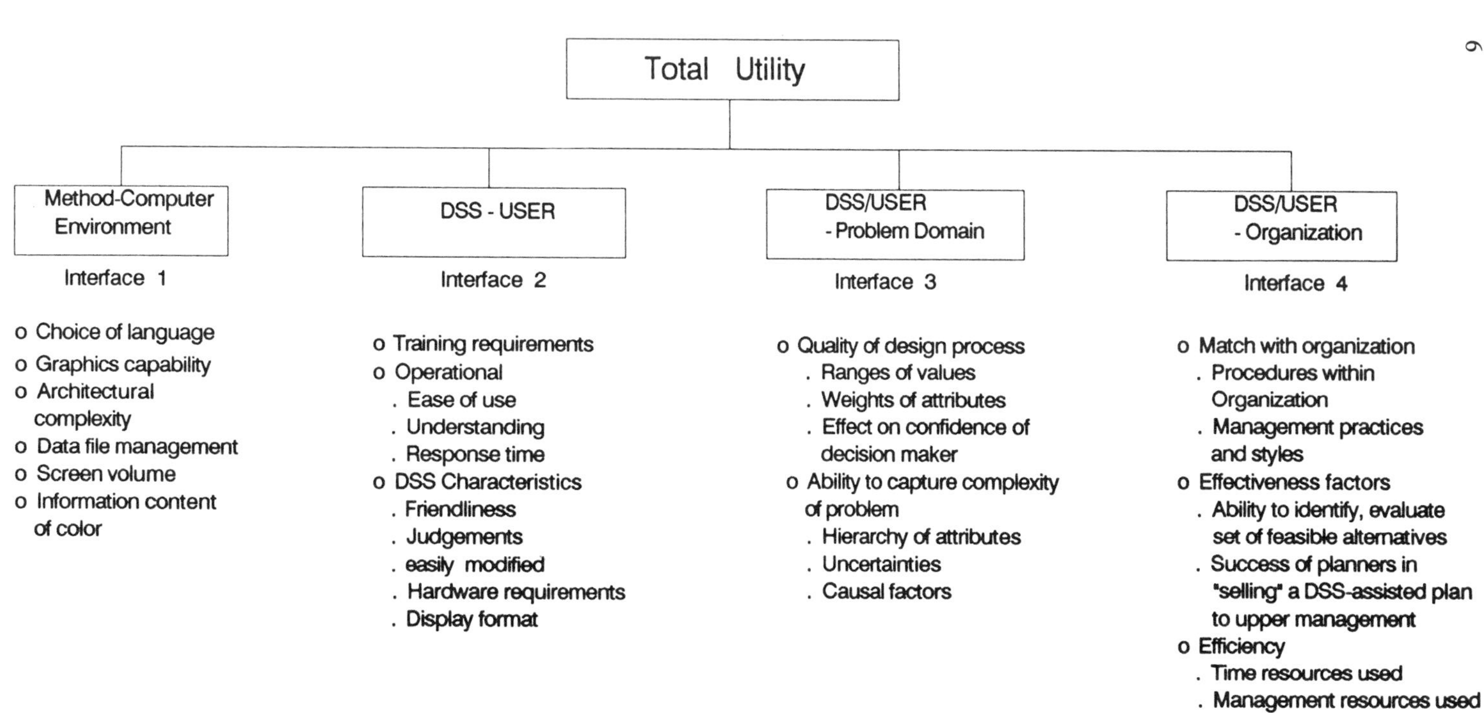

Figure 2. A Hierarchy of Evaluation Criteria at the Four Interfaces

4. EXPERIMENT RESULTS

Worth function values are presented on Table 1 and Table 2 for the Corps Group and the Student Group, respectively. Corps Evaluator 2, for instance, ranked EXPERT CHOICE first (0.63), followed by ARIADNE (0.47), ELECTRE (0.39), and MATS-PC (0.38).

Shown in Figure 5 is a distribution of the same worth function values, in an attempt to provide additional insight into the preferences of the two study groups. The distribution in Figure 5 is obtained directly from Tables 1 and 2, of course. Several observations are in order. Visual inspection of this distribution reveals that the Corps Group identified EXPERT CHOICE as a first choice, primarily, when compared to the other three DSS. That is, five of the Corps evaluators ranked EXPERT CHOICE highest, three evaluators ranked ELECTRE highest, two evaluators ranked ARIADNE first, and only one evaluator ranked MATS-PC first. Similarly, ARIADNE, ELECTRE, and MATS-PC were identified as second, third and fourth choices, respectively.

The Student Group perceived the capabilities of the four DSS somewhat differently as noted in Figure 5. EXPERT CHOICE was again identified as their first choice, but the remaining three DSS were ranked differently, so that MATS-PC is identified as second choice, followed by ELECTRE and ARIADNE competing for third place.

In summary, then, the two groups, ranked the four decision support systems as follows:

	Corps Group:	Student Group:
(1st. Choice)	**. EXPERT CHOICE**	**. EXPERT CHOICE**
(2nd. Choice)	**. ARIADNE**	**. MATS-PC**
(3rd. Choice)	**. ELECTRE**	**. ELECTRE**
(4th. Choice)	**. MATS-PC**	**. ARIADNE**

These two rankings reflect, then, the characteristics of the four multiple criteria DSS as perceived by the two groups. Further analysis and evaluation of the DSS by the users in the practicing community for other applications will be required before reaching final conclusions. Individual analytical capability and skills undoubtedly play a role in how a user perceives the relative usefulness of these four methods. The "system-user" interface of MATS-PC, for example, was perceived to be quite adequate and friendly by the Student Group, but somewhat confusing as perceived by the Corps Group, particularly with regards to its weights assessment technique, i.e., cross-checking weights, viewing and changing weights.

5. LIMITATIONS OF THIS ANALYSIS

Several limitations of this analysis are noted. Only one real-world problem, namely that of a large-scale water supply system, was addressed in this experiement. Other problems in a variety of decision environments need to be addressed before reaching conclusions. Also, in addition to the "predictive validity" criterion which was applied in our analysis, other critieria (e.g., inter-expert validity, inter-temporal, normative validity, etc.) need to be applied and investigated, as Hamedeh et al. (1990) have done in their

INTERFACE 3: Issues Between the DSS/User and to Problem Domain

(3.1) Do you feel that EXPERT CHOICE is a useful tool in the analysis of alternative project plans?

	low rating				high rating
(a) assessment of weights	1	2	(3)	4	5
(b) assessment of attributes	1	2	(3)	4	5
(c) confidence and insight gained into problem	1	2	(3)	4	5

(3.2) How do you feel about EXPERT CHOICE's ability to help you represent the complexity of the problem?

(a) hierarchy of attributes	1	2	(3)	4	5
(b) representation of uncertainty	(1)	2	3	4	5
(c) identification of potential impacts of uncertainty on plan attributes	(1)	2	3	4	5
(d) clarify your own values and thinking	1	(2)	3	4	5

(3.3) Do you have confidence in the results (the final ranking of those alternative plans) presented by EXPERT CHOICE i.e.,

none	perhaps	some	a lot
1	(2)	3	4

Figure 3. A Sample of Responses to the DSS Evaluation Questionnaire for Questions 3.1, 3.2 and 3.3

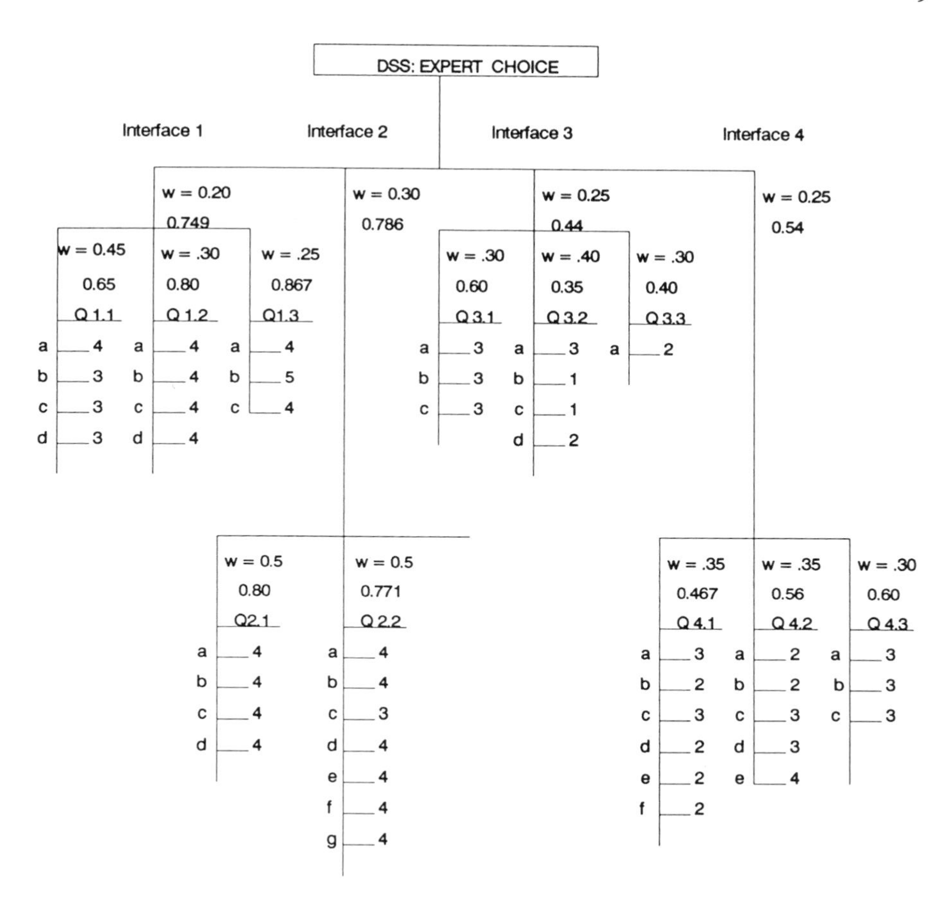

Figure 4. Distribution and Ranking of the Four DSS

	EVALUATORS										
	1	2	3	4	5	6	7	8	9	10	11
MATS-PC	.34	.38	.29	.46	.61	.53	.62	.49	.26	.36	.27
EXPERT CHOICE	.42	.63	.61	.56	.51	.73	.61	.62	.55	.64	.59
ARIADNE	.51	.47	.60	.50	.50	.66	.34	.82	.48	.69	.41
ELECTRE	.57	.39	.70	.48	.47	.35	.69	.50	.28	.56	.32

Table 1. Worth Scores by Corps Group

	EVALUATORS								
	1	2	3	4	5	6	7	8	9
MATS-PC	.65	.92	.79	.71	.61	.55	.72	.71	.79
EXPERT CHOICE	.89	.86	.86	.91	.54	.86	.80	.77	.84
ARIADNE	.43	.68	.65	.73	.64	.59	.57	.87	.68
ELECTRE	.59	.71	.51	.52	.58	.39	.61	.68	.58

Table 2. Worth Scores by Student Group

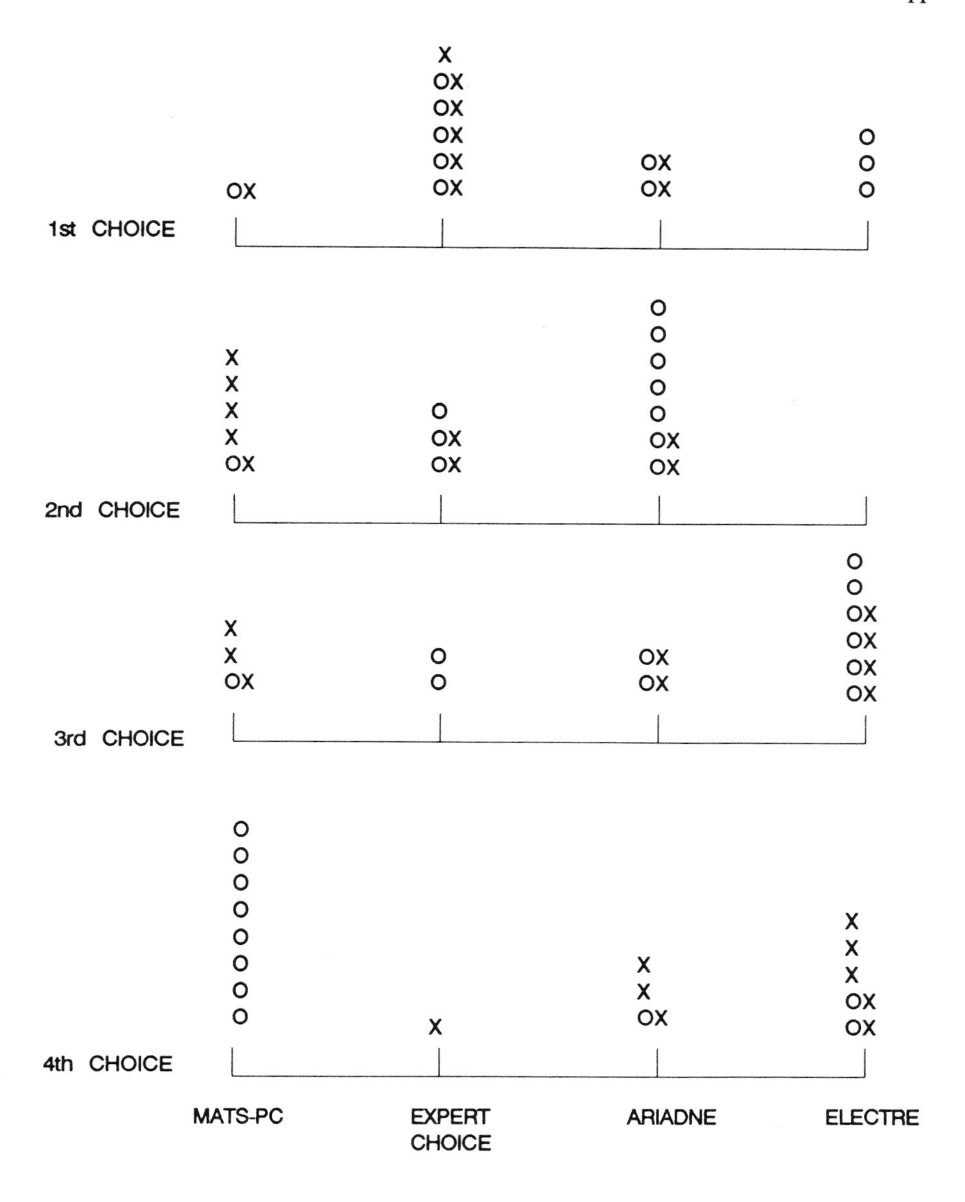

Figure 5. Distribution and Ranking of the Four DSS

study. Additionally, "individual differences", although recognized and measured, were not adequately controlled in the design of the experiment and later related to evaluation results.

The analysis, however, did consider the "impact" of the DSS on the ranking of the alternative water plans, and this was addressed in great detail via non-parametric statistical testing in Tasks I and II, and these results have been presented in Goicoechea et al. (1991).

ACKNOWLEDGEMENTS

The authors wish to thank Mr. Michael Krouse, Chief of Research Division, Institute for Water Resources, for his continued support and funding of this research project. Special thanks to Mr. Hal Kinback, Director of the Planning Associates Program, Washington Level Review Center, and specially the eleven Corps planners and our students, for giving generously of their time, patience and expertise, to test the four decision systems as reported in this paper.

APPENDIX

EVALUATION QUESTIONNAIRE FOR MATS-PC

Evaluator's Name:______________________

MATS-PC

The purpose of this experiment (Experiment 3) is to obtain your opinion and input to help evaluate MATS with your responses to the following questions. We would like to evaluate this tool in four major areas, also called Interfaces:

. Interface 1: [Method] - [Computer Environment]
. Interface 2: [Decision-Support System (DSS)] - [User]
. Interface 3: [DSS] - [Problem Domain]
. Interface 4: [DSS] - [Corps Organization]

Please see Figures 1 and 2 for a pictorial description of these four interfaces.

INTERFACE 1: [Method] - [Computer Environment]

(1.1) How comfortable are you with the amount (volume) of information provided on the screen to assist you with:

	not comfortable				very comfortable
(a) creating the data file	1	2	3	4	5
(b) weights elicitation	1	2	3	4	5
(c) generate hierarchical tree	1	2	3	4	5
(d) utility curves	1	2	3	4	5

Comments on any of the above?

____

____

(1.2) Were the graphics (e.g., tables, figures, menus, boxes, circles, arrows, etc....) helpful to you?

	not helpful				very helpful
(a) introduction to method	1	2	3	4	5
(b) execution of steps	1	2	3	4	5
(c) giving the responses that you wanted	1	2	3	4	5
(d) understanding results	1	2	3	4	5

Comments on any of the above?

____

____

(1.3) Was the use of color (blue, red, green) in the screen displays useful to you?

	little help				very helpful
(a) highlighting of MENU	1	2	3	4	5
(b) input of data	1	2	3	4	5
(c) presentation of results	1	2	3	4	5

Comments?____

____

(1.4) (System designer only) Choice of computer language to code the MATS methodology.

	difficult				easy
(a) coding of method	1	2	3	4	5
(b) incorporation of graphics	1	2	3	4	5
(c) error correction, debugging	1	2	3	4	5

(1.5) (System designer only) Architectural complexity required in software design.

	difficult				easy
(a) subroutine control	1	2	3	4	5
(b) file creation and management	1	2	3	4	5

INTERFACE 2: [DSS] - [User]

(2.1) How easy was it to understand MATS-PC with regards to the following?

	very difficult				very easy
(a) weights	1	2	3	4	5
(b) FACTORS	1	2	3	4	5
(c) FUNCTION FORMS	1	2	3	4	5
(d) IMPACTS	1	2	3	4	5
(e) RANKINGS	1	2	3	4	5

(2.2) How do you rate the MATS package with regards to the following?

	low rating				high rating
(a) easy to use in general	1	2	3	4	5
(b) helps you organize your thoughts (weights and preferences)	1	2	3	4	5
(c) helps you gain a better understanding of problem	1	2	3	4	5
(d) interface friendliness	1	2	3	4	5
(e) graphics and color environment	1	2	3	4	5
(f) explanation of results	1	2	3	4	5
(g) amount of effort,					

time required of you	1	2	3	4	5

INTERFACE 3: [DSS/User] - [Problem Domain]

(3.1) Do you feel that MATS is a useful tool in the analysis of alternative project plans?

	low rating				high rating
(a) assessment of weights	1	2	3	4	5
(b) assessment of attributes	1	2	3	4	5
(c) confidence and insight gained into problem	1	2	3	4	5

(3.2) How do you feel about MATS's ability to help you represent the complexity of the problem?

	low rating				high rating
(a) hierarchy of attributes	1	2	3	4	5
(b) representation of uncertainty	1	2	3	4	5
(c) identification of impacts of uncertainty on plan attributes	1	2	3	4	5
(d) clarify your own values and thinking	1	2	3	4	5

(3.3) Do you have confidence in the results (the final ranking of those alternative plans) presented by MATS?

none	perhaps	some	a lot
1	2	3	4

(3.4) How well, do you think, the MATS method gives a ranking of alternatives that is close to your own ranking (Experiment 1)

Totally different				identical
1	2	3	4	5

(3.5) If you think that MATS is helpful in the evaluation and ranking of alternative plans, how much is it due to:

	Percentage:				
(a) the methodology	20	40	60	80	100%
(b) the computer package itself	20	40	60	80	100%

INTERFACE 4: [DSS/User] - [Corps Organization]

(4.1) Match with organization: Is this tool potentially useful to the Corps of Engineers with regards to the following?

	not useful				very useful
(a) planning procedures	1	2	3	4	5
(b) engineering and economic analysis	1	2	3	4	5
(c) evaluation and ranking of					

alternative plans	1	2	3	4	5
(d) management practices and styles	1	2	3	4	5
(e) as a training tool for new planners	1	2	3	4	5
(f) to assist with the documentation and preparation of reports	1	2	3	4	5

(4.2) Effectiveness Factors: In your opinion, can this tool be useful during the planning process in any of the following areas?

	Not useful				Very useful
(a) identification of alternative project plans	1	2	3	4	5
(b) development of each plan (engineering & economic features)	1	2	3	4	5
(c) comparison and trade-off analysis among alternative plans	1	2	3	4	5
(d) in your capacity as a planner to present your ideas and recommendations to your supervisor in an effective way	1	2	3	4	5
(e) assist your group with the task of recommending a "best plan" to upper management	1	2	3	4	5

(4.3) Efficiency: Does this tool have the potential for any of the following?

	little potential				great potential
(a) saving (using less) planning resources: money, time, effort	1	2	3	4	5
(b) facilitating, expediting the plan evaluation process (economic and engineering factors)	1	2	3	4	5
(c) facilitating, expediting the Public Participation Process (show sensitivity of results to inputs by the various "publics")	1	2	3	4	5

THANK YOU

BIBLIOGRAPHY

Adelman, L., and M. Donnell, "Evaluating Decision Support Systems: A General Framework and Case Study", in Micrompuţer Decision Support Systems, by S.J. Andriole (Editor), QED, 1986.

Brown, C., D.P. Stinson, and R.W. Grant, Multiobjective Tradeoff System: Personal Computer User Manual (MATS-PC), Burau of Reclamation, Engineering and Research Center, Denver, Colorado, September 1986.

Forman, E.H., "The Analytic Hierarchical Process as a Decision Support System", Proceedings of the IEEE Computer Society, Fall 1983.

Goicoechea, A., D. Hansen and L. Duckstein, Multiobjective Decision Analysis with Engineering and Business Applications, Wiley, 520 pgs., 1982.

Goicoechea, A., Experimental Evaluation of decision Support Systems (DSS) and their Use in Situations Involving Multiple Criteria and Uncertainty, STATCOM Report 90-32, STATCOM, Inc., 7921 Jones Branch Drive, McLean, Virginia, October 24, 1990.

Goicoechea, A., E. Stakhiv and Fu Li, "Experimental Evaluation of Multiple Criteria Decision Support Systems for Application to Water Resources Planning", Paper submitted to Water Resources Bulletin, January 15, 1991.

Goicoechea, A., E.Z. Stakhiv, and F. Li., "Experimental Results in Multiple Criteria Decision Making With and Without the Aid of Decision Support Systems", Proceedings of the Engineering Foundation Conference on Risk-Based Decision Making, Santa Barbara, California, October 15-20, 1989.

Hamadeh, W., B.F. Hobbs, V. Chankong, and E.Z. Stakhiv, "Does Choice of Multiobjective Method Matter? An Experiment", Draft Report submitted to the U.S. Army Corps of Engineers, Institute for Water Resources, Ft. Belvoir, Virginia, December 1990.

Roy, B., "Classement et Choix en Presence de Criteres Multiples", Riro 2 Annee, No. 8, 1968.

Saaty, T.L., The Analytic Hierarchy Process, McGraw-Hill, 1980.

U.S. Army Corps of Engineers, Washington Metropolitan Area Water Supply Study Report, Institute for Water Resources, Ft. Belvoir, Virginia, November 1975.

White, C.C., A.P. Sage, S. Dozono, and W.T. Scherer, "Performance Evaluation of a Decision Support System", Large-Scale Systems, 6, 39-48, 1984.

A PROCEDURE FOR SELECTING MCDM TECHNIQUES FOR FOREST RESOURCES MANAGEMENT*

Aregai Tecle
Assistant Professor, School of Forestry
Northern Arizona University, Flagstaff, Arizona 86011-4098

and

Lucien Duckstein
Professor, Department of Systems and Industrial Engineering
University of Arizona, Tucson, Arizona 85721

ABSTRACT: A procedure for choosing a multicriterion decision-making (MCDM) technique is modeled as a multicriterion problem. The methodology consists of identifying a set of feasible MCDM techniques, and evaluating them with respect to four sets of choice criteria, namely (1) problem related criteria (2) DM/analyst related criteria (3) technique-related criteria, and (4) solution-related criteria. As an example of application, fifteen MCDM techniques are evaluated in terms of 24 criteria, forming an evaluation matrix of criteria versus alternatives. The evaluation matrix is then analyzed by means of a composite programming algorithm resulting in a preference ranking of the alternatives. The algorithm involves using weights and scales in Lp-norm distance measures. Extensive sensitivity analysis involving a total of 84 trials shows the algorithm to be robust with respect to small changes in criterion weights. Application of a subset of the techniques to a forest resources management problem illustrates the methodology throughout the paper.

INTRODUCTION

This paper presents a procedure to select a proper MCDM technique to model and solve a multicriterion forest resources management problem. Because of the large number of techniques available, an applied analyst can get confused in determining which techniques to select when confronted with a real problem. Such ambiguity can cause inappropriate selection of MCDM techniques resulting in a misleading solution and wrong conclusions. For example, Cohon and Marks (1975), in their classification of MCDM techniques that can be used to solve water resources problems, stated that ELECTRE was not applicable to these problems. Krysztofowicz et al. (1977), however, stated the opposite by pointing out that ELECTRE was successfully applied to water resources problems and that its use in this area should be continued. The latter was further proven to be true in subsequent studies by Gershon et al. (1982) and Tecle et al. (1988a, 1988b). Given such apparent contradictions, it is not surprising to see a mismatch in practice between a real world problem and the MCDM technique applied to solve that particular problem. There are four possible consequences of this kind of mismatch. First, as stated above, the solution resulting from poorly matched problem-MCDM technique situation will be misleading or unsatisfactory. Second, useful techniques may be judged inappropriately as in the case of ELECTRE. Third, the mismatch subsequently results in wrong decisions incurring losses in valuable time, energy, and money. Finally, it may discourage potential users from applying MCDM techniques to real world problems. To avoid these kinds of difficulties, a procedure for selecting the "best" solution technique for a particular problem may be of help.

As presented in this paper, the paradigm for selecting an appropriate MCDM technique for evaluating a multicriterion forest resources management problem can be briefly described in terms of the following steps: 1. Define the desired objectives or purposes that the MCDM techniques are to fulfill. 2. Select evaluation criteria that relate technique capabilities to objectives. 3. List and specify MCDM techniques available for attaining the objective of modeling the multicriterion problem on hand. 4. Determine technique capabilities or the levels of performance of a technique with respect to the evaluation criteria by setting up and solving a multicriterion problem. 5. Construct an evaluation matrix (techniques versus criteria array), the elements of which represent the capabilities of alternative techniques in terms of the selected criteria (obtained in step 4). 6. Analyze the merits of the alternative MCDM techniques and select the most satisficing technique.

Steps 1 through 5 constitute the problem formulation procedure, while step 6 is the implementation of the MCDM technique selection procedure. The roles of these steps in the selection process, that is, the steps of the technique selection algorithm, are described in the subsequent sections.

CRITERIA FOR SELECTING ALTERNATIVE TECHNIQUES AND THEIR EVALUATION

Tecle (1988) identified 49 different criteria upon which the choice of an appropriate MCDM technique for a particular problem can be based. These criteria can be categorized into four groups depending on whether they describe mainly (1) the characteristics of the decision maker (DM) and/or analyst involved, (2) the characteristics of the algorithm for solution, (3) the characteristics of the problem under consideration, or (4) the nature of the obtainable satisficing solution. In this paper only 24 selection criteria are used for evaluating the appropriateness of a set of 15 MCDM techniques to solve, as an example, a forest resources management problem. The limiting factor for utilizing only 24 criteria is the lack of adequate information on the actual use of the rest of the criteria to evaluate the performance of the techniques. The role of the 24 criteria in the evaluation process will be discussed later. In the meantime, a list of feasible MCDM techniques is provided from which the best technique for a forest watershed resources management is selected. More than 70 MCDM techniques are identified in Tecle (1988). However, for practical reasons only a few MCDM techniques are considered for evaluation in this study. These are the set of alternative techniques in Table 1 with which the authors have some familiarity. Here, they are evaluated for their suitability with respect to a multicriterion forest resources management problem such as the one described in Tecle (1988).

An approach for evaluating the performance of the alternative MCDM techniques under consideration is now provided in terms of the four different groups of criteria. This procedure lays the basis for the development of four evaluation matrices each of which consist of techniques versus criteria array. The array elements are the evaluation scores of all the

Table 1.--Alternative MCDM techniques considered for selection

1. **Analytic Hierarchy Process (AHP) (Saaty 1977, 1980)**
2. **Composite Programming (CTP) (Bardossy et al. 1985)**
3. **Compromise Programming (CP) (Zeleny 1973, 1982)**
4. **Cooperative Game Theory (CGT) (Nash 1953, Szidarovszky et al. 1984)**
5. **Displaced Ideal/DISID) (Zeleny 1974, Nijkamps 1979)**
6. **ELECTRE (ELEC) (Benayoun et al. 1966, Roy 1968)**
7. **Evaluation and Sensitivity Analysis Program (ESAP) (Mumpower and Bollacker 1981)**
8. **Goal Programming (GP) (Charnes and Cooper 1961, Ignizio 1976)**
9. **Multiattribute Utility Theory (MAUT) Keeney and Raiffa 1976)**
10. **Multicriterion Q-Analysis (MCQA) (Hiessl et al. 1985)**
11. **Probabilistic Tradeoff Development Method (PROTRADE) (Goicoechea et al. 1976)**
12. **Zionts-Wallenius (Z-W) (Zionts and Wallenius 1976)**
13. **Step Method (STEM) (Benayoun et al. 1971)**
14. **Surrogate Worth Trade-off (SWT) (Haimes and Hall 1974)**
15. **PROMETHEE (PRM) (Brans and Vincke 1985)**

techniques with respect to each criterion in each criterion group. The level of these scores is based on the authors' experience in applying the techniques as well as the evaluation results of previous works such as those of Khairullah and Zionts (1979), Ozernoy (1987), and Rietveld (1980). The arrays so formed provide in turn the information necessary to select a technique. After constructing these arrays, the DM's preference structure over the set of criteria in each matrix is determined to complete the problem formulation. Every criterion in any one of the four categories is assigned a weight relative to its importance in that group. For example, when considering the technique-related criteria, "ease of coding" is weighted much higher than the "number of parameters required" in a technique. Based on the authors' experience, it is more difficult to understand and use a technique characterized by coding complexity than one which requires a lot of parameters, all other characteristics remaining the same.

Evaluation Using DM or Analyst Related Criteria

The DM or analyst-related criteria are meant to reflect the DM's or analyst's level of knowledge and willingness to use these criteria and are given as row entries in Table 2. These criteria can be evaluated irrespectively of the characteristics of the problem under consideration. Evaluation with respect to these criteria is made using a subjective scale with a value ranging from one to ten, where one represents the worst case, that is, a task the particular technique does not fulfill and ten indicates the best possible performance that can be attributed to the particular technique. Alternatively the rating of the MCDM technique may be a fuzzy set membership function. In any case, the evaluation matrix of Table 2 is constructed.

Table 2.--Evaluation matrix of techniques versus DM or analyst-rated criteria

Criteria	Weight	Alternative Techniques														
		CP	CGT	CTP	GP	STEM	MAUT	ESAP	MCQA	ELEC	AHP	Z-W	PROTR	SWT	DISD	PRM
DM's level of knowledge	4	4	3	6	5	6	3	4	7	9	5	9	2	4	4	9
DM's desire to interface	3	9	9	10	8	5	7	8	9	8	6	3	2	5	5	8
Time avail. of DM	3	10	10	8	8	5	3	2	9	9	7	7	5	6	6	8
DM's actual knowledge	2	9	7	8	10	9	6	5	10	8	7	7	4	7	9	7
Analysts skill	1	10	9	8	7	6	7	6	4	3	6	6	7	5	8	6

Table 3.--Evaluation matrix of techniques versus technique-related criteria

Criteria	Weight	Alternative Techniques														
		CP	CGT	CTP	GP	STEM	MAUT	ESAP	MCQA	ELEC	AHP	Z-W	PROTR	SWT	DISD	PRM
CPU time required	3	6	4	4	7	7	4	3	8	10	6	7	7	4	5	8
No. of param. required	2	9	9	7	7	6	8	7	4	5	7	7	6	5	9	7
Ease of use	4	8	7	6	8	1	9	7	9	7	7	8	8	2	6	8
Computational burden	4	6	5	7	7	7	4	4	9	8	8	7	5	2	6	7
Ability to get eff. points	5	8	9	9	1	5	8	6	6	4	5	8	8	10	7	4
Ease of Coding	4	8	9	6	7	7	8	7	9	7	7	6	3	1	8	7

Solving the MCDM Technique Selection Problem

The sequential steps taken during a multicriterion decision making process can be lumped into two complementary stages: the problem formulation stage and the problem solution one. The problem formulation stage culminates in the formation of the evaluation matrices. The problem solution stage, on the other hand, constitutes the application of a MCDM technique on the matrices to determine the desired solution. According to Wymore (1976), the problem formulation stage is as important, if not more so, than the solution stage. This is particularly true for problems concerning large scale systems where, at least partly, the manner in which the problem is formulated helps in determining the most suitable MCDM technique to be used (Gershon and Duckstein 1984).

Now, since the problem of selecting MCDM techniques is itself a multicriterion problem, the MCDM technique selection algorithm can be applied to select the technique for this purpose also. However, since this procedure would lead to a cyclical process, a technique must be arbitrarily chosen and applied on the MCDM choice evaluation matrices to select the most satisficing technique for the problem at hand. This choice, however, must take into account the type of the problem to be solved. In this case, for example, since the MCDM technique selection problem is defined by a discrete set of systems and must be analyzed at two levels that consist of (1) solving each evaluation matrix separately and then (2) combining the results obtained to form a new evaluation matrix from which the final compromise solution is determined, the technique choice cannot be just arbitrary. This is because the discrete formulation and the two-level solution procedure requirements of the problem limit the selection of the choice algorithm.

An algorithm which fulfills this requirement and is used for the purpose of selecting the best MCDM technique is composite programming (CTP) (Tecle 1988). This algorithm, which is an extension of compromise programming, (Bardossy et al. 1985) is adapted here to perform the two level trade-off analysis required in the MCDM technique choice problem. At the first level, different L_p-norms are applied to seek a compromise within each of the four criterion groups, and then a different L_p-norm is applied to produce a trade-off among these four groups. A final preference ranking of the alternative MCDM techniques under consideration is thus reached. The L_p-norms, Φ_k for each of the k (k=1,...,4) aggregated criterion groups are:

$$\Phi_k = \left[\sum_{i=1}^{I_k} \alpha_{ki} \left(\frac{d^*_{ik} - d_{ijk}}{d^*_{ik} - d^{**}_{ik}} \right)^{P_k} \right]^{1/P_k} \tag{1}$$

where d^*_{ik} is the maximum of d_{ijk} over alternatives j = 1,...,J, and d^{**}_{ik} is the minimum of d_{ijk}, α_{ki} is the weight associated with criterion i (i=1,...,I_k) in group k (k = 1,...,4), and p_k is the associated compromise programming parameter with criterion k.

The overall composite goal function, **G** for the MCDM technique selection problem can be written as:

$$G = \left(\sum_{k=1}^{K} \beta_k \Phi_k^q \right)^{1/q}, \text{ for } k = 1,...,4 \quad (2)$$

and

$$\sum_{k=1}^{K} \beta_k = 1, \quad \text{for } k = 1,...,4 \quad (3)$$

where β_k is the weight of criterion group k, and q is the inter-criterion group trade-off parameter. The MCDM technique selection problem was solved for the set of weights and CP parameters given in Table 6 and 84 sets of β_k values provided in Tecle (1988). The results are discussed in the next sections.

Table 6.--Weights α_{ki} and CP Parameters, p_k in each Criterion Group, k

Criterion Group k	Criteria in Each Criterion Group							CP Parameter p_k
	1	2	3	4	5	6	7	
1	4	3	3	2	1			2
2	3	2	4	4	5	4		3
3	5	4	3	3	4	3	2	2
4	2	2	3	3	1	1		4

APPLICATION

The solution to the MCDM technique selection problem is obtained in terms of a L_p-distance from an ideal point. The ideal point in this case is considered to be the point consisting of the highest criterion score in each row of the evaluation matrix. Since composite programming algorithm was used to solve the problem, the results for the first level of analysis are presented first. Table 7 represents the compromise solutions corresponding to the four evaluation matrices of Table 2 through 5. It is interesting to note in Table 7 that the preference ranking of the alternative techniques differ from one criterion group to the other. If the status of ELECTRE is examined, for example, it is observed to be ranked first when evaluated using the DM/analyst-related group of criteria but falls to tenth, sixth, and fourth ranks when evaluated using the technique, problem, and solution-related groups of criteria, respectively. There are at least two possible reasons for this phenomenon. One reason could be due to biases attributable to the way the subjective performance evaluation scores and weights are obtained. The performance evaluation scores are determined on the basis of the experience of individuals in using the techniques while the criterion weights are meant to represent the DM's preference structure. Among the techniques considered for discrete action problems, ELECTRE is one of the most widely used. This is particularly true when the European experience with MCDM is taken into consideration. The other reason is due to the fact that the different criterion groups evaluate quite different characteristics of the techniques. Thus, since ELECTRE is taken to

Table 7.--Compromise ranking of alternatives and their respective L_p-distance

	Criteria Groups							
	DM/Analyst		Technique		Problem		Solution	
Alts.	Value	Rank	Value	Rank	Value	Rank	Value	Rank
CP	2.340	4	2.555	1	0.000	1	5.117	7
CGT	3.594	10	4.818	5	3.000	2	5.147	8
CTP	3.042	6	6.031	7	0.000	1	0.584	1
GP	2.556	5	11.413	14	7.071	8	9.831	15
STEM	3.226	8	9.580	13	7.348	9	6.287	10
MAUT	4.677	14	6.472	9	5.196	5	9.019	13
ESAP	4.564	13	7.988	12	5.196	5	9.256	14
MCQA	1.524	2	4.428	3	5.831	6	4.472	5
ELEC	1.465	1	6.582	10	5.831	6	4.375	4
AHP	3.173	7	5.401	6	5.000	4	7.295	12
Z-W	3.285	9	3.000	2	6.856	7	4.903	6
PROTR	5.718	15	6.655	11	7.937	10	2.367	2
SWT	3.929	12	13.855	15	7.937	10	5.618	9
DISID	3.758	11	4.797	4	3.606	3	2.887	3
PRM	1.938	3	6.387	8	5.831	6	6.910	11

represent all four versions (ELECTRE I, II, III, and IV), the evaluation scores used are representative of the average performance levels of all four types. Besides, there are some limitations to using ELECTRE. It can only be applied to discrete problems that have finite numbers of alternatives (Duckstein and Bogardi 1988). It is no wonder then to see ELECTRE having very low preference ranking, tenth, when evaluated with respect to the technique-related criteria. The discrepancies in the rankings of the other alternatives with respect to the different groups of criteria can be explained similarly.

But since our interest is in the final preference ranking, the results at the second level of evaluation become more important. Table 8 shows the overall preference rankings and the associated L_p-distances of the alternatives under consideration.

Of the 15 techniques examined, compromise programming (CP), composite programming (CTP), the method of the displaced ideal, and cooperative game theory, in that order, are found to be the most preferred techniques. Since all of these techniques are of the distance-based type, the concept of distance may have helped to make these techniques be easily understood and become popular. Further reasons for their high ranking may be found by reexamining the results of the first stage of evaluation in Table 7. All four techniques are highly rated with respect to the problem-related criteria as shown in the two columns under problem in Table 7. These techniques can handle all the tasks ascribed by this group of criteria except the task of directly solving problems having non-numerical data. Problems of the latter type are usually handled by a heuristic scaling procedure which consists of converting the data into numerical form (Gershon et al. 1982, Hiessl et al 1985, Tecle 1987, Tecle et al. 1988b).

The situation, however, is quite different when the techniques are evaluated with respect to the other criterion groups. With respect to the DM/analyst criterion group, compromise programming and composite programming are ranked fourth and sixth while cooperative game theory (CGT) and the method of the displaced ideal fall to the tenth and eleventh ranks, respectively. The reason for this is that CGT and the method of the displaced ideal are rated low with respect to some of the criteria in this group. Both techniques, for example, require the DM to have some knowledge of utility theory, the concepts of the ideal point in the displaced ideal case, and status quo point in the CGT case, and then of some axioms to understand and use the

Table 8.--Final overall ranking of alternatives for a particular inter-criterion group trade-off parameters

Technique	Distance Value	Rank
Compromise programming	3.0074	1
Composite programming	3.9427	2
Displaced ideal	4.0227	3
Cooperative game theory	4.3099	4
Multicriterion q-analysis	4.7194	5
Zionts-Wallenius method	4.8567	6
Analytic hierarchy process	5.5514	7
ELECTRE methods	5.6188	8
PROMETHEE methods	6.0370	9
Probabilistic trade-off development method	6.4037	10
Multiattribute utility theory	6.5810	11
Evaluation and sensitivity analysis program	7.2692	12
Step method	7.8649	13
Goal programming	9.3320	14
Surrogate worth trade-off method	10.1758	15

techniques (Szidarovszky et al. 1984; Zeleny 1982). For this reason the evaluations with respect to the "level of knowledge required of the DM" criterion produce low scores as shown in column 2 of Table 2. Other factors contributing to the low ranking of these alternatives are the levels of the criterion. As shown in Table 2, the "level of knowledge required of the DM" is given the highest weight in the group. This supplements the criterion scores in determining the ranking of the alternatives. An alternative having a high evaluation score with respect to an important criterion (that is, one with high criterion weight) will be ranked high, and the converse is true. The latter contributes to the low ranking of CGT and the method of the displaced ideal with respect to the DM/analyst-related criteria shown in columns 2 and 3 of Table 7. Similar reasoning can be given to the average ranking of CTP with respect to the technique-related criteria and CP and CGT with respect to the solution-related criteria (Table 7).

There is also high variability in the ranks of the lowest ranked alternatives across the criterion groups. Goal programming which is ranked fourteenth overall, for example, is ranked fairly high (fifth) with respect to the DM/analyst-related criteria, low with respect to the problem-related criteria and very low with respect to both technique-related (fourteenth) and solution-related (fifteenth) criteria. The concept of GP is easily understood and the technique is widely accepted for real-world application; for these reasons it is ranked above average with respect to the DM/analyst-related criteria. GP is, however, ranked eighth, only two notches higher than the lowest rank with respect to problem-related criteria because of its inability to handle dynamic problems as well as problems with finite number of alternatives and non-numerical data. Likewise GP is ranked very low with respect to both the technique-related and the solution-related groups of criteria. This low ranking is mainly due to the propensity of GP to provide a dominated solution to a multicriterion problem. Because the technique practically solves any problem as a single objective one, that is, minimizing the sum of the deviations of the objective function values from a goal point, it cannot, in its original version, produce a strongly efficient solution that simultaneously satisfies the different objectives. Besides, some of the solution points in GP may be dominated points. Note that the consideration of a modified GP algorithm, such as a direction-based one (projecting the current point on the Pareto optimum) does alleviate this problem (Kananen et al. 1990). Other factors that contribute to the ranking level of alternatives are the range of the values of the parameters used during the two-level evaluation scheme (Tecle 1988).

DISCUSSION AND CONCLUSIONS

Equation (1) shows the number and type of the different constants required to use the composite programming algorithm. The different constants used to carry out the selection procedure in this study include four sets of criterion weights, a set of optimal (maximal) values, a set of worst (minimal) values, four compromise parameters p_k such that $1 \leq p_k < \infty$, k=1,...,4 a set of criterion group weights and an inter-criterion group trade-off parameter, q where $1 \leq q < \infty$. Given this multitude of parameters it would be quite cumbersome, time consuming and expensive to carry out sensitivity analysis on every possible combination of parameter values. Thus, here, a sensitivity analysis is carried out only with respect to group of parameters β_k, k=1,...,4. Such sensitivity analyses were made for 84 sets of β_k parameters. Table 8 is one of the 84 ranking tables determined for β_1=0.1, β_2=0.3, β_3=0.2, β_4=0.4 and q=2. The 84 sets of β_k values, given in Tecle (1988), yield 84 preference ranking tables which are compared with each other; the rankings of each alternative MCDM technique in all of these tables are then enumerated to obtain the ranking frequency of Table 9. It is observed in this table that out of a possible first to 15th rank in each of the 84 trials, compromise programming and composite programming are found to be ranked first 50 and 34 times out of 84, respectively. The displaced ideal, cooperative game theory and multicriterion Q-analysis are respectively next in line. The ELECTRE, AHP and Zionts-Wallenius technique are mostly ranked in the upper half, ranging from 5th to 10th. The lowest ranked alternative techniques are goal programming and the surrogate worth trade-off methods which are mostly ranked 14th and 15th. The rest of the techniques are found to be ranked a little higher than the last two, having ranks between 8th and 12th in the case of PROTRADE and the PROMETHEE methods and between 10th and 13th ranks in the case of STEM, MAUT and ESAP. By comparing the results in Tables 8 and 9, it can be said that the MCDM technique choice algorithm is fairly robust with respect to the β_k parameters; namely, changes in β_k values cause only small changes in ranking of the alternatives.

Table 9.--Ranking Frequency of Alternative Techniques for 84 Sets of β_k (k=1,...,4) Values

Alter-native	Rank														
	1	2	3	4	5	6	7	8	9	10	11	12	13	14	15
CP	50	20	13	0	1	0	0	0	0	0	0	0	0	0	0
CGT	0	0	9	31	20	18	6	0	0	0	0	0	0	0	0
CTP	34	30	7	4	2	7	0	0	0	0	0	0	0	0	0
GP	0	0	0	0	0	0	0	0	0	0	0	0	2	53	29
STEM	0	0	0	0	0	0	0	0	0	5	24	16	39	0	0
MAUT	0	0	0	0	0	0	0	0	0	29	28	19	6	0	0
ESAP	0	0	0	0	0	0	0	0	0	1	14	32	27	10	0
MCQA	0	6	10	24	41	3	0	0	0	0	0	0	0	0	0
ELEC	0	0	0	2	6	27	31	18	0	0	0	0	0	0	0
AHP	0	0	0	0	1	6	22	25	20	9	1	0	0	0	0
Z-W	0	6	5	3	10	21	22	11	4	1	1	0	0	0	0
PROTR	0	0	1	0	0	1	3	16	6	22	16	11	6	2	0
SWT	0	0	0	0	0	0	0	0	0	0	0	6	4	19	55
DISID	0	22	39	20	3	0	0	0	0	0	0	0	0	0	0
PRM	0	0	0	0	0	1	0	14	52	17	0	0	0	0	0

To sum up this study, a possible algorithm has been developed for selecting, from among a set of feasible MCDM techniques, the most appropriate technique for application. The algorithm is based on a two level process. The first level consists of an evaluation of the techniques with respect to four different groups of criteria: (a) DM/analyst-related, (b) technique-related, (c) problem-related, and (d) solution-related criteria. This results in the ranking of the techniques with respect to each criterion group. Then an overall ranking of the techniques is obtained at a second level through a combination of the first stage results. For a forest resources management problem such as the one described in Tecle (1988), the MCDM technique choice algorithm used in this study shows compromise programming and composite programming to be the most preferred techniques and surrogate worth trade-off method and ordinary goal programming to be the least preferred ones. The other techniques are ranked in-between these extremes. For other types of problems, ranking of the techniques could be quite different. As a final remark, it should be pointed out that the MCDM technique choice algorithm discussed in this paper can be applied not only for selecting techniques to solve natural resources management problems, but also to any kind of MCDM technique choice problems (Duckstein et al. 1989). The aim should be toward the development of a handy MCDM choice algorithm for matching a particular MCDM technique to any real world multicriterion problem.

REFERENCES

Bardossy, A.; Bogardi, I.; Duckstein, L. 1985. Composite programming as an extension of compromise programming. In: Serafini, P., ed. Mathematics of multiobjective optimization. New York, NY: Springer-Verlag: 375-408.

Benayoun, R.; de Montgolfier, J.; Tergny, J.; Lariche, O.I. 1971. Linear programming with multiple objective functions: STEP methods (STEM). Mathematical Programming. 1(3):366-375.

Brans, J.P.; Vincke, P. 1985. A preference ranking organization method (the PROMETHEE method for multiple criteria decision-making). Management Science. 31(6):647-656.

Charnes, A.; Cooper, W.W. 1961. Management models and industrial applications of linear programming, vols. I and II. New York, NY: John Wiley and Sons.

Cohon, J.L.; Marks, D.H. 1975. A review and evaluation of multiobjective programming techniques. Water Resources Research. 11(2):208-220.

Duckstein, L., Bobee, B. and Ashkar, F. 1989. Multicriterion choice of estimation techniques for fitting extreme floods. Proceedings Intern. Conf. in Multiple Criterion Decision Making: Applic. in Industry and Service, Bangkok, Thailand. pp. 1069-1083. December.

Duckstein, L.; Bogardi, I. 1988. Multi-objective approaches to river basin planning. Handbook of Civil Engineering. Hasbrouck Heights, NJ; Technomic Publishing. pp. 415-452.

Duckstein, L.; Gershon, M.; McAniff, R. 1982. Model selection in multiobjective decision making for river basin planning. Advances in Water Resources. 5(3):178-184.

Gershon, M. 1981. Model choice in multiobjective decision making in water and mineral resources systems. Natural Resource Systems Tech. Rep. Series. No. 37. Tucson, AZ: Department of Hydrology and Water Resources, University of Arizona.

Gershon, M.; Duckstein, L.; McAniff, R. 1982. Multiobjective river basin planning with qualitative criteria. Water Resources Research. 18(2):193-202.

Gershon, M.; Duckstein, L. 1984. A procedure for selection of a multiobjective technique with application to water and mineral resources. Applied Mathematics and Computation. 14:245-271.

Goicoechea, A.; Duckstein, L.; Fogel, M.M. 1976. Multiobjective programming in watershed management: a case study of the Charleston Watershed. Water Resources Research. 12(6):1085-1092.

Haimes, Y.Y.; Hall, W.A. 1974. Multiobjectives in water resources systems analysis: the surrogate worth trade-off method. Water Resources Research. 10(4):615-624.

Hiessl, H., L. Duckstein, and E.J. Plate. 1985. Multiobjective analysis with concordance and discordance concepts. Applied Mathematics and Computation 17:107-122.

Ignizio, J.P. 1976. Goal programming and extensions. New York, NY: Lexington Books.

Kananen, I., P. Korhonen, J. Wallenius, and H. Wallenius. 1990. Multiple-objective analysis of input-output models for emergency management. Operations Research 38(2):193-201.

Keeney, R.L.; Raiffa, H. 1976. Decisions with multiple objectives: preferences and value tradeoffs. New York, NY: John Wiley and Sons.

Khairullah, Z.; Zionts, S. 1979. An experiment with some approaches for solving problems with multiple criteria. Working Paper No. 442. Third international conference on multiple criteria decision making. Konigswinter, West Germany.

Khalili, D. 1986. A decision methodology for the resource utilization of rangeland watersheds. Tucson, AZ: School of Renewable Natural Resources, University of Arizona. Unpublished Ph.D. dissertation.

Krzystofowicz, R., E. Castano, and R. Fike. 1977. "Comment on 'A review and evaluation of multi-objective programming techniques' by J.L. Cohon and D.H. Marks," Water Resources Research 13(3):690-692.

Mumpower, J., and L. Bollacker. 1981. User's Manual: Evaluation and Sensitivity Analysis Program (ESAP). U.S. Army Corps of Engineers, Waterways Experiment Station, Technical Report E-81-4, 279p.

Nash, J.F. 1953. Two-person cooperative games. Econometrica. 21:128-140.

Nijkamp, P. 1979. A theory of displaced ideals: an analysis of interdependent decisions via nonlinear multiobjective optimization. Environment and Planning. 11:1165-1178.

Ozernoy, V.M. 1987. Some issues in mathematical modelling of multiple criteria decision-making problems. 5th International Conference on Mathematical Modelling. Berkeley, CA: University of California: 212-215.

Rietveld, P. 1980. Multiple objective decision methods and regional planning. Amsterdam, Holland: North Holland Publication Co. 330 p.

Roy, B. 1968. Classement et choix en presence de points de vue multiple (La Methode ELECTRE). Revue d'Informatique et de Recherche Operationelle, No. 8:57-75.

Saaty, T.L. 1977. A scaling method for priorities in hierarchical structures. Journal of Mathematical Psychology. 15(3):234-281.

Saaty, T.L. 1980. The analytic hierarchy process: planning, priority setting, resources allocations. New York, NY: McGraw-Hill International Book Co. 187 p.

Szidarovszky, F.Duckstein, L.; Bogardi, I. 1984. Multiobjective management of mining under water hazard by game theory. European Journal of Operation Research. 15(2):251-258.

Tecle, A. 1987. Multiobjective forest watershed management. In: Proceedings, Seventh annual AGU Front Range branch "Hydrology Days;" April 21-23. Fort Collins, CO: Colorado State University: 110-125.

Tecle, A. 1988. Choice of multicriterion decision making techniques for watershed management. Tucson, AZ: School of Renewable Natural Resources, University of Arizona. Unpublished Ph.D. dissertation.

Tecle, A.; Fogel, M.; and Duckstein, L. 1988a. Multicriterion selection of wastewater management alternatives. ASCE J. Water Resour. Plan. & Manag. 114(4):383-398.

Tecle, A.; Fogel, M.; and Duckstein, L. 1988b. Multicriterion analysis of forest watershed management alternatives. Water Resource Bulletin 24(6): 1169-1178.

Wymore, A.W. 1976. Systems engineering methodology for interdisciplinary teams. New York, NY: Wiley. 481p.

Zeleny, M. 1973. Compromise programming. In: Cochrane, J.L.; Zeleny, M., eds. Multiple criteria decision making. Columbia, SC: University of South Carolina Press: 263-301.

Zeleny, M. 1974. A concept of compromise solutions and the method of the displaced ideal. Computers and Operations Research. 1(14):479-496.

Zeleny, M. 1982. Multiple criteria decision making. New York, NY: McGraw-Hill. 563 p.

Zionts, S.; Wallenius, J. 1976. An interactive programming method for solving the multiple criteria problem. Management Science. 22(6):652-663.

THE STATE OF MULTIPLE CRITERIA DECISION MAKING: PAST, PRESENT, AND FUTURE

STANLEY ZIONTS

ALUMNI PROFESSOR OF DECISION SUPPORT SYSTEMS

SCHOOL OF MANAGEMENT
STATE UNIVERSITY OF NEW YORK AT BUFFALO
BUFFALO, NEW YORK 14260

ABSTRACT

This paper presents an overview of multiple criteria decision making, both from historical as well as present and future perspectives. It divides the field into four subareas: multiple criteria mathematical programming; multiple criteria discrete alternatives; multiattribute utility theory; and negotiation theory. It reviews developments in each of the areas. It speculates on future developments, and suggests topics for future research. A substantial number of references are included.

I. Introduction

I was originally asked to talk about the history of Multiple Criteria Decision Making (MCDM) at this meeting. That was some time ago. When I sat down to prepare my paper for this meeting, I felt that to spend the entire time talking about the history would be dull and boring, as well as backward looking. Therefore, I felt at liberty to deviate from and broaden my original charge to include not only a bit of history, but also an appraisal of the present and the future. Though I have attempted to include representative approaches that were developed early on, I have covered only a small portion of the work that has been done. There is an enormous amount of work. See proceedings from earlier meetings, special issues of journals, books, as well as journal articles. Please excuse me if I've omitted your favorite approaches or, even worse, your work.

The paper consists of seven sections. In this section I have presented an introduction. In Section 2 I overview the field. I present a brief history of the field in Section 3, emphasizing early developments. In Section 4 I make a few comments about the future, and in Section 5 I discuss what I think are some fruitful research ideas. I overview my own current activities in Section 6, and conclude the paper in Section 7.

II. What is Multiple Criteria Decision Making (MCDM)?

In this presentation, some knowledge of the field is assumed. For definitions of some of the terms that I'm not defining, see any good introduction to MCDM, such as Steuer (1986).

Just what is the field of multiple criteria decision making? It means different things to different people. The following is my view of the field. A general definition of MCDM is the solving of decision problems that involve multiple (generally conflicting) objectives. Broadly speaking, we may divide MCDM into four subareas:

A. Multiple Criteria Mathematical Programming
B. Multiple Criteria Discrete Alternatives
C. Multiattribute Utility Theory
D. Negotiation Theory

Unfortunately, these categories need not be mutually exclusive and collectively exhaustive, though I define them to be so. I now define each of them more precisely, distinguish between them, and indicate possible overlap.

The first category, multiple criteria mathematical programming, refers to the solving of mathematical programming problems that have multiple objectives. The solution of such problems, as I define them, is generally carried out without first assessing the utility or value function (utility functions are normally defined for stochastic problems, whereas value functions are usually limited to deterministic problems - I shall use utility function in both cases) of a decision maker. It may be solved by enumerating all or a subset of nondominated (either extreme point or other) solutions, or it may be solved by helping a user to identify a specific solution.

The second category of problems involves multiple criteria discrete alternatives. It consists of problems which have a set of discrete alternatives that may be represented as a matrix in which the rows constitute alternatives and the columns constitute objectives. The entries in the matrix represent the performance of each objective for a given alternative. Here too, the problem is generally solved without first assessing the utility function of a decision maker.

The third category of MCDM problems involves multiattribute utility theory, usually done for probabilistic outcomes. It is defined as the assessing and fitting of utility functions and probabilities. Utility functions are assessed by giving a decision maker a choice between alternatives or between alternative lotteries. The responses are used to generate functions. The utility functions are in turn used to rank alternatives. Thus the user can explore alternative decisions and choose one. The utility functions may also be used as objective functions for solving mathematical programming problems.

Usually the first three topics are decisions involving one decision maker, or possibly a group consisting of members who all have similar interests. Negotiation theory consists of the study of negotiation behavior. Simply, it may be thought of as multiple criteria decision making where there are two or more decision makers or disputants who reach agreement employing some kind of voting mechanism that may be used to effectuate an agreement. One possible -- and common -- mechanism is unanimity! In that case all parties to the negotiation must agree on a solution. In many negotiations there are two negotiators (e. g., a buyer and a seller of a commodity), but there are negotiations involving many negotiators, such as the U. S. Congress.

III. A Brief History of MCDM

A. Multiple Criteria Mathematical Programming and Multiple Criteria Discrete Alternatives

Multiple criteria mathematical programming and multiple criteria discrete alternatives are grouped together here because their development is perhaps more intertwined than the other two areas. Probably the oldest widely recorded method of solving multiple criteria problem was developed by Benjamin Franklin in the eighteenth century. It is known as Franklin's moral or prudential algebra: "It was to Dr. (Joseph) Priestly that Franklin imparted the well-known expedient which he called moral or prudential algebra. Priestly asked him in one of his letters how he contrived to make up his mind, when strong and numerous arguments presented themselves for both of two proposed lines of conduct? 'My way is,' replied Franklin, 'to divide half a sheet of paper by a line into two columns; writing over the one pro, and over the other con; then, during three or four days' consideration, I put down under the different heads short hints of the different motives that at different times occur to me, for or against the measure. When I have thus got them all together in one view, I endeavor to estimate the respective weights...[to] find at length where the balance lies. (I)f, after a day or two of farther (sic) consideration, nothing new that is of importance occurs on either side, I come to a determination accordingly.' He added that he derived great help from equations of this kind; which, at least, rendered him less liable to take rash steps." (Parton, 1864) Franklin describes just how he finds the balance, by striking off reasons that balance each other.

Franklin's approach constitutes a way of balancing or trading off objectives against each other by cancelling one or more arguments in favor of a proposition against some number of arguments against the proposition. What is implied in Franklins's moral algebra is an additive utility function.

The first modern treatment of multiple criteria decision making, though not widely viewed as such, is probably that of Herbert A. Simon (1958). His work on satisficing is very much a multiple criteria method. It involves setting targets or aspiration levels, and then searching until one finds a solution that achieves the target or aspiration levels. If many solutions exist, then the aspiration levels are tightened; if no solutions exist, then the aspiration levels are relaxed.

Charnes and Cooper's (1961) development of goal programming is one of the early modern technical approaches. It appears to build on Simon's work. Goal Programming is one of the best known approaches to MCDM. It works with linear programming models, and involves choosing targets or goals for each objective. It uses weights to construct an objective function that minimizes the "distance" from the target or goals from the actually achieved solution. In their work, Charnes and Cooper refer to the term vector optimization, a term that has since been widely accepted. See also Koopmans (1951).

A well-known approach of Kepner and Tregoe (1965) includes categorizing each objective as a must, a want, or an ignore objective. The must objectives are used as filters for eliminating alternatives; the ignore objectives are sufficiently unimportant that they are ignored. The want objectives are used for making choices among the alternatives that have survived the filter process. Alternatives are compared, according to the objectives, and a solution is taken according to a seven-step decision making procedure that uses an arbitrary additive value function. According to the Kepner and Tregoe book (ibid., p. 1), the approach has been widely and successfully used.

The methods Electre (see Roy, 1968 for an early article) was intended to structure the set of alternatives using constructs known as concordance and discordance. The idea is to provide a partial order of alternatives as a graph and then use the kernel of the graph as a set of outranking or "best" solutions. Several versions of the method were developed, and a number of applications have been reported.

The Step Method, (Benayoun et al., 1971), also known as STEM, is one of the earlier interactive methods. It identifies an ideal solution, and then finds the "closest" nondominated solution to the ideal solution. Then the user specifies which objectives he is willing to relax from the nearest solution, and then calculates a new ideal solution given the restrictions, and continues until the user is satisfied with the "closest" solution.

Enumeration of all nondominated solutions may be thought of as an outgrowth of the work on efficiency and vector optimization. Several researchers seriously investigated exploring all nondominated extreme point solutions. Evans and Steuer (1973) and Yu and Zeleny (1975) both followed this line of activity. The investigation of all nondominated extreme point solutions was not very successful; the number of such solutions is enormous. The method did lead to subsequent work, however, which has been much more successful.

Work of Geoffrion, Dyer, and Feinberg (1972) used a method of nonlinear programming to solve multiobjective linear programming problems. In the method the user is asked to provide tradeoffs between objectives that are used to establish weights. The weights are then used to solve a linear programming problem, and a binary search is then used to find the best solution between the old solution and a new solution. The process is then repeated until the method converges. The method is like a method of steepest ascent used to solve a multiple objective linear programming problem. An application is described in the paper.

An interactive approach for solving multiple objective linear programs was proposed by Zionts and Wallenius (1976). It assumes a linear utility function and generates a starting nondominated extreme point solution (the incumbent). The user is then asked to compare the incumbent with another specially defined nondominated solution point, an adjacent efficient extreme point. Based on the user's response, an inequality constraint on weights is generated. The constraint restricts the weights on the objectives to be consistent with a linear utility function. Then a new set of consistent weights is identified and a new nondominated solution is found, and the procedure is continued. The procedure converges to an optimal solution. The method is defined quite similarly for a more general class of problems.

B. Multiattribute Utility Theory (MAUT)

Among the early work on MAUT, the work of von Neumann and Morgenstern (1944) is classic. It did not explicitly address multiattribute problems, but implicitly one can argue that utility theory is by nature a multiattribute theory. It uses lotteries to assess utility functions, by having, at least in theory, a decision maker try to find lotteries equivalent to fixed outcomes. The lotteries typically involve two outcomes, the best and the worst possible overall. Most if not all of MAUT utilizes aspects of utility theory.

An additive value model for multiple objectives was proposed by Churchman and Ackoff (1954). It is a rather primitive model, and it is proposed with some variations in the article. To give some feel for how the fields overlap, the work by Kepner and Tregoe, though less mathematical, is probably just as much a primitive MAUT method as is the work of Churchman and Ackoff.

Ten years later Fishburn (1964) published a volume that explored multiattribute models in expected utility theory. The volume is oriented toward decision analysis, but it does have a multiattribute flavor.

Based on earlier work, Keeney and Raiffa (1976) produced a state-of-the-art presentation of the world of multiattribute utility theory that includes most of the areas considered in this paper. It includes both the theory of MAUT as well as many of the practical aspects of applying MAUT, such as the assessment of utilities and probabilities, the fitting of functions to data, some aspects of negotiation, and other relevant materials. It is a classic.

The Analytic Hierarchy Process (AHP) (Saaty, 1980) may be thought of as an MAUT approach, though some MAUT aficionados disagree. It involves an importance-ratio assessment procedure. It uses a hierarchy to establish preferences and orderings. A linear model is then derived and used to rank alternatives. Sensitivity analysis is possible by changing weights. The method is embodied in software known as Expert Choice (Forman et al., 1983).

Other work may be seen as common to both MCDM and MAUT. For example, the French school developed several approaches, including the Electre method discussed above that employ outranking relations (Roy, 1968) as well as methods that use piecewise linear approximations to utility functions (see, for example, Jacquet-Lagreze and Siskos, 1982).

C. Negotiation Theory

I view negotiations as a natural extension of MCDM and MAUT. Negotiation theory has its roots in several different areas. It may be viewed as a branch of game theory, or it may be viewed as related to utility theory, or it may be viewed as part of psychology, organizational behavior, or political science. As with multiattribute utility theory, the work of von Neumann and Morgenstern (1944) is very important to the area. Negotiation, as defined by Pruitt (1981), "is a process by which a joint decision is made by two or more parties". Each party usually presents a proposal that initiates negotiation. The parties then engage in what Raiffa (1982) calls a negotiation dance that progresses from initial proposals to a final agreement or stalemate.

Some early but more recent relevant references from an economic perspective are Cross (1965), Harsanyi (1965), and Schelling (1960). Contini and Zionts (1968) explore the impact of a known imposed solution in a simple bargaining scheme. An imposed solution will be enforced upon the disputants if no agreement can be reached. The imposed solution influences the results of the bargaining.

Much of the work is oriented toward helping the negotiators "find" the Pareto-optimal frontier. The process of finding the frontier involves squeezing out joint gains, or at least being sure that no unused gains are left on the table at the end of the negotiations. There is a difficulty in using "conventional" negotiations, followed by a phase in which joint gains are identified and then gleaned. Often the parties are sufficiently relieved when a satisfactory decision has been found that they do not want to negotiate further. They are pleased to have found a solution and don't want to risk losing what they have achieved.

In addition to the Raiffa (1982) reference, other contemporary references are Fisher and Ury (1981) and Lax and Sebenius (1986). These are particularly good references. The field of group decision and negotiation support has blossomed with research as well as application. In addition to software and methodology for helping to resolve negotiations, special hardware has been developed and used to assist in this task. Though perhaps not considered as hardware by some, decision rooms are a particularly impressive kind of hardware. See also DeSanctis and Gallupe (1987), Fraser and Hipel (1984), Jelassi, Kersten, and Zionts (1990), Nunamaker et al. (1988), Read and Gear (1989), and Shakun (1988).

IV. Where are we going?

Where are these fields going? There is no question that they are maturing. Work is progressing from theory to practice, and in some respects, vice versa. Methods are being used more in applications, and the microcomputer has made the use of the methods feasible for a multitude of users.

Researchers are always interested in productive areas for future research as well as areas that have been extensively mined. I'll return to some tradeoffs between the two. My opinion is that we should look at two levels: the macro level and the micro level.

On the macro level, I see two important problems:

1. Negotiation. Though much has been done and much has been written about negotiation, we still have a long way to go. Our models continue to be extremely primitive. Some think we may not be able to develop a theory of negotiation that has a

solid foundation. Such a foundation would help negotiators in disputes move toward the Pareto-optimal frontier. However, I am stubborn enough that I think we can develop such a theory. I think that in our current research we are trying to take steps that are simply too big. We should work on smaller tasks more thoroughly until we can proceed with the larger tasks. For example, it may be useful to develop protocols (or expert systems or decision support systems) for helping negotiators understand and structure their own positions, and to help them determine a utility function to be used in evaluating alternative decisions.

2. MCDSS methods and microcomputer implementations of MCDM methods. To apply methods applicable to MCDM, MAUT, and negotiations, it is extremely useful to have a decision support system (DSS). It is not difficult to understand the general concepts of a methodology and then be hand-held through the use of the method, as with a DSS, whereas without a DSS, the process is extremely cumbersome. The DSS carries out the dirty work of handling the steps of the procedure, performing the calculations and logic. Doing this without a DSS requires either intensive prior study of a method or someone to hold one's hand throughout the procedure.

Further, I think it is important enough to carry out the development of appropriate DSS's that the theory can be left to later. This may sound cavalier. Raiffa (1982) has stated that some of his concerns about MAUT, in particular, that certain assumptions be satisfied before a methodology be used, seem not very important in practice. For example, in order to use an additive utility function, it is theoretically necessary to determine that the attributes are mutually preferentially independent. However, even if the attributes are not mutually preferentially independent, an additive utility function may be used as an approximation. In such cases the function may yield incorrect results, but often it is more than adequate for most purposes. The theory can always be developed later, even though we may not be able to develop the precise theory we desire. We can instead develop something approximating the theory.

On the micro level, I see four aspects of research that I believe are very useful:

1. The use of a Tchebycheff or L_∞ norm as a proxy utility function. The Tchebycheff or L_∞ norm when maximized (more precisely, maximizing it as a quasiconcave function or minimizing it as a quasiconvex function) can generate any nondominated solution point. This need not be the case with other quasiconcave functions such as linear functions, and some higher order functions. Linear functions are effectively restricted to extreme point solutions, and cannot identify convex dominated solutions, and higher order functions may "miss" some nondominated or Pareto-optimal solutions.

2. Cone dominance. Cone dominance is a concept developed by Korhonen, Wallenius, and Zionts (1984) and independently by Hazen (1983). It is a way of using expressed preferences to eliminate alternatives. The cone structure is straightforward and rigorous, but there have been interesting extensions of it, with such ideas as using nonexistent or dummy solutions for comparison purposes. See, for example, Köksalan et al. (1984), Breslawski (1986) and Prasad (1990). What cone dominance does is to increase the information learned in the asking of preference questions of users.

3. Computer graphics in MCDSS. In multiple criteria decision support systems, there is increasing use of computer graphics. This is in the domain of so-called user interfaces, but the use of computer graphics in these systems has become very important. The value of graphics has added greatly to the impact of computers.

4. Eclectic Approaches. As the field continues to mature, the number of "new" ideas may diminish. Both on the macro level, as well as on the micro level, I expect that eclectic approaches will be more and more utilized. Rather than come up with new ideas, we will refine, repackage, and reuse old ideas.

Many academic fields are prone to fads. Someone gets an idea, espouses it, and then others are attracted to do research in the area. As new fads replace old, researchers are attracted to the new areas and leave the old. Are fads bad? I don't think so, as long as we don't get carried away by them, and flit from one area to another like a bee flying from flower to flower. Moreover, by tackling something new, I don't mean that we should abandon our other work. The value of the new work is promising, but there may be many valuable projects that may come from older projects that are worthwhile. There is nothing wrong with continuing the older projects. Though the ex ante expectations may not be so high, the ex post results may just be fine.

V. Some Sources of Research Topics

In addition to my observations regarding research topics, I believe that there are a number of sources of MCDM research topics. Some of these ideas would generally, I imagine, apply to any field. One idea is to look to related fields and to research with which you are not familiar. Though the idea of looking to related fields and to research with which one may not be familiar may seem obvious, it is amazing how resistant we are to doing that. It's a question of laziness, or inertia, or both. However, to try and overcome that, I try to include research with which I am not familiar in an MCDM course that we offer every two years, as well as have Ph. D. students working independently study and report on research with which I am not familiar. People who do research in different areas should be aware of research in other areas. Other ideas are to attend seminars and conferences in related areas, and to study work that may be similar sounding, yet quite different.

One such neighboring area that I have heard of, but have not done work in is Data Envelopment Analysis (DEA) (See Banker, Charnes, and Cooper, 1984). DEA is a way of comparing different organizations that operate with the same kinds of inputs and outputs. The organizations are compared by relating a weighted sum of outputs to inputs. By using a linear programming formulation derived from a linear fractional program, organizations are classified either as efficient or inefficient. A numerical measure of efficiency is found in the solution of the linear programming problem.

Some aspects of zero-based budgeting have elements in common with MCDM. The idea of zero-based budgeting is to remove every expenditure from an organization. Then every possible expenditure is ranked, from the most valuable to the least. The ranking is then used to choose among competing activities.

Some ideas of fuzzy set theory may prove useful in multiple criteria decision making. Though some such work has been done, I expect that much more remains to be done.

Another way of finding research of which one may not be aware is to look at the work of researchers in other countries. Even though there is quite a bit of cross fertilization among the researchers of different countries, not all work is universally known. To the contrary, much of the work is not known outside of the country. With the International Society for Multiple Criteria, the European Working Group on MCDM, ESIGMA, and the various other organizations, there is considerable interchange among researchers. Nonetheless, there is considerable work in other countries that many of us aren't aware of. Some good sources of this research are the respective national journals, and national and international meetings. Until recently, work done in Eastern Europe was not widely known in the west and vice versa. That was particularly true of developments in the Soviet Union. Developments in the east parallel those in the west. Because of limited cross fertilization of ideas, researchers may benefit by studying approaches developed in the Soviet Union. (See

Lieberman, 1990 and Ozernoy, 1988), as well as translations of Soviet journals.

Another way of stimulating research is to look to the areas of applications. What are important current problems, locally, nationally, and internationally? What problems are important to the individual and society? Some current problems of which I am aware include water resource management, solid and dangerous waste disposal, protection of the environment, consumption and replenishment of our natural resources, pollution of our planet, and so on. Yet another important problem is helping some of the newly opened countries of Eastern Europe and other rapidly developing countries to develop and interact with the rest of the world. Some of these countries have grave economic problems. All of these problems are complicated multiple criteria problems. Further, any one of the problems I identified may in fact consist of multiple problems. There may be grants available for the study of some of them through national and international agencies.

VI. One Approach to Doing MCDM Research

One approach that I like draws on the old adage, "Make new friends, but keep the old. One is silver; the other is gold." I try to continue on old projects, so long as it makes sense. On the other hand, I also try to work on new projects.

My current major activities include negotiation decision support. We have undertaken several projects at Buffalo, and I currently have two Ph. D. students S. Prasad and J. Teich who are completing dissertations. Though their topics are superficially similar, they are quite different. Prasad, who is working with M. Karwan, M. Sudit, and me, uses expanded dominance cones to eliminate alternatives and a network method to order the alternatives for each participant to the negotiation. He has measures of possible errors in the ordering process, and the method asks additional pairwise comparisons until a certain confidence in the ordering is achieved. He then combines the different orders to identify the Pareto-optimal solutions to the problem.

Teich uses a procedure to identify an approximate contract curve or set of Pareto-optimal solutions to the problem. He does this by asking users to respond to budget appropriation questions that involve assigning a fixed budget to competing activities. He reports the results of some experimental tests, and an application of the approach to a negotiation in Finland.

We are continuing our work on AIM, the Aspiration-Level Interactive Method for discrete MCDM. (See Lotfi, Stewart, and Zionts 1990.) It is an eclectic approach for solving discrete alternative multiple criteria decision problems, implemented on an IBM-compatible personal computer. In it, the user determines levels of aspiration for different objectives. The user is given considerable feedback as to the feasibility of his aspirations. The closest nondominated solution to the solution specified by the levels of aspiration is provided, as are other useful outputs. The method is easy to use and easy to understand.

Vahid Lotfi and I are working with a Ph. D. student, Yong-Seok Yoon, who recently defended his doctoral dissertation. Yoon's work (Yoon 1990), initially intended as an extension of AIM to continuous problems using linear programming, draws on the philosophy of AIM, and on the work of Steuer (1986). Yoon has the user choose among p + 1 solutions at each stage, where p is the number of objectives. Yoon reports favorable comparisons of his approach with others.

I am also working on a number of other projects, some that are new, and some that draw on earlier work I have done. I try to be eclectic and draw from others' ideas as well. At forums such as these meetings, I like to hear what others have to say, and draw from the ideas I like.

VII. Conclusion

In this paper I have overviewed the field of multiple criteria decision making and then presented a brief history of the field. I then forecasted something about the future, though it is always risky to speculate where a field is going. Macro areas that I highlighted as promising were negotiations, multiple criteria decision support systems and eclectic approaches. Micro areas that I feel are important are the use of the Tchebycheff or L_∞ norm, cone dominance, and computer graphics. Then I briefly outlined some research currently in progress. Though the future work I outlined as promising is a speculation on my part, I am confident that future developments in the field will be as exciting if not more exciting as past developments.

REFERENCES

Banker, R. D., A. Charnes, and W. W. Cooper, "Some Models for Estimating Technical and Scale Inefficiencies in Data Envelopment Analysis," Management Science, Vol. 30, No. 9, 1984, 1078-1092.

Benayoun, R., J. de Montgolfier, J. Tergny, and O. Larichev, "Linear Programming with Multiple Objective Functions: Step Method (STEM)," Mathematical Programming, 1, pp. 366-375, 1971.

Breslawski, S., Investigations in Multiple Objective Linear Programming, Doctoral Dissertation, School of Management, State University of New York at Buffalo, 1986.

Charnes, A., and W. W. Cooper, Management Models and Industrial Applications of Linear Programming, New York: John Wiley, 1961.

Churchman, C. W., and R. L. Ackoff, "An Approximate Measure of Value," Operations Research, 2, 1954, 172-187.

Contini, B. and Zionts, S., "Restricted Bargaining for Organizations with Multiple Objectives," Econometrica, 1968, Vol. 36 No. 2, pp. 397-414.

Cross, J., "A Theory of the Bargaining Process," American Economic Review, March, 1965, p. 67-72.

DeSanctis, G. and B. Gallupe (1987) "A Foundation for the Study of Group Decision Support Systems", Management Science, Vol. 33, No. 5, 589-609.

Evans, J. P. and R. E. Steuer, "Generating Efficient Extreme Points in Linear Multiple Objective Programming: Two Algorithms and Computing Experience," in Cochrane and Zeleny (1973), 349-365.

Fishburn, P. C., Decision and Value Theory, John Wiley, New York, 1964.

Fishburn, P. C., R. E. Steuer, J. Wallenius, and S. Zionts, "Multiple Criteria Decision Making; Multiattribute Utility Theory -The Next Ten Years", Working paper No. 747, School of Management, State University of New York at Buffalo, Buffalo, N. Y. 14260, April 1990.

Fisher, R., and W. Ury, Getting to Yes, Boston: Houghton Mifflin, 1981.

Forman, E. H., T. L. Saaty, M. N. Selly, and R. Waldron, Expert Choice, Decision Software, Inc., 1983.

Fraser, N. M., and K. W. Hipel (1984) Conflict Analysis: Models and Resolutions, New York, New York, North-Holland.

Geoffrion, A. M., J. S. Dyer, and A. Feinberg, "An Interactive Approach for Multicriterion Optimization with an Application to the Operation of an Academic Department," Management Science, 19, 1972, 357-368.

Harsanyi, J. C., "Bargaining and Conflict Situations in the Light of a New Approach to Game theory," American Economic Review Proceedings, May 1965.

Hazen, G. B., "Preference Convex Unanimity in Multiple Criteria Decision Making," Mathematics of Operations Research, Vol. 8, No. 4, 1983, pp. 505-516.

Jacquet-Lagrèze, E., and J. Siskos, "Assessing a Set of Additive Utility Functions for Multicriteria Decision-Making, The UTA Method", European Journal of Operations Research, 10, 2 (1982), 151-164.

Jelassi, T., G. Kersten, and S. Zionts, "An Introduction to Group Decision and Negotiation Support", in Bana e Costa, C. A. (ed.), Readings in Multiple Criteria Decision Aid, Springer-Verlag, Berlin, 1990 (Forthcoming).

Keeney, R. L., and H. Raiffa, Decisions with Multiple Objectives: Preferences and Value Tradeoffs, New York: John Wiley, 1976.

Kepner, C. H., and B. B. Tregoe, The Rational Manager: A Systematic Approach to Problem Solving and Decision Making, New York, McGraw-Hill, 1965.

Köksalan, M., Karwan, M. H., and Zionts, S., "An Improved Method for Solving Multiple Criteria Problems Involving Discrete Alternatives," IEEE Transactions on Systems, Man, and Cybernetics, Vol. 14, No. 1, January 1984, pp. 24-34.

Koopmans, T. C., "Analysis of Productionas an Efficient Combination of Activities." in Koopmans, T. C. (ed.), Activity Analysis of Production and Allocation, Cowles Commission Monograph No. 13 (New York: John Wiley and Sons, Inc., 1951).

Korhonen, P., J. Wallenius, and S. Zionts, "Solving the Discrete Multiple Criteria Problem Using Convex Cones," Management Science, 30, 1984, 1336-1345.

Lax, D. A., and J. K. Sebenius, The Manager as Negotiator, The Free Press, New York, 1986.

Lieberman, E. R., Multi-Objective Programming in the USSR, unpublished manuscript, State University of New York at Buffalo, School of Management, 1990. 444 pp.

Lotfi, V., T. Stewart, and S. Zionts, "An Aspiration-level Interactive Model for Multiple Criteria Decision Making," Unpublished working paper, Revised May 1990.

Nunamaker, J. F., L. M. Applegate, and B. R. Konsynski, "Computer-Aided Deliberation: Model Management and Group Decision Support," Operations Research, Vol. 36, No. 6, 1988, pp. 826-848.

Ozernoy, V., "Multiple Criteria Decision Making in the USSR: A Survey," Naval Research Logistics, Vol. 35, No. 6, 1988, pp. 543-566.

Parton, J., (1864). Life and Times of Benjamin Franklin, Vol. I, New York: Mason Brothers, p. 547.

Prasad, S., unpublished dissertation, Department of Industrial Engineering, State University of New York at Buffalo, 1990, in preparation.

Pruitt, D., Negotiation Behavior, New York: Academic Press, 1981.

Raiffa, H., The Art and Science of Negotiation, Cambridge: Harvard University Press, 1982.

Read, M. J., and A. E. Gear, "Interactive Group Decision Support", Paper presented at the IIASA International Workshop on MCDSS, Helsinki School of Economics, August 7-11, 1989.

Roy, B., "Classement et Choix en Présence de Points de Vue Multiples (La Méthode ELECTRE)," Revue d'Informatique et de Recherche Opérationelle, 8, pp. 57-75, 1968.

Saaty, T. L., The Analytic Hierarchy Process, New York: McGraw-Hill, 1980.

Schelling, T., The Strategy of Conflict, McGraw-Hill, New York, 1960.

Shakun, M., (1988) Evolutionary Systems Design, Oakland, CA, Holden-Day, Inc.

Simon, H. A., Administrative Behavior, MacMillan: New York, 1958.
Steuer, R. (1986) Multiple Criteria Optimization: Theory, Computation and Application, Wiley, New York.

von Neumann, J. and O. Morgenstern, Theory of Games and Economic Behavior, New York: John Wiley, 1944.

Yoon, Y.-S., An Interactive Aspiration-Level Multiple Objective Linear Programming Method: Theory and Computational Tests, Unpublished doctoral dissertation, School of Management, State University of New York, Buffalo, N. Y., 1990.

Yu, P. L. and M. Zeleny, "The Set of All Nondominated Solutions in the Linear Cases and a Multicriteria Simplex Method," Journal of Mathematical Analysis and Applications, 49, 1975, 430-468.

Zionts, S. and Wallenius, J., "An Interactive Programming Method for Solving the Multiple Criteria Problem," Management Science, 1976, Vol. 22, No. 6, pp. 652-663.

AN INTERACTIVE INTEGRATED MULTIOBJECTIVE OPTIMIZATION APPROACH FOR QUASICONCAVE / QUASICONVEX UTILITY FUNCTIONS

Jumah E. Al-alwani*
Ph.D. Candidate
Benjamin F. Hobbs*
Associate Professor
Behnam Malakooti*
Associate Professor
Department of Systems Engineering
Case Western Reserve University
Cleveland, Ohio 44106

Abstract

In this paper, a new interactive integrated approach for solving multiobjective optimization problems is presented. The approach is very flexible and general in that it handles two broad classes of implicit utility functions: quasiconcave and quasiconvex. The first step is to use preference comparison-based tests to determine the class of utility function that is consistent with the decision maker (DM's) underlying preferences with respect to a sample of nondominated alternatives. Then one of the imbedded algorithm that is appropriate for the selected utility function is used. In the case of quasiconcave utility, a modified Geofferion-Dyer-Feinberg algorithm is developed by projecting the gradient-based improvement direction on the nondominated frontier as well as providing an interactive termination criterion. The quasiconvex utility-based algorithms are briefly discussed due to the space limitations. The demands upon the DM are kept to a minimum in the sense that only paired comparisons of alternatives and tradeoff evaluations are elicited. Example application is presented for the quasiconcave utility -based algorithm.

1. Introduction

In this paper, an interactive integrated approach for solving general multiobjective optimization problems is presented. The multiobjective optimization problem (MOP) is expressed as follows, MOP: maximize $\{f_1(\mathbf{x}), f_2(\mathbf{x}), \ldots\ldots, f_k(\mathbf{x})\}$

$$\text{s. t. } \mathbf{x} \in X = \{\mathbf{x} \in R^n \mid g_i(\mathbf{x}) \leq 0,\ i = 1, \ldots, m\},$$

where $\mathbf{x}$ is the set of the decisions variables, $f_1(\mathbf{x}), f_2(\mathbf{x}), \ldots\ldots, f_k(\mathbf{x})$ are the objective functions and $g_1(\mathbf{x}), g_2(\mathbf{x}), \ldots\ldots, g_m(\mathbf{x})$ are the problem constraints. The approach has the unique feature of not only accommodating a decision maker (DM) with a quasiconcave utility function but also those that may exhibit a quasiconvex utility preference structure. This

*Names are in alphabetical order.

important feature is not present in any previous multiobjective methodology. The approach's imbedded interactive algorithms is able to handle the nonlinear MOP case. It also takes into consideration certain properties of nondominated alternatives which may arise due to nonlinearities in the feasible set. Further, the imbedded algorithms minimize the burden of the preference information demanded at each iteration of the solution process.

In the next section, the general framework of the integrated approach is outlined together with utility class testing theory. The quasiconcave utility-based interactive algorithm is discussed in detail in Section 3. A brief summary of the quasiconvex utility-based interactive algorithms is provided in Section 4. Conclusions are presented in Section 5.

2. The Interactive Integrated Approach

In this section, the integrated approach is presented. It considers the decision maker that might exhibit either a quasiconcave or quasiconvex preference structure behavior and provides a default step in case that the underlying preference is not consistent with any of these classes. The underlying utility function U(.), which is implicitly known only to the decision maker, is assumed to be a nondecreasing continuous and differentiable Quasiconcave or Quasiconvex utility function on a convex set V that contains the feasible set X. The decision maker would be able to provide local preference information, based on his/her implicit utility function, such as a paired comparisons between two nondominated alternatives and simple tradeoff questions. The objectives are continuous and differentiable functions of the decision variables. The constraint set is compact (i. e., closed and bounded) with continuous and differentiable function constraints. Next, definitions for quasiconcave as well as quasiconvex functions are presented. In the next section, the steps of the integrated approach are outlined. The utility class testing theory is discussed in subsection 2.2.

Definition

Let U:V-->R, where V is a nonempty convex set in R^k. The function U(.) is said to be quasiconcave if, for each $\mathbf{f}^i$ and $\mathbf{f}^j \in$ V, the following inequality is true :

$$U[\alpha\mathbf{f}^i + (1-\alpha)\mathbf{f}^j] \geq \text{minimum } \{U(\mathbf{f}^i),U(\mathbf{f}^j)\} \text{ for each } \alpha \in (0,1).$$

Definition

Let U:V-->R, where V is a nonempty convex set in R^k. The function U(.) is said to be quasiconvex if, for each $\mathbf{f}^i$ and $\mathbf{f}^j \in$ V, the following inequality is true :

$$U[\alpha\mathbf{f}^i + (1-\alpha)\mathbf{f}^j] \leq \text{maximum } \{U(\mathbf{f}^i),U(\mathbf{f}^j)\} \text{ for each } \alpha \in (0,1).$$

2.1 The Steps of the Approach

The approach includes four main steps. The first step includes two tasks. The first task is to generate a representative sample of (e.g. 2k+2) nondominated alternatives. This task can be accomplished using the Lambda and Filter routines (Steuer 1986). However, the second task requires interaction with the DM. The interaction results in a complete pairwise-comparison preference information set. The second step of the approach checks whether the set constructed above is consistent with the class of the quasiconcave utility functions or not. If it is, the quasiconcave-based algorithm is used to solve the problem. If it is not, the third step of the approach is invoked. This step checks whether the set, constructed in the first step, is consistent with the class of the quasiconvex utility functions or not. If it is consistent, the quasiconvex-based algorithm that is appropriate to the structure of the MOP is used to solve the problem. If the set is not consistent with the quasiconcave utility class or the quasiconvex utility class, the fourth step of the approach is used. This step is a default which includes two tasks. In the first task, the quasiconcave-based algorithm is executed for several times with a different starting point at each time and their final alternatives are accumulated in a list. In the second task, all of the alternatives that were accumulated in the list are presented to the DM for evaluation. He/she identifies the most-preferred alternative for the whole problem. Next, the main steps of the approach are presented.

Step 1 Initialization

(i) Generate a representative sample of the nondominated frontier.

(ii) Construct a complete pairwise-comparison preference set.

Step 2 Check for the Quasiconcave Utility Class

(i) Test for the Quasiconcave utility class.

(ii) If the test is positive execute the quasiconcave-based algorithm, else go to step 3.

Step 3 Check for the Quasiconvex Utility Class

(i) Test for the Quasiconvex utility class.

(ii) If the test positive, execute the appropriate quasiconvex-based algorithm, else go to step 4.

Step 4 Default Case

(i) Execute the quasiconcave-based algorithm for several times with a different starting point each time and accumulate their final alternatives in a list.

(ii) Present this list to the decision maker to select the most-preferred alternative.

2.2 Utility Class Testing Theory

The concept of testing for the form of the utility function was suggested by Korhonen, Moskowitz & Wallenius (KMW) (1986), where necessary tests for detecting linearity as well as quasiconcavity of the utility function were provided with respect to a pairwise preference information of a set of nondominated alternatives. In this subsection, some elements of our testing theory for two broad classes of utility functions are presented (see Al-alwani 1990 for details). The testing theory which is an extension of the KMW's concept includes mathematical programs that provide diagnostic tests which are necessary as well as sufficient for two classes of functions in the sense they test for quasiconcavity as well as quasiconvexity of the decision maker's underlying utility function. Furthermore, in contrast to the KMW's test in which indifferent responses were not used, those responses will be utilized as well.

Let Z be a subset of nondominated alternatives generated from the nondominated frontier. Also let the set P represent the pairwise preference data, where P Z ZxZ and the set P is defined as:

$P = \{<f^i, f^j> \mid f^i, f^j \in Z, f^i$ is preferred or indifferent to f^j, $i \neq j$, $i,j = 1,\ldots\ldots,z\}$.

From the above set P, we identify all subsets that have a common least-preferred alternative as well as a common most-preferred one. These subsets can be defined as:

$PL^j = \{<f^i, f^j> \mid f^i$ is preferred or indifferent to f^j, $i \neq j$, for all $i\}$ $j = 1,\ldots\ldots,z$.

$PM^i = \{<f^i, f^j> \mid f^i$ is preferred or indifferent to f^j, $i \neq j$, for all $j\}$ $i = 1,\ldots\ldots,z$.

Next, definition 2.1 provides a condition for a pairwise-comparison preference set P to be consistent with a certain class of utility functions. Based on this definition as well as properties of the quasiconcave and quasiconvex functions, a necessary mathematical programming-based tests (programs 2.1 and 2.2) are derived for the two classes respectively. Each test utilizes a different grouping of the elements of the preference set P.

Definition 2.1

Let U* be a class of utility functions. A pairwise-comparison preference set P is U*-consistent if and only if there exists a utility function $U \in U^*$ such that $U(f^i) \geq U(f^j)$ for all $<f^i, f^j> \in P$, $i \neq j$.

Lemma 2.1

Let U(.) be a quasiconcave function on a convex set V Z R^n. Let $f^i \in V$, $i = 1,\ldots..,m$. If $U(f^i) \geq U(f^j)$ for all i and some j, $i \neq j$, then there exists a set of multipliers $\lambda = (\lambda_1,\ldots\ldots.,\lambda_k)$, $\lambda \geq 0$, $\Sigma^k_{r=1}\lambda_r = 1$ such that, for all i, $\lambda^t(f^i - f^j) \geq 0$, $i = 1,\ldots..,m$, $i \neq j$.

Proof (by construction)

Since U(.) is a quasiconcave function on V Z R^n and $f^i \in V$, $U(f^i) \geq U(f^j)$ implies that $\nabla U(f^j)(f^i - f^j) \geq 0$. Now, set $\lambda = \nabla U(f^j)/[\Sigma^k_{r=1}\partial U(f^j)/\partial f_r]$, where $\lambda \geq 0$ and $\Sigma^k_{r=1}\lambda_r = 1$. Then $\lambda^t(f^i - f^j) \geq 0$, $i = 1,\ldots..,m$, $i \neq j$, Q.E.D.

Theorem 2.1

Let U_{qv} be the class of quasiconcave utility functions and $U \in U_{qv}$. Given a pairwise-comparison preference set P, if P is U_{qv}-consistent then $\varepsilon^* \geq 0$ in the following program,

Program 2.1: ε^* = Maximize ε

s. t. $(\lambda^j)^t(f^i - f^j) \geq \varepsilon$ for all $<f^i,f^j> \in P$, $i, j = 1,.....z$, $i \neq j$.

$\lambda^j \geq 0$, $\Sigma^k_{r=1}\lambda^j_r = 1$, for all $PL^j \neq \emptyset$, $j = 1,........,z$.

Proof (by contradiction)

Let $\varepsilon^* < 0$ in the above program. Since the set P is U_{qv}-consistent then there exists $U \in U_{qv}$ such that $U(f^i) \geq U(f^j)$ for all $<f^i,f^j> \in P$. From the definition of the subsets PL^j and by Lemma 2.1, there exists a set $\lambda^j = (\lambda^j_1,.......,\lambda^j_k)$, $\lambda^j \geq 0$ such that $(\lambda^j)^t(f^i-f^j) \geq 0$ for all $<f^i,f^j> \in PL^j$, $PL^j \neq \emptyset$, for all j, $j = 1,.......,z$, $i \neq j$. (1)

Now in the above program, $\varepsilon^* = \text{minimum}_{j=1,.....,z} \{(\lambda^j)^t(f^i-f^j)\}$. This implies that there exist a j such that $(\lambda^j)^t(f^i-f^j) = \varepsilon^* < 0$, which contradicts (1). Therefore $\varepsilon^* \geq 0$, Q.E.D.

Lemma 2.2

Let U(.) be a quasiconvex function on a convex set $V \; Z \; R^n$. Let $f^j \in V$, $j = 1,....,m$. If $U(f^i) \geq U(f^j)$ for all j and some i, $i \neq j$, then there exists a set of multipliers $\lambda = (\lambda_1,......,\lambda_k)$, $\lambda \geq 0$, $\Sigma^k_{r=1}\lambda_r = 1$ such that, for all j, $\lambda^t(f^i-f^j) \geq 0$, $j = 1,.....,m$, $i \neq j$, Q.E.D.

Proof (by construction)

Since U(.) is quasiconvex function on $V \; Z \; R^n$ and $f^j \in V$, $U(f^i) \geq U(f^j)$ implies that $\nabla U(f^i)(f^i-f^j) \geq 0$. Now, set $\lambda = \nabla U(f^i)/[\Sigma^k_{r=1}\partial U(f^i)/\partial f_r]$, where $\lambda \geq 0$ and $\Sigma^k_{r=1}\lambda_r = 1$. Then $\lambda^t(f^i-f^j) \geq 0$, $j = 1,..,m$, $i \neq j$.

Theorem 2.2

Let U_{qx} be the class of quasiconvex utility functions and $U \in U_{qx}$. Given a pairwise-comparison preference set P, if P is U_{qx}-consistent then $\varepsilon^* \geq 0$ in the following program,

Program 2.2: ε^* = Maximize ε

s. t. $(\lambda^i)^t(f^i - f^j) \geq \varepsilon$ for all $<f^i,f^j> \in P$, $i, j = 1,.....z$, $i \neq j$.

$\lambda^i \geq 0$, $\Sigma^k_{r=1}l^i_r = 1$, for all $PM^i \neq \emptyset$, $i = 1,........,z$.

Proof (by contradiction)

Let $\varepsilon^* < 0$ in the above program. Since the set P is U_{qx}-consistent then there exists an $U \in U_{qx}$ such that $U(f^i) \geq U(f^j)$ for all $<f^i,f^j> \in P$. From the definition of the subsets PM^i and by Lemma 2.2, there exists a set $\lambda^i = (\lambda^i_1,......,\lambda^i_k)$, $\lambda^i \geq 0$ such that $(\lambda^i)^t(f^i-f^j) \geq 0$ for all $<f^i,f^j> \in PM^i$, $PM^i \neq \emptyset$, for all i, $i =1,.......,z$, $i \neq j$. (2)

Now in the above program, $\varepsilon^* = \text{minimum}_{i=1,.....,z}\{(\lambda^i)^t(f^i-f^j)\}$. This implies that there exists an i such that $(\lambda^i)^t(f^i-f^j) = \varepsilon^* < 0$, which contradicts (2). Therefore $\varepsilon^* \geq 0$, Q.E.D.

In summary, the above utility class tests are executed by generating a representative set Z of the nondominated frontier first using a weighted Tchebycheff-based scheme (Steuer 1988). Then, through interaction with the DM, the preference set P is constructed. Finally, using the above two groupings PL^j and PM^i testing programs 2.1 and 2.2 provide a sufficient condition for the set P of nondominated alternatives to be <u>inconsistent</u> with the quasiconcave and quasiconvex classes, respectively.

3. The Quasiconcave Utility-based Interactive Multiobjective Algorithm

Several multiple criteria decision making (MCDM) approaches have been developed for the pseudoconcave and quasiconcave utility function. The pseudoconcave utility-based methods include the method of Korhonen and Laakso (1986). However, the quasiconcave utility-based methods include heuristic as well as exact methods for discrete MCDM (Malakooti 1988a, 1989 and 1990) and multiobjective linear programming (MOLP) (Malakooti 1988b). In this section, an interactive multiobjective nonlinear programming algorithm for the quasiconcave utility function is presented. The developed interactive algorithm builds on the Frank & Wolf (1956) (FW) as well as the Geofferion-Dyer-Feinberg (1972) (GDF) algorithms. It differs from the GDF approach in that it does not perform a one dimensional line search. Rather, it searches a nondominated curve. This process will eliminate the possibility of providing the DM with dominated outcomes. It also differs in that it provides an interactive tradeoff-based termination criterion. This will give the decision maker the control over the final outcome of the process. Further, although our algorithm is similar to the method of Korhonen & Laakso (1986) in the sense that a certain direction is projected on the nondominated frontier, we believe that the gradient-based improvement direction is easier to obtain from the DM's point of view. This stems from the fact that it is based on eliciting local paired-comparison preference information rather than a global reference point estimate. Experiments revealed that asking the decision maker to provide point estimates is a difficult question (Wallenius 1975). The algorithm includes five main steps. The first step (Step 0) is the initialization of the algorithm, where any of the previous solutions can be used. The next step includes assessing the gradient of the utility function as well as finding an improving direction based one the assessed gradient. In the third step, a sample of points along the direction is generated and projected on the nondominated frontier. The projection process is accomplished using a weighted Tchebycheff or lexicographic-based scheme. The fourth and the fifth step include the search process and checking for termination. Next, the steps of the quasiconcave utility-based algorithm are outlined. Because of the space limitation, only the interaction termination criterion (ITC) is discussed in subsection 3.2 (see Al-alwani 1990 for the details of the algorithm).

3.1 The Steps of the Algorithm

Step 0 : (Initialization)

(a) Generate an initial efficient solution. Denote it $\mathbf{x}^1$.

(b) Set i = 1.

Step 1 : (Direction Finding)

(a) Assess DM's utility function gradient at the $\mathbf{f}(\mathbf{x}^i)$ using paired comparison-based scheme.

(b) Determine the point $\mathbf{f}(\mathbf{y}^i)$ such that $\mathbf{y}^i$ solves the following direction problem:

$$\text{maximize } \nabla_{\mathbf{x}} U(f_1(\mathbf{x}^i), f_2(\mathbf{x}^i), \ldots\ldots, f_k(\mathbf{x}^i)).\mathbf{y}$$

$$\text{s. t. } \quad \mathbf{y} \in X.$$

Set $\mathbf{d}^i = \mathbf{f}(\mathbf{y}^i) - \mathbf{f}(\mathbf{x}^i)$

Step 2 : (Projection)

Generate $\mathbf{M}^i$ nondominated alternatives along a nondominated curve by projecting the gradient-based improvement direction $\mathbf{d}^i$ on the nondominated frontier, where $\mathbf{M}^i$ is predefined. This can be done using a weighted Tchebycheff or lexicographic-based scheme.

Step 3 : (One-dimensional Curve Search)

Interact with the decision maker to select the best among the generated $(\mathbf{M}^i\mathbf{+2})$ alternatives; denote its associated efficient solution as $\mathbf{y}^*$.

Step 4 : (Check for Termination)

(a) **IF** $\mathbf{f}(\mathbf{y}^*)$ is the same as $\mathbf{f}(\mathbf{x}^i)$, then check the interactive termination criterion. If the conditions are satisfied declare $\mathbf{f}(\mathbf{x}^i)$ as the most-preferred alternative and $\mathbf{x}^i$ as the most-preferred decision, STOP; Otherwise, determine any alternative that is better than $\mathbf{f}(\mathbf{x}^i)$, call it $\mathbf{f}(\mathbf{y}^*)$ and go to (b). **ELSE** go to (b).

(b) set $\mathbf{x}^{i+1} = \mathbf{y}^*$, $i = i + 1$, go to Step 1.

3.2 The Interactive Termination Criterion (ITC)

In order to insure that decision maker has the control over the conclusion of the solution process, an interactive termination criterion for the solution process is clearly needed. The desired criterion ought not only to avoid asking for unreasonable amounts of preference information but it should also be able to provide a new improvement direction. The concept of tradeoff in the objective space will play a key role in our termination method. Since every tradeoff in the objective space is induced by a direction in the decision space, there are advantages to working in the decision space because it is already constructed through its function constraints. This will be crucial when the model is linearly constrained.

Let the set $I(\mathbf{x}^*) = \{1,.....,r\}$ be the set of binding constraints at point $\mathbf{x}^*$ (i.e. $g_i(\mathbf{x}^*) = 0$); then, the set of the feasible directions $D(\mathbf{x}^*)$ is contained in the cone $C(\mathbf{x}^*) = \{ \mathbf{d} \mid \nabla g_i(\mathbf{x}^*)\mathbf{d} \leq 0,\ i \in I(\mathbf{x}^*)\}$ (Bazaraa & Shetty 1979). Since the cone $C(\mathbf{x}^*)$ is a convex polyhedral cone with a vertex $\mathbf{x}^*$, it can be represented using its minimal generators and can be written as follows:

$$C(\mathbf{x}^*) = \{ \mathbf{x} \mid \mathbf{x} = \mathbf{x}^* + \Sigma^q_{i=1} \mu_i \mathbf{d}^i ,\ \mu_i \geq 0 \},$$

where $\mathbf{d}^i$, $i = 1,...., q$ are generation directions. Next, theorem 3.1 establishes a sufficient condition for point $\mathbf{x}^*$ to be a global optimal solution using the enclosing cone $C(\mathbf{x}^*)$, assuming the utility function "U(.)" is quasiconcave on the convex constraint.

Theorem 3.1

Let the feasible region X be convex, the utility function $U'(\mathbf{x}) = U(\mathbf{f}(\mathbf{x}))$ be quasiconcave on X and $\mathbf{x}^*$ be an efficient point. Furthermore, consider the cone $C(\mathbf{x}^*) = \{ \mathbf{x} \mid \mathbf{x} = \mathbf{x}^* + \Sigma^q_{i=1} \mu_i \mathbf{d}^i,\ \mu_i \geq 0 \}$. If $\nabla U'(\mathbf{x}^*)\mathbf{d}^i \leq 0$ for all $\mathbf{d}^i$, $i = 1,....,q$ with at least one strict inequality (<), then $\mathbf{x}^*$ is a strict global optimal solution.

Proof (by contradiction)

It is sufficient to show that $\mathbf{x}^*$ is a strict local optimum (i.e. $U'(\mathbf{x}^*) > U'(\mathbf{x})$ for all $\mathbf{x}$ in the neighborhood of $\mathbf{x}^*$) (Avriel 1976). Therefore for the contradiction, let $\mathbf{x}'$ be a feasible point such that $U'(\mathbf{x}') \geq U'(\mathbf{x}^*)$. Since $\mathbf{x}'$ is a feasible point, then $\mathbf{d}^* = (\mathbf{x}'-\mathbf{x}^*)$ is a feasible direction of the convex set X. This implies $\mathbf{d}^* \in D(\mathbf{x}^*)$ and hence $\mathbf{d}^* \in C(\mathbf{x}^*)$. Therefore it could be expressed as a linear combination of C's generators. Then there exist $\mu_i \geq 0$ such that $\mathbf{d}^* = \Sigma^q_{i=1}\mu_i\mathbf{d}^i$. Therefore $\nabla U'(\mathbf{x}^*)\mathbf{d}^* = \nabla U'(\mathbf{x}^*)[\Sigma^q_{i=1}\mu_i\mathbf{d}^i] = \mu_1\nabla U'(\mathbf{x}^*)\mathbf{d}^1 + ... + \mu_q\nabla U'(\mathbf{x}^*)\mathbf{d}^q < 0$ since there at least one "<" (1). Now, since $U'(\mathbf{x})$ is quasiconcave on X, $U'(\mathbf{x}') \geq U'(\mathbf{x}^*)$ implies $\nabla U'(\mathbf{x}^*)(\mathbf{x}'-\mathbf{x}^*) \geq 0$, therefore $\nabla U'(\mathbf{x}^*)\mathbf{d}^* \geq 0$ contradicting (1). Thus, $\mathbf{x}^*$ is a strict local optimal solution and therefore $\mathbf{x}^*$ is also a strict global optimal solution, Q.E.D.

The main issue in the above generators-based representation of the enclosing cone $C(\mathbf{x}^*)$ is the efficient construction of the generation directions $\mathbf{d}^i$, $i = 1,........,q$. For this reason, we introduce the concept of the enclosing polyhedral set next. Theorem 3.2 provides a way of constructing an enclosing set .

Definition 3.1

A set $X^* = \{A\mathbf{x} \leq \mathbf{b},\ \mathbf{x} \geq \mathbf{0}\}$ is an enclosing polyhedral set of the set $X = \{ \mathbf{x} \mid g_i(\mathbf{x}) \leq 0,\ i = 1,......,m \}$ if X is a subset of X^*.

Theorem 3.2

Let $\mathbf{x}^* \in$ boundary(X) and $X^* = \{\mathbf{x} \mid g_i(\mathbf{x}^*) + \nabla g_i(\mathbf{x}^*)(\mathbf{x}' - \mathbf{x}^*) \leq 0, i \in I(\mathbf{x}^*), \mathbf{x} \geq \mathbf{0}\}$. Then $X^* \supset X$.

Proof

Since $\mathbf{x}^* \in$ boundary(X), the hyperplane $g^*_i(\mathbf{x}) = g_i(\mathbf{x}^*) + \nabla U(\mathbf{x}^*)(\mathbf{x} - \mathbf{x}^*)$ supports X at $\mathbf{x}^*$ for each $i \in I(\mathbf{x}^*)$. Then $\nabla g_i(\mathbf{x}^*)(\mathbf{x} - \mathbf{x}^*) \leq 0$ for all $\mathbf{x} \in X$. Since $g_i(\mathbf{x}^*) = 0$ then $g^*_i(\mathbf{x}) = g_i(\mathbf{x}^*) + \nabla U(\mathbf{x}^*)(\mathbf{x} - \mathbf{x}^*) \leq 0$ for all $\mathbf{x} \in X$. Now consider the set $X^i = \{ \mathbf{x} \mid g_i(\mathbf{x}^*) + \nabla U(\mathbf{x}^*)(\mathbf{x} - \mathbf{x}^*) \leq 0\}$. Therefore $\mathbf{x} \in X$ implies $\mathbf{x}^* \in X^i$ for each $i \in I(\mathbf{x}^*)$, and hence $\mathbf{x} \in X$ implies $\mathbf{x}^* \in \cap^r_{i=1} X^i = X^*$. Then $X^* \supset X$, Q.E D.

The advantage of working with the set X* is that the number of constraints is much smaller that the ones for the set X and hence the computational saving will be substantial. Furthermore, the enclosing set X* should be constructed even though the set X is a polytope. This is due to the fact that we need only binding constraints at the point of interest. The crucial idea needed to characterize to the generator directions is the concept of extreme directions (Bazaraa & Shetty 1979). Let the set of decision variables at an extreme point $\mathbf{x}^* \in X^*$ be partitioned into two subsets, dependent (basic) variables subset $\mathbf{y}^*$ and independent (nonbasic) variables subset $\mathbf{z}^*$, such that matrix A is partitioned into two submatrices B and N where B is an rxr nonsingular matrix and N is an rx(n-r) matrix. Therefore $\mathbf{x}^* = (\mathbf{y}^*,\mathbf{z}^*)^t$, where the basic variables $\mathbf{y}^* = (y^*_1, y^*_2, \ldots\ldots, y^*_r)^t$ and the nonbasic variables $\mathbf{z}^* = (z^*_1, z^*_2, \ldots\ldots, z^*_{n-r})^t$. Based on this partition, the extreme direction $\mathbf{d}^j$ is constructed for each nonbasic variable z_j, $j = 1,\ldots\ldots, n-r$, and is expressed as follows,

$$\mathbf{d}^j = \begin{vmatrix} -\mathbf{B}^{-1}\mathbf{N}_j \\ \cdots\cdots \\ \mathbf{I}_j \end{vmatrix} \quad \text{for each } z^*_j \in \mathbf{z}. \qquad \text{Eq. (1)}$$

However in the case that $\mathbf{x}^* \in X^*$ is not an extreme point, some of the nonbasic variables values will be positive and therefore each nonbasic variable with a positive value can be increased or decreased in order to generate a feasible direction in the decision space (a feasible tradeoff in the objective space) (Malakooti 1988b). This implies that for each such nonbasic variable, two directions opposite to each other, should be considered. Therefore in order to construct the desired set of generators, let $s = n-r$ and let the sets $\mathbf{z}^{*+} = \{z_1,\ldots\ldots,z_p\}$ and $\mathbf{z}^{*o} = \{z_{p+1},\ldots\ldots,z_s\}$ represent the sets of positive and zero nonbasic variables respectively. Then the generators set $\{\mathbf{d}^1, \mathbf{d}^2,\ldots\ldots, \mathbf{d}^p, \mathbf{d}^{p+1},\ldots\ldots, \mathbf{d}^s, \mathbf{d}^{s+1}, \ldots\ldots, \mathbf{d}^q\}$ follows, where $q = s+p$.

$$\mathbf{d}^j = \begin{vmatrix} -\mathbf{B}^{-1}\mathbf{N}_j \\ \cdots\cdots\cdots \end{vmatrix} \quad \text{for each } z^*_j \in \mathbf{z}^*.$$

$$\text{and} \qquad \mathbf{d}^{s+j} = \begin{vmatrix} \\ -\mathbf{d}^{j} \\ \mathbf{I}_j \end{vmatrix} \qquad \text{for each } z^*_j \in \mathbf{z}^{*+}.$$

Because of the construction of the above directions in the sense that the directions in two sets $\{\mathbf{d}^1,\mathbf{d}^2,.......,\mathbf{d}^p\}$ and $\{\mathbf{d}^{s+1},......,\mathbf{d}^{s+p}\}$ are opposite to each other, the cone $C(\mathbf{x}^*)$ can better be represented using the set $\{\mathbf{d}^1, \mathbf{d}^2,.......,\mathbf{d}^p, \mathbf{d}^{p+1},......,\mathbf{d}^s\}$ and can be expressed as follows,

$$C(\mathbf{x}^*) = \{ \mathbf{x} \mid \mathbf{x} = \mathbf{x}^* + \Sigma^p_{j=1} \mu_j \mathbf{d}^j + \Sigma^s_{j=P+1} \mu_j \mathbf{d}^j,\ \mu_j \geq 0, j = P+1,......,s \}$$

where the μ_i, $i = 1,.....,p$ are unrestricted. In this representation, only "s" directions are used. The above set of directions are sufficient to span the enclosing cone $C(\mathbf{x}^*)$ that contains the feasible directions set $D(\mathbf{x}^*)$ (see Al-alwani 1990 for details). However, in the case that the multiobjective model is linearly constrained, those directions are also feasible directions $D(\mathbf{x}^*)$ and consequently, every direction spanned by them is feasible.

Tradeoffs in the objective space is determined with respect each independent (nonbasic) variables z_j. This can be done using the expression of the reduced gradient for the objective functions at the point $\mathbf{x}^* = (\mathbf{y}^*,\mathbf{z}^*)$,

$$\mathbf{t}^j = \nabla_{zj}\mathbf{f}(\mathbf{y}^*,\mathbf{z}^*) - \nabla_{\mathbf{y}}\mathbf{f}(\mathbf{y}^*,\mathbf{z}^*)B^{-1}N_j \qquad \text{Eq. (2)}$$

Next, we establish the relationship between the above directions and their associated tradeoffs through the utility gradient in the decision space as well as in the objective space. This relation is important to transform DM's tradeoff responses to the decision space where termination is checked.

Lemma 3.1

Let $\mathbf{x}^* \in X$ be a convex set. Consider the convex cone $C(\mathbf{x}^*) = \{ \mathbf{x} \mid \mathbf{x} = \mathbf{x}^* + \Sigma^p_{j=1} \mu_j \mathbf{d}^j + \Sigma^s_{j=P+1} \mu_j \mathbf{d}^j,\ \mu_j \geq 0, j = P+1, ..., s \}$. Let the directions $\mathbf{d}^j$, $j=1,......., s$ be characterized as follows:

$$\mathbf{d}^j = \begin{vmatrix} -\mathbf{B}^{-1}\mathbf{N}_j \\ .... \\ \mathbf{I}_j \end{vmatrix} \qquad \text{for each } z_j \in \mathbf{z}.$$

Then $\nabla_{\mathbf{x}}U(\mathbf{f}(\mathbf{x}^*))\mathbf{d}^j = \nabla_{\mathbf{f}}U(\mathbf{f}(\mathbf{x}^*))\mathbf{t}^j$ for all $j = 1,........, s$.

Proof

Consider the directions $\mathbf{d}^j$, $j = 1,......., s$. $\nabla_{\mathbf{x}}U(\mathbf{f}(\mathbf{x}^*))\mathbf{d}^j = \nabla_{\mathbf{f}}U(\mathbf{f}(\mathbf{x}^*)\nabla_{\mathbf{x}}\mathbf{f}(\mathbf{x}^*)\mathbf{d}^j = \nabla_{\mathbf{f}}U(\mathbf{f}(\mathbf{x}^*)\nabla_{\mathbf{y},\mathbf{z}}\mathbf{f}(\mathbf{y}^*,\mathbf{z}^*)\mathbf{d}^j$ by the chain rule of calculus. Let $C_{\mathbf{y}^*} = \nabla_{\mathbf{y}}\mathbf{f}(\mathbf{y}^*,\mathbf{z}^*)$ and $C_{\mathbf{z}^*} = \nabla_{\mathbf{z}}\mathbf{f}(\mathbf{y}^*,\mathbf{z}^*)$, then $\nabla_{\mathbf{x}}U(\mathbf{f}(\mathbf{x}^*))\mathbf{d}^j = \nabla_{\mathbf{f}}U(\mathbf{f}(\mathbf{x}^*)[\ C_{\mathbf{y}^*} | C_{\mathbf{z}^*}\ ]\mathbf{d}^j = \nabla_{\mathbf{f}}U(\mathbf{f}(\mathbf{x}^*)\ [\ -C_{\mathbf{y}^*}\mathbf{B}^{-1}\mathbf{N}_j + I_j C_{\mathbf{z}^*}\ ] = \nabla_{\mathbf{f}}U(\mathbf{f}(\mathbf{x}^*)\mathbf{t}^j$, $j = 1,....., s$, Q.E.D.

Definition 3.2

(1) $\nabla_f U(\mathbf{f}(\mathbf{x}^*))\mathbf{t}^* \leq 0$ if $\mathbf{t}^*$ is unpreferred (i.e. DM either declines or is indifferent to $\mathbf{t}^*$)

(2) $\nabla_f U(\mathbf{f}(\mathbf{x}^*))\mathbf{t}^* < 0$ if the decision maker declines tradeoff $\mathbf{t}^*$.

(3) $\nabla_f U(\mathbf{f}(\mathbf{x}^*))\mathbf{t}^* = 0$ if the decision maker is indifferent to tradeoff $\mathbf{t}^*$.

Theorem 3.3

Let the utility function $U'(\mathbf{x}) = U(\mathbf{f}(\mathbf{x}))$ be quasiconcave on the convex feasible set X and $\mathbf{x}^*$ be an efficient point. If (1) the decision maker is indifferent to every tradeoff of the set $T^{*+} = \{\mathbf{t}^1, \mathbf{t}^2, \ldots\ldots, \mathbf{t}^p\}$ and (2) every tradeoff of the set of $T^{*0} = \{\mathbf{t}^{p+1}, \ldots\ldots, \mathbf{t}^s\}$ is unpreferred with at least one of them being rejected, then $\mathbf{x}^*$ is the most-preferred solution.

Proof

Since the decision maker is indifferent to each one of the tradeoffs $\{\mathbf{t}^1, \mathbf{t}^2, \ldots\ldots, \mathbf{t}^p\}$ then, by definition 3.1, $\nabla_f U(\mathbf{f}(\mathbf{x}^*))\mathbf{t}^j = 0$ for all j = 1, ..., p and therefore $\nabla_\mathbf{x} U'(\mathbf{x}^*)\mathbf{d}^j = \nabla_f U(\mathbf{f}(\mathbf{x}^*))\mathbf{t}^j = 0$ by lemma 3.1. It is also true that for the opposite directions $\mathbf{d}^{s+j} = -\mathbf{d}^j$, i = 1,......., p, $\nabla_\mathbf{x} U'(\mathbf{x}^*)\mathbf{d}^{s+j} = 0$. Furthermore, since each one of the tradeoffs $\{\mathbf{t}^{p+1}, \ldots\ldots, \mathbf{t}^s\}$ is unpreferred with one of them being rejected then $\nabla_f U(\mathbf{f}(\mathbf{x}^*))\mathbf{t}^j \leq 0$ for all j = p+1, ..., s with one of the inequalities being strict. Thus, by lemma 3.1, $\nabla_\mathbf{x} U'(\mathbf{x}^*)\mathbf{d}^j = \nabla_f U(\mathbf{f}(\mathbf{x}^*))\mathbf{t}^j \leq 0$, with one being strict inequality. Finally, since the set $\{\mathbf{d}^1, \ldots, \mathbf{d}^p, \mathbf{d}^{p+1}, \ldots, \mathbf{d}^s, \mathbf{d}^{s+1}, \ldots, \mathbf{d}^{s+p}\}$ span the cone $C(\mathbf{x}^*)$, then $\mathbf{x}^*$ is optimal by theorem 3.1. Hence $\mathbf{x}^*$ is the most-preferred solution, Q.E.D.

The execution of the interactive termination criterion (ITC) at a certain solution, say $\mathbf{x}^*$, includes several tasks. First, the enclosing set X^* as defined by theorem 3.2 is constructed. Then, the decision variables at $\mathbf{x}^*$ are partitioned into basics and nonbasics, where some of the nonbasics might be positive. Using the nonbasic variables, the associated extreme directions are generated via Eq. (1). Using Eq. (2), tradeoffs induced by these directions are generated. Tradeoffs associated with positive nonbasic variables are efficient tradeoffs (Malakooti 1988b), wheres the ones associated with zero nonbasic variables might not be efficient. Therefore, the next task is to screen all of the zero nonbasic tradeoffs for efficiency. This process can accomplished using a tradeoff efficiency routine (Malakooti 1988b). Finally, all of the resulted efficient tradeoffs are presented to the decision maker for evaluation.

Next, an example to illustrate the steps of the algorithm is presented.

Example 3.1

$$\text{maximize } \{ f_1 = x_1 , f_2 = x_2 , f_3 = x_3 \}$$
$$\text{s. t.} \quad x_1^2 + x_2^2 + x_3^2 \leq 225$$
$$x_i \geq 0 \quad \text{for } i = 1,2,3.$$

Assume that the implicit DM's utility function

$$U(f_1,f_2,f_3) = -(f_1-16)^2-(f_2-16)^2-(f_3-16)^2$$

Solution: Initialization

The ideal point $\mathbf{f}^* = (f_1^*,f_2^*,f_3^*) = (15,15,15)$. Assuming the ε vector is (1,1,1), the Utopian point $\mathbf{f}^{**} = \mathbf{f}^* + \varepsilon$ is (16,16,16). Let the initial solution and alternative be $\mathbf{x}^1$ = (15,0,0) and $\mathbf{f}(\mathbf{x}^1)$ = (15,0,0). Set i = 1 .

Iteration(1)

Step(1): $\nabla U(\mathbf{f}) = 2(1,16,16)$, the direction finding problem is, maximize $y_1 + 16y_2 + 16y_3$ s.t. $\mathbf{y} \in X$. Using GINO (Liebman, Scharge, Lasdon & Waren 1986), $\mathbf{y}^1$ = (0.66,10.60,10.59)' = $\mathbf{f}(\mathbf{y}^1)$.

Step(2): generate three (M^1=3) convex combination points of $\mathbf{f}(\mathbf{x}^1)$ and $\mathbf{f}(\mathbf{y}^1)$. These are (11.42,2.65,2.65)', (7.83,5.30,5.30)' and (4.25,7.95 ,7.44)'. To project these points using the weighted Tchebycheff-based scheme, their associated Tchebycheff weights are calculated. So, $w_i = [1/(f_i^{**}-f'_i)]/[\Sigma_{i=1}(1/(f_i^{**}-f'_i))]$, for i = 1,2,3, where f_i^{**} and f'_i are the ith level of the Utopia point and generated point on the line respectively. For this iteration, we have the following table:

point on the direction line	projection	utility value	
(15.00,0.00,0.00)	(15.00,0.00,0.00)	-513.00	$\mathbf{f}(\mathbf{x}^1)$
(11.42,2.65,2.65)	(12.51,5.89,5.89)	-216.60	A
(7.83,5.30,5.30)	(9.88,7.98,7.98)	-166.10	B*
(4.25,7.95,7.95)	(6.57,9.54,9.53)	-172.52	C
(0.66,10.60,10.59)	(0.66,10.60,10.59)	-294.07	$\mathbf{f}(\mathbf{y}^1)$

Step(3): Decision maker prefers alternative "B".

Iteration(2): Assuming $\mathbf{M}^2 = 2$, the following table results:

point on the direction line	projection	utility value	
(9.88,7.98,7.98)	(9.88,7.98,7.98)	-166.10	$\mathbf{f}(\mathbf{x}^2)$
(8.96,8.43,8.43)	(9.01,8.48,8.48)	-161.96	D *
(8.04,8.88,8.88)	(8.10,8.93,8.93)	-162.38	E
(7.12,9.33,9.33)	(7.12,9.33,9.33)	-167.83	$\mathbf{f}(\mathbf{y}^2)$

Decision maker prefers alternative "D".

Iteration(3): Assumes $M^3 = 1$, the following table results:

point on the direction line	projection	utility value	
(9.01,8.48,8.48)	(9.01,8.48,8.48)	-161.96	$\mathbf{f}(\mathbf{x}^3)$
(8.63,8.67,8.67)	(8.63,8.67,8.67)	-161.77	F *
(8.25,8.86,8.86)	(8.25,8.86,8.86)	-162.02	$f(\mathbf{y}^3)$

Decision maker prefers alternative "F".

Iteration(4): repeating the process again using $\mathbf{f}(\mathbf{x}^4) = (8.63, 8.67, 8.67)$, the direction finding problem yields the same point. This implies that this point is a candidate for checking the optimality condition. Assuming the termination criterion is satisfied, the solution process is concluded with the most-preferred alternative as (8.63,8.67,8.67) and the most-preferred decision as (8.63, 8.67, 8.67) for the whole problem assuming the utility function U(.) is quasiconcave. In order to see how close this alternative to the true one, the simulated utility function is explicitly maximized using GINO (Liebman, Scharge, Lasdon & Waren 1986). It turns out that the utility of the final alternative given by our algorithm is 99.99% of the utility of the true final alternative given by GINO.

4. Quasiconvex Utility-based Multiobjective Algorithms: A Brief Summary

In this section, we assume that the decision maker's underlying preference is represented by a quasiconvex utility function. It was not, until recently, that this class of functions has been suggested to be viable (Malakooti 1990). For instance, certain classes of risk seeking multiattribute utility functions are convex (Keeney & Raiffa 1976), (Harrison & Rosenthal 1988). Furthermore, convex as well as quasiconvex single utility functions have been used to represent DM's preference (Duckstein, Bobee & Ashkar 990).

The quasiconvex utility-based algorithms depend on the structure of constraint set of the model under consideration. It turns out that in the case of the multiobjective linear programming, the most-preferred solution occurs at an extreme point of the feasible set. The developed interactive MOLP algorithm guarantees a global optimal solution, involves only finite number of pivoting operations and requires only paired comparison-based questions (see Al-alwani 1990 or Al-alwani & Malakooti 1990 for details). Through a branch and bound framework, the algorithm extends the tree of solutions at each iteration. At each iteration, a subset of solutions is examined by the decision maker, through their associated alternatives, to decide on the one that is most-preferred. Having identified the most-preferred alternative among the subset, its associated terminal tree node is declared for the next expansion. Unless the termination criterion is satisfied, the algorithm iterates by executing the expansion process.

In the case of multiobjective nonlinear programming, it turns out that, under mild conditions such as the utility function being strict quasiconvex or having a positive gradient everywhere, the most-preferred solution is a supported or proper alternative, respectively. Due to the these results, a heuristic algorithm can be designed, (see Al-alwani 1990 for details). It uses the fact that the simple scalarization mechanism of the weighted objectives can be used in this case. Interestingly enough, the newly generated alternative at the end of each iteration through this scheme is guaranteed to have a higher utility value than the one at its beginning unless the scheme returns the same alternative. The needed weights is provided by assessing the gradient of the utility function.

5. Conclusions

The developed integrated approach exhibits the following appealing features. First, it is flexible, since two general classes of utility functions will be handled. The least restrictive form of generalized concavity or convexity (i.e. quasiconcave or quasiconvex) is assumed to represent the DM's preference. This feature has not been provided by any previous research. We believe it would be an effective concept for enhancing the overall DM's satisfaction of the final recommended solution. It is general, since it handles nonlinear multiobjective optimization problems. Furthermore, it could be extended to handle multiobjective integer as well as discrete multicriteria decision problems cases. Also, it is easy to use, since every algorithm of the approach includes easy steps. This feature is necessary in order for users (especially naive ones) to understand the underlying problem-solving process. Consequently, they are likely to be more satisfied with it. It places minimal cognitive load, since the user will be faced only with pairwise preference questions as well as simple tradeoff questions. These type of questions are considered to be easy relative to other types such as marginal rate of substitution and reference values. Consequently, the cognitive burden would be minimal. Finally, all of alternatives presented to the DM during the process are nondominated and consequently the final solution must also be nondominated.

References

Al-alwani, J. E., Multiobjective Systems: An Interactive Integrated Optimization Approach, Ph.D. Dissertation, Department of Systems Engineering, Case Western Reserve University, Cleveland, Ohio, forthcoming August, 1990.

Al-alwani, J. E. and Hobbs, B. F. and Malakooti, B., "A Quasiconcave Utility-based Interactive Multiobjective Optimization Approach", Department of Systems Engineering, Case Western Reserve University, Cleveland, Ohio, working paper #1-90, 1990.

Al-alwani, J. E. and Malakooti, B., "A Quasiconvex Utility-based Interactive Multiobjective Optimization Approach", Department of Systems Engineering, Case Western Reserve University, Cleveland, Ohio, working paper #2-90, 1990.

Avriel, M., Nonlinear Programming: Analysis and Methods, Prentice-Hall, Englewood Cliffs, NJ, 1976.

Bazaraa, M. and Shetty, A., Nonlinear Programming: Theory and Algorithms, Wiley, New York, 1979.

Duckstein, L., Bobee, B. and Ashkar, F., "Mulitcriterion Choice of Estimation Techniques for Fitting Extreme Floods", in Tabucanon, M. and Chankong, V. (Eds.), MCDM: Applications to Industry and Service; Asian Institute of Technology, Bangkok, Tailand, 1989, 1069-1083.

Dyer, S., "The effects of errors in the estimation of the gradient on the Frank-Wolf's algorithm, with application to interactive programming", Operations Research, Vol. 22, pp. 160-174, 1974.

Frank, M. and Wolf, P."An Algorithm for Quadratic Programming", Naval Research Logistic Quarterly, Vol. 3, 1956.

Geofferion, A. M., Dyer, S. J. and Feinberg, A.,"An Interactive Approach for Multicriterion Optimization with An Application to The Operation of An Academic Department", Management Science, Vol. 19, No. 4, pp. 357-368, 1972.

Keeney, R. L. and Raffia, H., Decisions with Multiple Objectives: Preferences and Value Tradeoffs, John Wiley & Sons, New York, 1976.

Korhonen, P. and Laakso, K., "A Visual Interactive Method for Solving the Multiple Criteria Problem", European Journal of Operational Research, Vol. 24, No. 2, pp. 277-287, 1986.

Korhonen, P., Moskowitz, H. and Wallenius, J.,"A Progressive Algorithm for Modelling and Solving Multiple Criteria Decision problems", Operations Research, Vol. 34, No. 5, pp. 726-731, 1986.

Liebman, J., Scharge, L., Lasdon, L. and Waren, A., Modeling and Optimization with GINO, Scientific Press, Palo Alto, CA., 1986.

Malakooti, B., "A Decision Support System and Heuristic Interactive Approach for solving Discrete Multiple Criteria Problems", IEEE Transaction on Systems, Man and Cybernetics, Vol. 18, No. 2, pp. 787-801, 1988a.

Malakooti, B., "An Exact Interactive Method for Exploring the Efficient Facets of Multiple Objective Linear Programming Problems with Quasiconcave Utility Functions", IEEE Transaction on Systems, Man and Cybernetics, Vol. 18, No. 5, pp. 787-801, 1988b.

Malakooti, B., "Theories and An Exact Interactive Paired-Comparison Approach for Discrete Multiple-Criteria Problems", IEEE Transaction on Systems, Man and Cybernetics, Vol. 19, No. 2, pp. 787-801, 1989.

Malakooti, B., "Ranking Multiple Criteria Alternatives with Half-Space, Convex and Nonconvex Dominating Cones : Quasiconcave and Quasiconvex Multiple Attribute Utility Functions", Computers and Operations Research, Vol. 16, No. 2, pp. 117-127, 1990.

Steuer, R. E., Multiple Criteria Optimization: Theory, Computation, and Application, John Wiley & Sons, 1986.

Wallenius, J., "Comparative Evaluation of Some Interactive Approaches to Multicriteria Optimization", Management Science, Vol. 21, No. 12, pp. 326-344, 1975.

THE ANALYTIC HIERARCHY PROCESS WITH INTERVAL JUDGMENTS

Ami Arbel
Tel-Aviv University
Tel-Aviv, Israel

Luis G. Vargas
University of Pittsburgh
314 Mervis Hall. Pittsburgh, PA 15260. USA.

ABSTRACT

We show how the AHP can be extended to consider interval judgments. The approach is based on a set of optimization models that are solved to obtain the upper and lower bounds of the components of the eigenvector. The method is developed for both single reciprocal matrices and a hierarchic structure. The validity of the approach is inferred from the convergence of the models to the traditional AHP approach when the measure of the interval judgments converges to zero.

1. Introduction

The Analytic Hierarchy Process (AHP) [3, 4, 5] is based on the concept of paired comparison. It is a multicriteria, multiobjective, multilevel and multiparty approach to decision making, and unlike Multiattribute Utility Theory it can easily handle several levels of complexity structured either hierarchically or in the form of a network.

When paired comparisons (or judgments) are elicited in the AHP they are point estimates, because although time is always involved, these judgments are not usually represented as a function of it. Instead, time is dealt with by partitioning the time horizon it into several stages, and eliciting judgments for each of them. The uncertainty surrounding some decisions can be taken into account in this manner. However, for each stage, decision makers could still be unsured of their judgments and hence they may be unwilling to provide point judgments. At this stage, it is more likely that a range of values (interval) of the judgments could be more appealing and capture their uncertainty in preferences. As the process evolves they will sharpen their judgments and point judgments may be acceptable.

For example, consider R&D projects among which we must distribute limited resources. When a project is funded, we are not always sure if it is going to be successful. The validity of our judgments can be better established when the project is reviewed some time later. The problem is that some of the projects may be too costly, and one may not be able to afford a mistake. If this is the case, how does one decide which project to implement? One way could be to use benefit/cost ratios. However, what should one do if in addition the information about both the benefits and the costs is also uncertain? Should one use a probability distribution along with the benefits and the costs and make the decision based on the average benefit/cost ratio? Again, what if one cannot afford to make decisions on the average because a single mistake is too costly and we do not have the necessary resources?

There are two types of uncertainties [6] encountered in decision making: uncertainty about the consequences of our actions (*environmental uncertainty*) and uncertainty about the information or knowledge one has of a problem and its consequences (*judgmental uncertainty*). Utility theory attempts to deal with our perception of environmental uncertainty. Fuzzy set theory attempts to deal with judgmental uncertainty. The problem with both theories is that they do not reproduce known scales in situations for which scales have been constructed. All they provide is a tool to capture the preferences of decision makers. But, do they? The problems with Utility Theory have already been discussed elsewhere [2]. Fuzzy set, on the other hand, is too abstract and removed from reality. In other words, we must be able to reproduce known results obtained with other scales before using a given measurement theory.

The Analytic Hierarchy Process with interval judgments (AHPIJ) is an extension of the AHP developed to address the problem of judgmental uncertainty. The judgments are now intervals rather than single numbers. When the interval degenerates into a point, we obtain the traditional AHP approach [5].

In the AHP a matrix of paired comparisons is given by:

$$A = \{a_{ij} \,|\, a_{ji} = a_{ij}^{-1},\ a_{ij} > 0\}.$$

A ratio scale is recovered from this matrix by solving the eigenvalue problem $A\mathbf{w}=\lambda\mathbf{w}$ for the right eigenvector associated with the principal eigenvalue λ_{max} of A (see [3], [4]). When the entries of the matrix are intervals, then the coefficients of $A\mathbf{w}=\lambda\mathbf{w}$ are intervals and the traditional approach is not applicable.

We found two studies of this problem in the literature: a simulation approach [6] and a linear programming approach [1]. The simulation approach is always applicable, while the LP approach can only be used when A is a consistent matrix, i.e., $a_{ij}\, a_{jk} = a_{ik}$, for all i, j and k.

In this paper we extend the approach developed by Arbel [1] to deal with inconsistent matrices. The problem is formulated as a set of optimization models whose solutions provide the ranges of variation of the weights of the alternatives. We also extend the approach to hierarchic structures where all the judgments are of the interval type.

2. Reciprocal Matrices with Interval Judgments

A reciprocal matrix with interval judgments is a matrix $A=(I_{ij})$ whose entries satisfy:

(i) $I_{ij}= \{x| l_{ij} \leq x \leq u_{ij},\}$, and
(ii) For all $x \in I_{ij}$ and $y \in I_{ji}$, then $xy = 1$.

The problem is to find the range of values of each component of the principal right eigenvector. Let $I(\mathbf{w})$ the set of all principal right eigenvectors corresponding to the matrices constructed with values in the intervals I_{ij}, i.e.,

$$I(\mathbf{w}) = \{\mathbf{w}|A\mathbf{w}=\lambda_{max}\mathbf{w},\ A=(a_{ij}),\ a_{ij} \in I_{ij}\}.$$

In [6] we studied this problem using simulation. Arbel [1] has solved this problem using linear programming but only in the case in which the matrices are consistent.. He considers a consistent matrix $A = (w_i/w_j)$, and formulates the following problem:

$$\begin{aligned} &\text{Min } w_0 \\ \text{s.t.} \quad & \\ &\Sigma w_i = 1 \\ &l_{ij} \leq \frac{w_i}{w_j} \leq u_{ij},\ i,j = 1, 2, ..., n \\ &w_i > 0, \text{ for all } i \end{aligned} \tag{1}$$

or

$$\begin{aligned} &\text{Min } w_0 \\ \text{s.t.} \quad & \\ &\Sigma w_i = 1 \\ &l_{ij}w_j - w_i \leq 0,\ i,j = 1, 2, ..., n \\ &- u_{ij}w_j + w_i \leq 0,\ i,j = 1, 2, ..., n \\ &w_i > 0, \text{ for all } i. \end{aligned} \tag{2}$$

The solutions of (2) are the eigenvectors of the consistent matrices in terms of which all other consistent matrices can be generated. The following two results follow from this formulation [1]:

(a). If all vertices of the solution space exhibit the same rank order, then any interior point exhibits the same rank order.

(b). Let $\{\mathbf{w}_1, ..., \mathbf{w}_q\}$ be the solutions of (2). If we generate points $\alpha_1\mathbf{w}_1 + ... + \alpha_q\mathbf{w}_q$ by selecting $\alpha_1, \alpha_2, ..., \alpha_q$ as a random partition of [0,1], and hence $\Sigma\alpha_i = 1$, the point generated converges to the arithmetic mean of the points as the number of sampled points increases.

Example

Consider the following matrix of paired comparisons:

$$\begin{bmatrix} 1 & [2,5] & [2,4] & [1,3] \\ & 1 & [1,3] & [1,2] \\ & & 1 & [\frac{1}{2},1] \\ & & & 1 \end{bmatrix}$$

The LP-Approach [1]:

The solutions of (2) are convex combinations of the following six vertices:

$$\mathbf{w}_1^T = (0.5217, 0.1739, 0.1304, 0.1739)$$
$$\mathbf{w}_2^T = (0.5000, 0.1667, 0.1667, 0.1667)$$
$$\mathbf{w}_3^T = (0.4000, 0.2000, 0.2000, 0.2000)$$
$$\mathbf{w}_4^T = (0.4444, 0.2222, 0.1111, 0.2222)$$
$$\mathbf{w}_5^T = (0.4800, 0.2400, 0.1200, 0.1600)$$
$$\mathbf{w}_6^T = (0.4615, 0.2308, 0.1538, 0.1538)$$

and the average of these vertices is given by:

$$\bar{\mathbf{w}}^T = (0.4679, 0.2056, 0.1470, 0.1794).$$

The Simulation Approach [6]:

The simulation results for a sample of size n=1000 are given by:

Minimum	Average	Maximum
0.3694	0.4702	0.5517
0.1501	0.2138	0.2895
0.0929	0.1318	0.1891
0.1332	0.1842	0.2600

Note that (2) does not have a feasible solution for some inconsistent matrices, because the inconsistency may yield an eigenvector for which (w_i/w_j) does not belong to the interval I_{ij}.

3. Preference Programming

To deal with inconsistent matrices we develop a more general model of which (2) will be a particular case. We first reformulate (2) by adding two new constraints:

$$\sum_{j=1}^{n} x_{ij} w_j = \lambda w_i, \; i=1, 2, ..., n$$
$$x_{ij} x_{ji} = 1, \; i,j = 1, 2, ..., n.$$

The extreme points of the feasible region are given by the solutions of a set of optimization problems. We need two optimization problems for each component:

$$\begin{aligned} & \text{Min } w_i \\ \text{s.t.,} \quad & \Sigma w_i = 1 \\ & l_{ij} w_j - w_i \leq 0, \; i,j = 1, 2, ..., n \\ & -u_{ij} w_j + w_i \leq 0, \; i,j = 1, 2, ..., n \\ & \sum_{j=1}^{n} x_{ij} w_j = \lambda w_i, \; i=1, 2, ..., n \\ & x_{ij} x_{ji} = 1, \; i,j = 1, 2, ..., n \\ & w_i > 0, \text{ for all } i. \end{aligned} \tag{3}$$

and

$$\begin{aligned} & \text{Max } w_i \\ \text{s.t.,} \quad & \Sigma w_i = 1 \\ & l_{ij} w_j - w_i \leq 0, \; i,j = 1, 2, ..., n \\ & -u_{ij} w_j + w_i \leq 0, \; i,j = 1, 2, ..., n \\ & \sum_{j=1}^{n} x_{ij} w_j = \lambda w_i, \; i=1, 2, ..., n \\ & x_{ij} x_{ji} = 1, \; i,j = 1, 2, ..., n \\ & w_i > 0, \text{ for all } i. \end{aligned} \tag{4}$$

In the example, we have:

$$0.4000 \leq w_1 \leq 0.5217$$
$$0.1667 \leq w_2 \leq 0.2400$$
$$0.1111 \leq w_3 \leq 0.2000$$
$$0.1538 \leq w_4 \leq 0.2222$$

where w_1 attains its maximum and minimum at vertices $\mathbf{w}_1$ and $\mathbf{w}_3$, respectively; w_2 at vertices $\mathbf{w}_5$ and $\mathbf{w}_2$, respectively; w_3 at vertices $\mathbf{w}_3$ and $\mathbf{w}_4$, respectively; and w_4 at vertices $\mathbf{w}_4$ and $\mathbf{w}_6$, respectively.

We state without proof the following result.

Theorem 1: Problems (3) and (4) are equivalent to Problem (2).

Problems (3) and (4) only provide solutions for which the ratios (w_i/w_j) belong to the interval judgments. Eliminating this restriction we have:

$$\begin{aligned} &\text{Min } w_i,\ i = 1,2,\ldots,n \\ &\text{s.t.} \\ &\quad \Sigma w_i = 1 \\ &\quad l_{ij} \le x_{ij} \le u_{ij},\ i,j = 1, 2, \ldots, n \\ &\quad \sum_{j=1}^{n} x_{ij}\, w_j = \lambda\, w_i,\ i=1, 2, \ldots, n \\ &\quad x_{ij}\, x_{ji} = 1,\ i,j = 1, 2, \ldots, n \\ &\quad w_i > 0, \text{ for all } i. \end{aligned} \tag{5}$$

and

$$\begin{aligned} &\text{Max } w_i,\ i=1,2,\ldots,n \\ &\text{s.t.} \\ &\quad \Sigma w_i = 1 \\ &\quad l_{ij} \le x_{ij} \le u_{ij},\ i,j = 1, 2, \ldots, n \\ &\quad \sum_{j=1}^{n} x_{ij}\, w_j = \lambda\, w_i,\ i=1, 2, \ldots, n \\ &\quad x_{ij}\, x_{ji} = 1,\ i,j = 1, 2, \ldots, n \\ &\quad w_i > 0, \text{ for all } i. \end{aligned} \tag{6}$$

whose solutions provide the upper and lower bounds of the components of $\mathbf{w}^T = (w_1, \ldots, w_n)$.

Theorem 2: A feasible solution of (3) (or (4)) is also a feasible solution of (5) (or (6)).

Proof: Substituting $x_{ij} = w_i/w_j$ in (5) (or (6)), we obtain (3) (or (4)) and the result follows.

For the example given above, the solutions of (5) are given by:

Table 1
Lower Bounds of the Eigenvector Components

Min w_1 — Eigenvector

$$\begin{bmatrix} 1 & 2 & 2 & 1 \\ & 1 & 1 & 1 \\ & & 1 & \frac{1}{2} \\ & & & 1 \end{bmatrix} \begin{bmatrix} 0.3383 \\ 0.2046 \\ 0.1692 \\ 0.2879 \end{bmatrix}$$

Min w_2 — Eigenvector

$$\begin{bmatrix} 1 & 5 & 4 & 3 \\ & 1 & 1 & 1 \\ & & 1 & 1 \\ & & & 1 \end{bmatrix} \begin{bmatrix} 0.5667 \\ 0.1358 \\ 0.1429 \\ 0.1546 \end{bmatrix}$$

Min w_3 — Eigenvector

$$\begin{bmatrix} 1 & 5 & 4 & 1 \\ & 1 & 3 & 2 \\ & & 1 & \frac{1}{2} \\ & & & 1 \end{bmatrix} \begin{bmatrix} 0.4758 \\ 0.2229 \\ 0.0856 \\ 0.2157 \end{bmatrix}$$

Min w_4 — Eigenvector

$$\begin{bmatrix} 1 & 5 & 4 & 3 \\ & 1 & 3 & 2 \\ & & 1 & 1 \\ & & & 1 \end{bmatrix} \begin{bmatrix} 0.5656 \\ 0.2079 \\ 0.1036 \\ 0.1229 \end{bmatrix}$$

and the solutions of (6) are given by:

Table 2
Upper Bounds of the Eigenvector Components

Max w_1 — Eigenvector

$$\begin{bmatrix} 1 & 5 & 4 & 3 \\ & 1 & 1.55 & 1.61 \\ & & 1 & 1 \\ & & & 1 \end{bmatrix} \begin{bmatrix} 0.5686 \\ 0.1703 \\ 0.1253 \\ 0.1358 \end{bmatrix}$$

Max w_2 — Eigenvector

$$\begin{bmatrix} 1 & 2 & 2 & 1 \\ & 1 & 3 & 2 \\ & & 1 & 1 \\ & & & 1 \end{bmatrix} \begin{bmatrix} 0.3435 \\ 0.3111 \\ 0.1463 \\ 0.1990 \end{bmatrix}$$

Max w_3 — Eigenvector

$$\begin{bmatrix} 1 & 2 & 2 & 1 \\ & 1 & 1 & 1 \\ & & 1 & 1 \\ & & & 1 \end{bmatrix} \begin{bmatrix} 0.3465 \\ 0.2036 \\ 0.2036 \\ 0.2463 \end{bmatrix}$$

Max w_4 — Eigenvector

$$\begin{bmatrix} 1 & 2 & 2 & 1 \\ & 1 & 1 & 1 \\ & & 1 & \frac{1}{2} \\ & & & 1 \end{bmatrix} \begin{bmatrix} 0.3383 \\ 0.2046 \\ 0.1692 \\ 0.2879 \end{bmatrix}.$$

Summarizing, the intervals of variation of the components of the eigenvector are given by:

$$0.3383 \le w_1 \le 0.5686$$
$$0.1358 \le w_2 \le 0.3111$$
$$0.0856 \le w_3 \le 0.2036$$
$$0.1229 \le w_4 \le 0.2879$$

4. Preference Programming in Hierarchic Structures

Let us now formulate the problem for a hierarchic structure. We assume that all judgments are of the interval type . The objective is to determine the range of variation of the weights for the elements at the bottom of the hierarchy. There are four types of constraints in this problem:

(eigenvalue constraints)

$$\text{(i)} \ \sum_{j=1}^{n_h} x_{hij}^{(k)} w_{hj}^{(k)} = \lambda w_{hi}^{(k)}, \ i=1, 2, ..., n_h, \ h=1,2,...,L, \ k=1,2,...,n_{h-1}$$

where $x_{hij}^{(k)}$ is the pairwise comparison of the elements i and j of the h*th* level with respect to the k*th* element of the (h-1)*st* level.

(reciprocity constraints)

$$\text{(ii)} \ x_{hij}^{(k)} x_{hji}^{(k)} = 1, \ i,j=1,2,...,n_h, \ h=1,2,...,L, \ k=1,2,...,n_{h-1}$$

(interval judgments constraints)

$$\text{(iii)} \ l_{hij}^{(k)} \le x_{hij}^{(k)} \le u_{hij}^{(k)}, \ i,j=1,2,...,n_h, \ h=1,2,...,L, \ k=1,2,...,n_{h-1}$$

(Hierarchic Composition constraints)

$$\text{(iv)} \ W_L W_{L-1} \cdots W_1 W_0 = \mathbf{w}$$

where $W_h = (w_{hj}^{(k)})$, h=1,2,...,L, are matrices whose columns are the eigenvectors of the elements of the h*th* level with respect to the elements of the (h-1)*st* level, h=1,2,...,L, respectively; W_0 is the vector of weights of the elements at the second level with respect to the elements at the first level, and $\mathbf{w}^T = (w_1, ..., w_n)$ is the vector of weights of the elements at the bottom level of the hierarchy. The $2n_L$ optimization models are given by:

$$\text{Min(Max) } w_s,\ s=1,2,\ldots,n_L$$

$$\text{s.t.,}\quad \sum_{j=1}^{n_h} x_{hij}^{(k)} w_{hj}^{(k)} = \lambda\, w_{hi}^{(k)},\ i=1,2,\ldots,n_h,\ h=1,2,\ldots,L,\ k=1,2,\ldots,n_{h-1}$$

$$l_{hij}^{(k)} \leq x_{hij}^{(k)} \leq u_{hij}^{(k)},\ i,j=1,2,\ldots,n_h,\ h=1,2,\ldots,L,\ k=1,2,\ldots,n_{h-1} \qquad (7)$$

$$x_{hij}^{(k)} x_{hji}^{(k)} = 1,\ i,j=1,2,\ldots,n_h,\ h=1,2,\ldots,L,\ k=1,2,\ldots,n_{h-1}$$

$$W_L W_{L-1} \cdots W_1 W_0 = \mathbf{w},\ \mathbf{w} > 0$$

Consider the hierarchy of Figure 1 and the interval judgments given in Table 3.

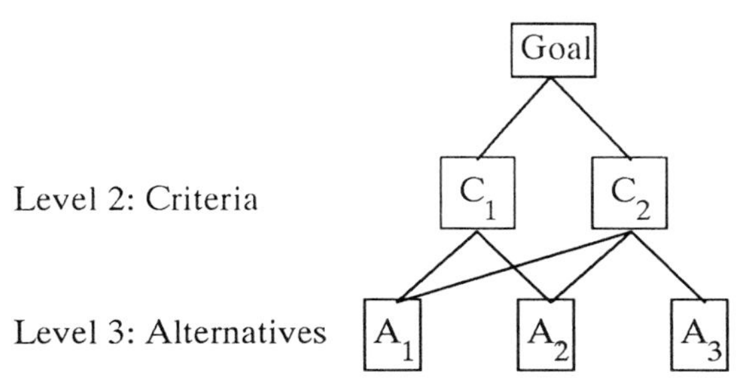

Figure 1. A hierarchy

Table 3
Interval Judgments in a Hierarchy

	C_1	C_2	Interval Eigenvector
C_1	1	[1,3]	(.50, .75)
C_2		1	(.25, .50)

	A_1	A_2	I.E.
A_1	1	[2,4]	(.67, .80)
A_2		1	(.20, .33)

	A_1	A_2	A_3	I.E.
A_1	1	[2,5]	[2,4]	(.327, .659)
A_2		1	[1,3]	(.122, .333)
A_3			1	(.185, .413)

The global interval priorities are given by:

$$0.4971 \le w_1 \le 0.7647$$
$$0.1610 \le w_2 \le 0.3333$$
$$0.0463 \le w_3 \le 0.2063$$

Note that A_1 dominates all other alternatives. Thus the logical choice would be A_1.

It is clear that if $l^{(k)}_{hij} \uparrow u^{(k)}_{hij}$ for all i, j , k and h, then, one would obtain the same results as in the traditional AHP. If the global interval priorities of two alternatives do not intersect, then the most preferred alternative is the one with the largest priority. However, if the global interval priorities intersect, then depending on the measure of the intersection, there maybe a chance that rank reversal takes place. In this case, one should also determine the most likely preference structure and select the alternative with the largest probability and priority. In all other cases it is not clear yet what criterion should be applied to determine the "best" alternative. This is a subject that needs further research.

5. Conclusions

We extended the traditional AHP approach to the case where judgments are of the interval type. This approach can be helpful in situations where the uncertainty of decision makers do not allow them to clearly state their preferences using point judgements. In addition, it could be used as a tool to perform sensitivity analysis and propagate changes throughout hierarchies carrying out the analysis in the traditional manner. The AHPIJ could also be employed when exploring problem formulations with different audiences where the intensity of judgments are not perceived to be the same. It is here, in this domain of uncertainty that we think this method will be more helpful.

REFERENCES

1. Arbel, A. , "Approximate Articulation of Preference and Priority Derivation," EJOR 43 (1989), pp. 317-326.
2. Colson, G. and C. De Bruyn, "Models and Methods in Multiple Objectives Decision Making," Math. Comput. Modelling 12 (1989) pp. 1201-1211.
3. Saaty, T.L., "Scaling Method for Priorities in Hierarchical Structure," J'l of Math. Psych. 15, 3 (1977), pp. 234-281.
4. Saaty, T.L., "Axiomatic Foundation of the Analytic Hierarchy Process," Management Science 32, 7 (1986), pp. 841-855.
5. Saaty, T.L., *Multicriteria Decision Making: The Analytic Hierarchy Process.*" RWS Pub., Pittsburgh, 1988. Original version published by McGraw-Hill, 1980.
6. Saaty, T.L. and L.G. Vargas, "Uncertainty and Rank Order in the Analytic Hierarchy Process," EJOR 32 (1987), pp. 107-117.

AN INTEGRATING DATA ENVELOPMENT ANALYSIS WITH MULTIPLE CRITERIA DECISION ANALYSIS

Valerie Belton
Department of Management Science
University of Strathclyde
Glasgow, Scotland

Abstract

In recent years, practical applications of both multiple criteria analysis and data envelopment analysis (DEA) have become more widespread. The approaches have much in common, both incorporating evaluation of the performance of a number of entities on multiple criteria, with the underlying objective of identifying those that perform better. Both approaches can be used to monitor the performance of similar entities. Both incorporate a model which features a process of assigning weights to criteria. The DEA approach is essentially to find a set of criteria weights which present each entity in the best possible light. Generally, a multiple criteria analysis would elicit weights from the decision makers which best represent their values. There are a number of ways in which those who use these approaches can learn from each other, to mutual benefit, and these will be discussed in this paper. The implementation of these ideas, using V.I.S.A linked with linear programming, is considered briefly.

Introduction - Why Write This Paper?

In recent years, practical applications of multiple criteria decision analysis (MCDA) and data envelopment analysis (DEA) have become more widespread. However, despite having much in common the two fields have developed almost entirely independently of each other. It is my belief, based on ten years experience in the application of multiple criteria decision analysis, and more recent experience in the application of data envelopment analysis, that the two approaches can be integrated to provide a more effective and easier to understand approach to performance measurement than is currently offered by DEA.

The principal aim of this paper is to introduce this suggested approach. There are also two secondary aims. Firstly, I would like to bring DEA to the attention of the MCDM community; I feel that there are further benefits to be gained by both MCDA and DEA as a result of cooperative work and integration of the techniques. Secondly, I would like to record some of the difficulties we have experienced in trying to use DEA. These difficulties are, I hope, largely overcome by the integrated approach which is suggested.

In the following section I will briefly introduce DEA, as most members of the MCDA community are unlikely to be familiar with it. In section 3, I describe some of the difficulties encountered in trying to apply DEA, and then go on to suggest, in section 4, a modified approach which maintains the important features of DEA and incorporates features of MCDA in an attempt to overcome these difficulties. In the final section I comment briefly on the use of DEA in practice.

Data Envelopment Analysis

Data envelopment analysis is a technique which is applied to the measurement of the relative efficiency of a number of similar units, for example, branches of a bank or building society, or University departments. These are referred to as decision making units (DMUs). Multiple, incommensurate, inputs are used to generate multiple, incommensurate outputs. Each DMU may have a pattern of inputs which is different from that of other DMUs, generating its own pattern of outputs. If the DMUs under consideration are branches of a bank, then the inputs might be numbers of counter staff, number of personal bankers, number of managers, number of automatic tellers, and so on; outputs might be number of personal accounts, number of corporate accounts, amount of lending, new accounts opened, number of mortgages, and so on. Defining appropriate inputs and outputs is in itself an important problem, but this paper does not address that issue. The aim of DEA is to discern, for each DMU, whether or not it is operating in an efficient manner, given its inputs and outputs, relative to all the other DMUs under consideration.

The measure of efficiency which is used is a ratio of a weighted sum of outputs to a weighted sum of inputs. For each DMU, a set of weights is determined to show it in the best possible light. The formulation of the problem for unit Q is as follows:

$$\text{Max} \quad H_Q = \Sigma w_i\, y_{iQ} \,/\, \Sigma v_j\, x_{jQ} \tag{1}$$

$$\text{Subject to} \quad \Sigma w_i\, y_{is} \,/\, \Sigma v_j\, x_{js} \leq 1 \qquad s = 1,\ldots Q \ldots S$$

$$w_i \,,\; v_j \geq \varepsilon \qquad i = 1,\ldots.I \quad j = 1,\ldots.J$$

Where
- S is the number of DMUs
- I is the number of outputs
- y_{is} is the amount of output i generated by unit s
- J is the number of inputs
- x_{js} is the amount of input j used by unit s
- w_i is the weight associated with output i
- v_j is the weight associated with input j
- ε is a small positive number

The above model is a fractional linear programme. It can be solved as a linear programme by setting the denominator in the objective function equal to an arbitrary constant (usually unity) and maximising the numerator, giving the following formulation:

$$\text{Max} \quad H_Q = \Sigma w_i \, y_{iQ} \tag{2}$$

$$\text{Subject to} \quad \Sigma v_j \, x_{jQ} = 1$$

$$\Sigma w_i \, y_{is} \, / \, \Sigma v_j \, x_{js} \leq 1 \qquad s = 1,...Q...S$$

$$w_i \, , \, v_j \geq \varepsilon \qquad i = 1,....I \quad j = 1,....J$$

In practice, for computational convenience, it is usually the dual of this problem which is solved. This is as follows:

$$\text{Min} \quad Z_Q - \varepsilon \, (\, \Sigma s_i^+ + \Sigma s_j^- \,) \tag{3}$$

$$\text{Subject to} \quad Z_Q x_{jQ} - \Sigma \alpha_s x_{js} - s_j^- = 0 \qquad j = 1,....J$$

$$\Sigma \alpha_s y_{is} - s_i^+ = y_{iQ} \qquad i = 1,....I$$

$$\alpha_s, \, s_i^+, \, s_j^- \geq 0$$

$$Z_Q \text{ unconstrained}$$

Where Z_Q, α_s, s_i^+, s_j^- are dual variables

A DMU is efficient if and only if the optimal solution to the above problem has $Z_Q = 1$ and all s_i^+, s_j^- equal to zero. A DMU is inefficient if the optimal value of Z_Q is less than 1, in which case the value of Z_Q gives an indication of the degree of inefficiency (distance from the efficient frontier). The non-zero α_s in the optimal solution indicate the set of DMUs with respect to which Q is inefficient, known as the reference set. A linear combination (with coefficients α_s) of the DMUs in this set defines a unit which produces a weighted combination of outputs equal to unit Q with a weighted combination of inputs which is a proportion Z_Q of that of unit Q.

A linear programme, as defined above, is solved for each DMU.

For details of the formulation of the problem, the derivation of the dual, and an explanation of the solution, see Lewin and Morey (1981) or Banker et al (1989).

The problem can be visualised in two dimensions in a number of ways. Figure 1 represents a set of DMUs producing differing combinations of two outputs, each for one unit of input. The DMUs which lie on the efficient frontier (A,B,C,D,E) will have relative efficiencies equal to 1. All other DMUs will have efficiency scores less than 1, the actual score being defined by their position along a ray from the origin to the efficient frontier. For example, the efficiency score of unit F is equal to the ratio of the distances OF and OZ. The reference set for unit F is units B and C.

The above is a brief description of DEA as originally defined and described by Charnes and Cooper (1978), although the ideas date back to Farrell (1957). There is a growing literature on DEA, a bibliography compiled by Seiford (1987) cites almost 200 papers written since 1978, the 1990 update cites 400 references. In recent years Conferences and Conference streams (e.g. IFORS 1990) have been dedicated to the subject. A number of alternative models have been developed, allowing for economies of scale (Banker et al, 1984), constrained weights (Thompson et al, 1986, Dyson and Thannassoulis, 1988) and additive, rather than ratio, models (see Banker et al, 1989). Banker et al (1989) give an up-to-date review of the different models and Charnes and Cooper (1990) review the interactions between research and application since 1978. There has been little interaction with MCDA, one exception being the work by Golany (1988).

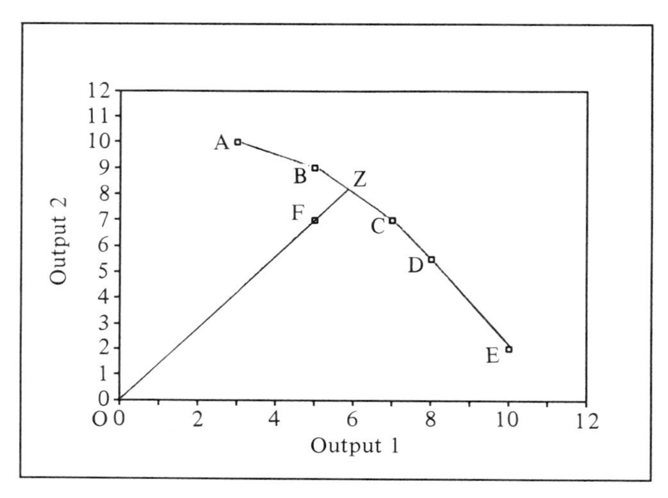

Figure 1 The Efficiency Frontier

Experiences with applying DEA

A number of attempts to apply DEA have not met with the resounding success one might expect from an approach enjoying its current level of popularity. I am not able to describe these attempts in detail here, as the permission of the organisations involved has not yet been obtained, however, they are real applications in large organisations. In each case we were working with managers in the organisation's OR group, people who are technically competent. A number of difficulties emerged, some of which related to the technical specification of the model, some of which were more general. The ones which I consider to be most important are discussed briefly below.

The concept of DEA, that of determining whether or not a unit is efficient in comparison with all other units under consideration, is easily understood. However, the weights which are determined in the solution of this problem are difficult to interpret. In the case of a single input and multiple outputs they can be interpreted as the amount of resource used to generate a unit of output (Dyson and Thanassoulis, 1988). Similarly, if there are multiple inputs, but only one output, the weights can be interpreted as the amount of each input needed to generate a unit of output. When there are multiple inputs and outputs then there is no intuitively appealing interpretation.

In the original formulation of DEA the only bounds on the weights are those constraining them to be greater than or equal to a small positive number. As a consequence, the analysis might indicate that a DMU is efficient but only as a result of an extreme distribution of the weights. For example, if a DMU uses relatively little of one input and/or produces more of one output than other DMUs then it may turn out to be efficient by allocating most, or all, weight to that input/output (see Sexton et al, 1986, for a discussion of this problem). This ignores the DMU's efficiency with respect to all other inputs and outputs. One should beware of jumping to immediate conclusions on the basis of such evidence, as for an efficient DMU the linear program outlined above will have at least two optimal solutions and possibly many more, each solution indicating a potentially different set of weights. Sexton et al (1986) describe an extension to DEA which allows to user to influence whether a more extreme, or a balanced set of weights should be selected if there is such a choice. Dyson and Thanassoulis (1988) approach the problem differently, by determining reasonable bounds on the weights (in the case of a single input) using regression analysis and incorporating these into the problem formulation to preclude extreme sets of weights.

It is reasonably straightforward to set bounds on the values of the weights and to interpret these in the case of either a single input or a single output. However, as before, the meaning and implication of such bounds is not clear in the general case of multiple inputs and outputs. This is partly a result of constraining the weighted sum of inputs for the DMU under consideration (target DMU) to equal unity. This means that the magnitude of the weights is affected by the scale of operation of the target DMU; if the target DMU has high levels of inputs then as a consequence the weights which can be assigned must be smaller than if the target DMU had low inputs. This has an effect on the values of the weights assigned to the outputs. As a consequence it is difficult to set meaningful bounds on the weights and to compare weights generated for different target DMUs, except in terms of ratios.

A consequence of these difficulties in interpreting weights is that the model is no longer transparent - it becomes a black box. The user is not in control of the analysis, and furthermore, much of the output cannot be readily interpreted. There is a feeling that one ought to be able, in some way, to constrain the weights allocated to inputs and outputs to prevent a DMU appearing to be efficient for the reasons given above, but an inability to do so because of the difficulty in interpretation.

The aim of the approach described in the next section is to overcome these limitations by redefining the underlying model and incorporating the revised version within a visual interactive model. This clarifies the definition of the weights, utilises a model which is transparent and easily understood by the user, incorporates effective graphical feedback and puts the user in control, allowing direct interaction with the model.

The proposed approach

The proposed approach is based on a simple hierarchical, weighted value function model. An alternative approach to the evaluation of the performance of similar DMUs would be to construct a value function model of inputs and outputs, determining a set of weights for each by direct questioning. The task of determining these weights is not an easy one and there may be many different views on what the weights should be. It is exactly this task which DEA is designed to avoid. However, the weights are well defined and have a clear meaning, namely that they define an acceptable trade-off between criteria in the model - in this case between inputs, or between outputs. In the proposed approach it is not required that an acceptable trade-off between an input and an output is defined. The proposed approach is a compromise between these two extremes.

The problem is viewed as the following hierarchical model:

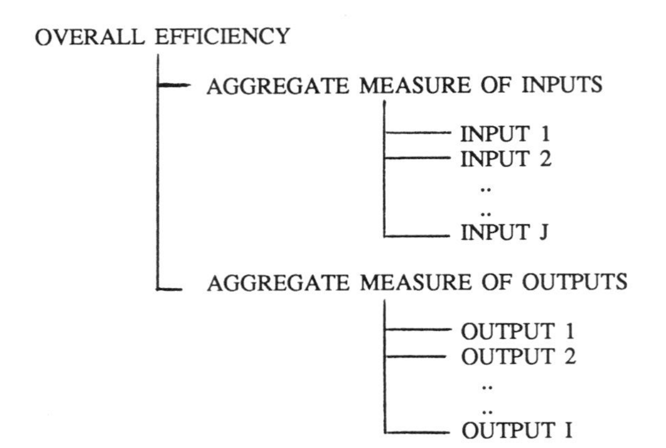

It is possible that the inputs and outputs could be further disaggregated, adding another layer to the hierarchy.

The inputs and outputs are weighted to give the aggregate measures of input and output. As is usual in a multiple criteria analysis of this kind the weights will be normalised so that the sum of input weights is 1 and similarly for output weights. The measurements of inputs and outputs may be rescaled so that the maximum value for each input or output is 100, although this is not necessary. The advantage of rescaling is simply that the aggregate scores are constrained to be no more than 100 (which facilitates automation of graph drawing). It is important to remember that the weights are scaling factors and are thus related to the scales in use.

For any set of input and output weights, we can plot aggregate output measure against aggregate input measure; efficient DMUs are easily identified.

Following the principle of DEA, we would like to know, for each DMU, if there is a set of weights which put it on the efficient frontier. This can be determined by solving the following problem:

Let $\Sigma w_i y_{is} = O_s$ (aggregate output for unit s)
$\Sigma v_j x_{js} = I_s$ (aggregate input for unit s)

β be a constant, $0 \leq \beta \leq 1$

$V_s = \beta O_s - (1-\beta) I_s$

If unit Q is efficient, then for some value of β, $V_Q \geq V_s$ for all units s (not including Q).

That is, if for some value of β, the optimal solution of the following linear programme has the value of the objective function equal to zero.

$$\text{Min } \Sigma\, D_s^- \tag{4}$$

Subject to: $V_Q - V_s + D_s^- - D_s^+ = 0$

$w_i,\ v_j\ ,\ D_s^-,\ D_s^+ \geq 0$

Where D_s^-, D_s^+ are deviation variables, as used in goal programming

Clearly this involves a search over values of ß, which slows down the solution process.[1]

If there is no set of weights which make unit Q efficient, then the value of the objective function at the optimum is non-zero and the solution gives a set of weights which minimise the extent to which other units perform better than Q (this may not be the set of weights which minimise the distance of Q from the efficient frontier, as all units are included in the measure).

The advantage of the above formulation over the original DEA formulation is that the interpretation of the weights is now clear, input weights and output weights are decoupled and it is possible to set bounds on the values of weights within the family of input weights or output weights, incorporating these in the above problem formulation. For example, upper and lower bounds may be set to ensure that a reasonably balanced view of performance is given. The formulation can be extended as mentioned above, and described in detail by Sexton et al (1986) to direct the choice of weights for efficient DMUs to the more extreme or the more balanced, as desired.

The output of the initial analysis is displayed graphically as shown in Figure 2. This presents the set of input weights, the set of output weights, and the position of all DMUs in the aggregate input/aggregate output plane given these weights. It can be seen whether or not the DMU of interest is on the efficient frontier, and if not, how far away it is.

The user can interactively investigate the effect of changing any of the input or output weights, seeing how the position of the target DMU changes relative to the efficient frontier. At any time a new target DMU may be selected, the analysis carried out to determine whether or not it can be made efficient, and the appropriate weights and plot of DMUs displayed.

It is intended to implement this system within V·I·S·A (Belton and Vickers, 1989). In addition to being able to investigate the efficiency of DMUs as described above, the user will be able to make use of V·I·S·A's facilities for interactive analysis of scores (corresponding to inputs and outputs in the DEA model) to investigate what changes are necessary to move an inefficient unit to the efficient frontier.

---

[1] I am currently looking for a more efficient way of solving this problem, and would be interested to hear from anyone who knows of one.

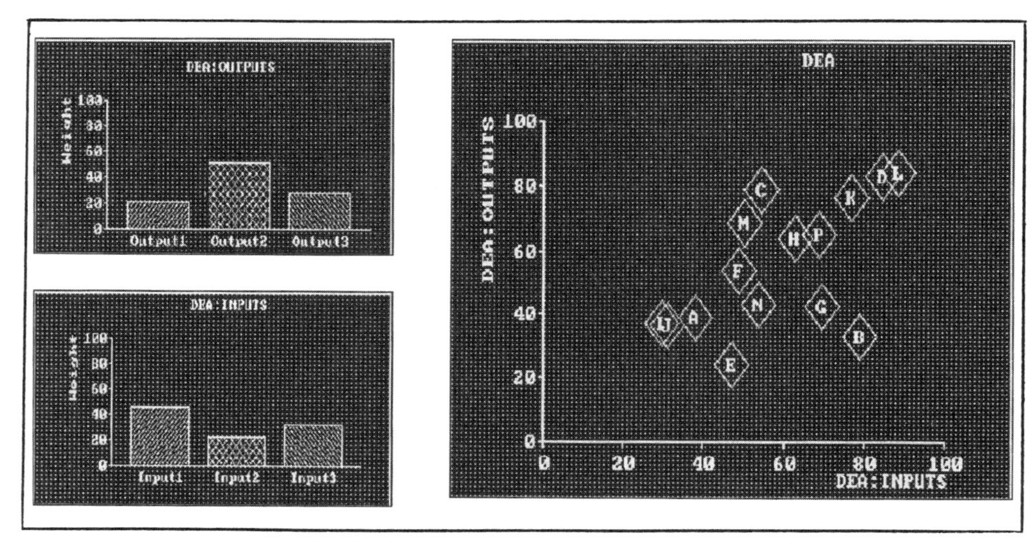

Figure 2 An integrated MCDA/DEA system - example of output

Discussion

As the title of the paper implies, the approach described above is, at the current moment, an idea for integrating DEA and MCDA. It is an extension of the principle of DEA to allow the user to better understand the analysis from the point of view of determining a weighted sum of inputs and outputs in the context of trying to identify efficient, or perhaps more pertinently, inefficient DMUs. In proposing an interactive model it puts control in the hands of the user and hopefully, as a consequence, will increase understanding of and confidence in the model.

The ideas are still at the research stage, but they will be put to the test in the future with the managers who rejected the use of DEA in its original form. Potential users of DEA, in whatever form, should take note that there are many other issues to be aware of in trying to implement such an approach. I have already mentioned that the definition of inputs and outputs is itself a complex area, but beyond the technical considerations it is of course of paramount importance that those who are to make use of the model are involved in its development and that a sense of ownership is established from the beginning.

References

RD Banker, A Charnes and WW Cooper (1984), Some Models for Estimating Technical and Scale Inefficiencies in Data Envelopment Analysis, Man Sci 30, 1078-1092.

RD Banker, A Charnes, WW Cooper, J Swarts and DA Thomas (1989), An Introduction to Data Envelopment Analysis with some of its Models and their Uses, Research in Governmental and Nonprofit Accounting (5), 125-163

V Belton and SP Vickers (1989), V·I·S·A - Visual Interactive Modelling for Multiple Criteria Decision Aid, in AG Lockett and G Islei (Eds), Improving Decision Making in Organisations, Springer-Verlag.

A Charnes, WW Cooper and E Rhodes (1978), Measuring Efficiency of Decision Making Units, Euro Jnl Opl Res, 429-444.

Charnes and Cooper (1990), DEA Usages and Interpretations, Graduate School of Business, University of Texas at Austin (presented at IFORS, Athens)

RG Dyson and E Thanassoulis (1988), Reducing Weight Flexibility in Data Envelopment Analysis, Jnl Opl Res Soc, 39(6), 563-576.

MJ Farrell (1957), The Measurement of Productive Efficiency, Jnl Royal Stat Soc (A) 120, 253-281.

B Golany (1988), An Interactive MOLP Procedure for the Extension of DEA to Effectiveness Analysis, Jnl Opl Res Soc, 39(8), 725-734

AY Lewin and RC Morey (1981), Measuring the Relative Efficiency and Output Potential of Public Sector Organisations: an Application of Data Envelopment Analysis, Int Jnl Pol Anal Info Syst 5(4), 267-285.

L Seiford (1988), A Bibliography of Data Envelopment Analysis (1978-1986), Working Paper, University of Massachusetts

TR Sexton, RH Silkman and AJ Hogan (1986), Data Envelopment Analysis: Critique and Extensions, in RH Silkman (Ed), Measuring Efficiency: An Assessment of Data Envelopment Analysis, Jossey-Bass, San Francisco.

RG Thompson, FD Singleton Jnr, RM Thrall and BA Smith (1986), Comparative Site Evaluations for Locating a High-Energy Physics Lab in Texas, Interfaces 16, 35-39.

INTERVAL VALUE TRADEOFFS: THEORY, METHODS, SOFTWARE, AND APPLICATIONS

Vadim P. Berman, Gennady Ye. Naumov,
ul. Socialisticheskaya, 76
Rostov-on-Don 344700 USSR
Vladislav V. Podinovskii
Rostovskaya nab., 3-14
Moscow 119121 USSR

Abstract. A new approach to the solution of multicriterial problems is suggested. It is based upon an interval estimation of tradeoffs. Such intervals can be found during an interrogation of the decision maker (DM). This information is easy enough for him/her. Interval rates of substitution are a base for approximation of the DM's preference structure and allows to reduce the set of Pareto-optimal alternatives to the set of nondominated alternative. In practical situation the nondominated alternative is unique quite often.

The methods are realised for IBM PC-compatible computers. A software is meant for a wide class of users and has a friendly interface.

1. Introduction

Let us consider a problem of choice of the best alternative from the given set of alternatives which are characterized by multiple criteria. We can assume that the first step in such choice is a selection of the set of Pareto-optimal alternatives.

This set has one remarkable feature. If one of the two Pareto-optimal alternatives is better than the other by one criterion it is worse inevitably by some other criterion. Hence if we want to choose only one alternative it is necessary to point out what failing off by one criterion we are ready to donate for the improvement by the other one. And so that is the analysis of tradeoffs.

In the classical theory the tradeoffs must be pinpointed in the form of rate of substitution (see, for example, Keeney and Raiffa [2]). But for two reasons it is very difficult, those are:

- the DM cannot determine the exact value of the rate of substitution;
- exact rate of substitution value varies when the criteria values change.

All these difficulties are surmounted in a new theory. It is just enough to point out the interval containing the rate of substitution value of one criterion for another. This interval may be "with a stock", i.e. wide enough. The lower bound of the interval is the maximal value of increment by one criterion such that any smaller number is certainly not acceptable for you in the exchange for decrement of another criterion by one unit. Similarly the upper bound of this interval is the minimal value such that any greater number is certainly acceptable.

Let us explain now how the comparison of alternatives on this basis can be accomplished. We will consider three successively complicating cases:

1) Two alternatives differ only by two criteria; the second alternative fails off by one unit for the first criterion against the first alternative but by the other criterion improvement equals value w.

2) The alternatives differ only by two criteria but in contrast to the first case the failing off by one criterion equals few units while improvement by the other criterion equals some value w.

3) Two alternatives differ by many criteria.

In case 1 it is enough to compare the value w

with the interval rate of substitution. If w is smaller than the lower limit of the interval then according to the definition of an interval the first alternative is better. If w is greater than the upper limit of the interval then the first alternative is worse. If the value of w is in this interval then these alternatives cannot be compared on the base of the given information on the rates of substitution and some more precise information is necessary.

Now let us consider case 2. For definiteness let us assume that the failing off is equal three units of the criterion. Then the following sequence of alternatives can be represented. The first and the last alternatives coincide with the given ones. The second one differs from the first one by failing off by one unit and improvement by $w/3$ and the third one differs by two units failing off and $2w/3$ improvement. Comparison of each pair of adjacent alternatives is realized as in the case 1. After all, the failing off is equal one unit. It is natural to suppose that if

the first alternative is better than the second
the second is better than the third
the third is better than the last one

then the first alternative is better than the last one.

Thus, if such sequence can be constructed then the first alternative is better than the last one. According to this example development of such sequence is quite evident.

Eventually let us consider case 3. The first alternative is better than the second one if there exists a sequence of alternatives where 1) the first and the last alternatives coincide with the given ones; 2) each adjacent pair differs only by values of two criteria; 3) it can be asserted (as in the case 2) that the first alternative of each pair is better than the second. (Note that such sequence can occur not a unique one, taking into account the order of criteria pairs, and the problem of existence of such sequence turns to be non-trivial). If such sequence can be spread from the second alternative to the first one then the second alternative is better. If none of these sequences can be constructed then it is impossible to compare the two alternatives on the basis of the given information on tradeoffs. It means that some more exact information is necessary.

2. Theoretical model of preferences

Interval approach supposes that the DM's preferences are modelled with the help of a preference relation with interval rates of substitution. Let us give the basic definitions and theoretical results.

Let S be a given set of available alternatives (or strategies), and $f=(f_1,\dots,f_n)$ be a vector criterion defined on S. Each alternative s is represented by its vector estimate $f(s)$ in a criterial space $\mathbb{R}^n$. The criteria f_i will be denoted by their numbers i, $i = 1,\dots,n$. For definiteness let us consider that all criteria are positively oriented, i.e. their larger values are more preferable than the smaller ones. Negatively oriented criterion smaller values of which are more preferable than larger ones can be formally transformed easily into the positively oriented one, for example, by changing the sign.

We'll mean a (strict) preference relation on $\mathbb{R}^n$ as a binary relation $P \subset \mathbb{R}^n\times\mathbb{R}^n$ which is irreflexive and transitive. We'll often use the Pareto relation P^0:

$xPy \iff (x_i \geq y_i,\ i=1,\dots,n \text{ and } x\neq y)$.

P^0 is an example of a preference relation.

The preference relations we use will be assumed to have certain properties. A preference relation P is:

1) continuous if an upper cut $P^+(x)=\{y\in\mathbb{R}^n \mid yPx\}$ and a lower cut $P^-(x)=\{y\in\mathbb{R}^n \mid xPy\}$ are open for every $x\in\mathbb{R}^n$;

2) monotone (with respect to P^0) if $xP^0y \Rightarrow xPy$ for every $x,y\in\mathbb{R}^n$;

3) radial if $xPy \Rightarrow xP[\alpha y+(1-\alpha)x]$ and $[\alpha x+(1-\alpha)y]Py$ for every $x,y\in\mathbb{R}^n$ and $\alpha>0$;

4) plane-decomposable if for every $x,y\in\mathbb{R}^n$: $xPy \Rightarrow$ { there exists a sequence $z^0=x, z^1,\dots, z^{m-1}, z^m=y$ such that z^{i-1} and z^i have not more than two different coordinates and $z^{i-1}Pz^i$ for all $i=1,\dots,m$}.

A cone $C\subseteq\mathbb{R}^n$ is plane-regeneratible (briefly PR-cone) if a convex hull of a union of all intersections $C_{ij} = C \cap S_{ij}$ where S_{ij} is a coordinate plane coincides with C: $C = \text{conv} \bigcup_{i>j} C_{ij}$.

A PR-cone is always polyhedral.

Theorem 1. A preference relation P is cone

$$P^-(x) = -P^+(x) = (x+C)\backslash\{x\},$$

where the dominance cone C is a PR-cone containing a negative orthant without origin $\{x\in\mathbb{R}^n | OP^O x\}\backslash\{O\}$ if and only if P has properties 1-4.

For $t>0$ the t-interval rate of substitution of a criterion j for a criterion i at a point $x\in\mathbb{R}^n$ for a continuous monotone preference relation P is the interval

$$\lambda_{ij}(x,t)=[\lambda^-_{ij}(x,t);\lambda^+_{ij}(x,t)]$$

where $\lambda^-_{ij}(x,t) = \sup\{\Delta/t \mid \Delta>0,\ xP(x\| x_i-t, x_j+\Delta)\}$,

$\lambda^+_{ij}(x,t) = \inf\{\Delta/t \mid \Delta>0,\ (x\| x_i-t, x_j+\Delta)Px\}$.

Here $(x\| x_i-t, x_j+\Delta)$ is a vector obtained from x by changing x_i for x_i-t and x_j for $x_j+\Delta$.

In accordance with this definition we have $(x\| x_i-t, x_j+\Delta)Px$ for any $\Delta\in(\lambda^+_{ij}(x,t),+\infty)$ and $xP(x\| x_i-t, x_j+\Delta)$ for any $\Delta\in(0,\lambda^-_{ij}(x,t))$. If $\Delta\in\lambda_{ij}(x,t)$ then x and $(x\| x_i-t, x_j+\Delta)$ are incompatible with respect to P.

If $\lambda_{ij}(x,t)$ is constant, i.e. not depending on values of x and t, than it will be called interval rate of substitution and denoted by λ_{ij}.

Theorem 2. The preference relation P satisfies the conditions of the theorem 1 if and only if all t-interval rates of substitution are constant and have properties:

1) $\lambda_{ij} = 1/\lambda_{ji}$; 2) $\lambda_{ij} \subseteq \lambda_{ik} \times \lambda_{kj}$.

In theorem 2 we used designations from the interval analysis taking into account that $0<\lambda^-_{ij}\leqslant\lambda^+_{ij}$:

$$1/\lambda_{ij} = [1/\lambda^+_{ji};\ 1/\lambda^-_{ji}];$$

$$\lambda_{ik} \times \lambda_{kj} = [\lambda^-_{ik} \cdot \lambda^-_{kj};\ \lambda^+_{ik} \cdot \lambda^+_{kj}].$$

In accordance with the theorem 1 and 2 a preference relation having properties 1-4 is called preference relation with interval rates of substitution (briefly IRS-relation).

Axiomatics and a structure of IRS-relation as well as properties of interval rates of substitution were given in Berman and Naumov [1]. Interval evaluation of substitution of a fixed criterion for each of other criteria was considered in Passy and Levanon[4], interval estimates were not constant in the criteria space.

It should be noted that there is a fundamental distinction between the interval approach and methods of separation of some part of the Pareto set with the help of a generalized criterion when the weight coefficients are restricted by intervals (Stuer [6]).

3. Interval approximation of DM's preferences

The interval approach to the solution of a multicriterial problem presupposes a reception of estimations of interval rates of substitution for some (not obligatory for all) pairs of criteria. To receive such estimation for a pair of criteria $\langle i,j \rangle$ the DM has to compare the following pairs of vector estimates

$x = (x_1, \ldots, x_i, \ldots, x_j, \ldots, x_n)$,

$y = (x_1, \ldots, x_i - 1, \ldots, x_j + \Delta, \ldots, x_n)$,

where all components $x_1, \ldots, x_n$ are fixed and $\Delta > 0$ is variable. As a result the positive numbers Δ^-_{ij} and Δ^+_{ij}, $\Delta^-_{ij} < \Delta^+_{ij}$ should be found, such that

a) if $\Delta \leqslant \Delta^-_{ij}$ then x is more preferable than y,

b) if $\Delta \geqslant \Delta^+_{ij}$ then y is more preferable than x,

c) if $\Delta^-_{ij} < \Delta < \Delta^+_{ij}$ then x and y are not comparable.

Then the DM increases the interval $(\Delta^-_{ij}, \Delta^+_{ij})$ while the conditions a)-c) are valid in any point $x \in \mathbb{R}^n$. Hence for a pair of criteria $\langle i,j \rangle$ we received from the DM intervals $\mu_{ij} = [\mu^-_{ij}; \mu^+_{ij}]$

having the following property: for every $x \in \mathbb{R}^n$, $t>0$ $(x \| x_i - t, x_j + \mu_{ij}^+ t)$ P x and x P $(x \| x_i - t, x_j + \mu_{ij}^- t)$.

A binary relation generated by μ_{ij} in accordance with this property will be denoted by $P(\mu_{ij})$. Let M be an information on DM's preferences formed by accumulated μ_{ij}. This information induces a relation P^M on $\mathbb{R}^n$ which is the transitive closure of the union of P^0 and all $P(\mu_{ij})$:

$$P^M = \mathrm{Tr}\left[P^0 \cup \left(\bigcup_{\mu_{ij} \in M} P(\mu_{ij}) \right) \right].$$

In accordance with definition xP^My is valid if and only if there exists a chain xP^1z^1, $z^1P^2z^2$, ..., $z^{m-1}P^my$, in which each $z^k \in \mathbb{R}^n$, P^k is P^0 or $P(\mu_{ij})$ for some $\mu_{ij} \in M$.

The information M is said to be consistent if P^M is irreflexive. Having the consistent information M we can consider the intervals μ_{ij} as estimates of interval rates of substitution for some IRS-relation.

A construction of preference relation using an information on constraints of rates of substitution (not only for single criteria but also for sets of criteria) was suggested in the framework of the theory of criteria importance in Podinovskii [5]. Methods for verificating consistency of such information and for constructing the corresponding preference relation were presented in Menshikova and Podinovskii [3].
Using these results we can construct the preference relation P^M basing on M directly.

For each μ_{ij} we introduce two row vectors $a^-(\mu_{ij})$ and $a^+(\mu_{ij})$ as follows:

$$a_k^-(\mu_{ij}) = \begin{cases} -\mu_{ij}^-, & k=j, \\ 1, & k=i, \\ 0, & k \neq i,j \end{cases},$$

$$a_k^+(\mu_{ij}) = \begin{cases} \mu_{ij}^+, & k=j, \\ -1, & k=i, \\ 0, & k\neq i,j. \end{cases}$$

Using all these row vectors we form a $2q\times n$-matrix A^M where q is a number of μ_{ij} in M. We'll use the following designations (for $x,y\in\mathbb{R}^n$): $x\geqq y \Leftrightarrow x_i\geqslant y_i,\ i=1,\dots,n;\ \ x>y \Leftrightarrow x_i>y_i,\ i=1,\dots,n.$

Let us introduce a set $B^M = \{\ \beta\in\mathbb{R}^n \mid \beta>0,\ \ A^M\beta>0\ \}$.

From the point of view of the theory of criteria importance M is the information on criteria importance of a special kind (proportional importance) and elements of B^M are vectors of importance coefficients.

Theorem 3. The information M is consistent if and only if B^M is non-empty.

Theorem 4. Given any $x,y\in\mathbb{R}^n$ and $x\neq y$, xP^My is valid if and only if there exists a vector $u\in\mathbb{R}^{2q}$ with nonnegative components such that the following vector inequality is satisfied:

$$x-y \geqq uA^m.$$

Theorem 5. xP^My if and only if $(x-y)\beta>0$ for each $\beta\in B^M$.

Let $Y=f(S)$ be a set of vector estimates of alternatives. The relation P^M determines a subset of nondominated points Y^M in M.

As a result the Pareto set of choice S^O is reduced to a set of nondominated alternatives S^M (vector estimates of which belong to Y^M).

In practical situations when the set S was finite and included several dozens of alternatives we had only a few nondominated ones. Quite often there was the only nondominated alternative.

4. Decision support system MCITOS

The interval approach to the solution of multicriterial problems is now supported by a modern software. The authors of the article have created a system MCITOS (MultiCriteria Interval Trade-Offs System). Input data for MCITOS are the next:

- set of criteria;
- set of alternatives with estimates of them by all criteria.

MCITOS asks DM about his preferences and chooses all non-dominated alternatives out of the given set quickly.

MCITOS can be used in any IBM PC-compatible computers with memory no less than 512 KByte. Color graphics monitor is recommended.

The number of criteria and alternatives is limited only by the accessible memory (approximately 1000 alternatives with 15 criteria).

System's interface is based on windows, menu, spreadsheets and is friendly. All the data are in easy form to examine and modify.

Features of MCITOS:

- availability of English and Russian versions;
- spreadsheet in the table of alternatives that allows to solve easily such problems as "what if ...";
- possibility to employ files created by dBase (Clipper, FoxBase).

5. Applications

The system MCITOS was successfully used for solving several practical problems including a problem of choice of equipment for automatic stores, problems of constructing a power gearing, and some problems of personal choice.

New type of gearing was invented by G.Juravlev (Mixed gear, PCT/SU N 86/00110). Let us consider an example of problem of design of mixed gearing.

Design versions (alternatives) are evaluated by four criteria: 1) strenth resource, 2) interaxial distance (mm), 3) ring width (mm), 4) class index n^o. The criteria 1 and 4 are positively oriented. The criteria 2 and 3 are negatively oriented. Table 1 shows vector estimates of all forty-two alternatives.

This problem was analysed by the designer with help of MCITOS. Three intervals were received:

$\mu_{12}=(10;70)$, $\mu_{42}=(10;120)$, $\mu_{31}=(0.005;0.03)$.

The information $M=\{\mu_{12},\mu_{42},\mu_{31}\}$ is consistent. The only alternative whose vector estimate is nondominated is the fourth one. Hence the design version 4 of gearing is optimal.

Table 1. Design versions of mixed gearing

Alter-natives	Criteria			
	1	2	3	4
1	1.02197	202.194	60	5
2	1.43539	202.194	70	5
3	1.36653	202.194	70	6
4	1.26218	202.194	70	7
5	1.13743	212.836	60	5
6	1.04408	212.836	60	6
7	1.59756	212.836	70	5
8	1.52093	212.836	70	6
9	1.40478	212.836	70	7
10	1.0439	223.477	50	5
11	1.24806	223.477	60	5
12	1.14563	223.477	60	6
13	1.05874	223.477	60	7
14	1.75295	223.477	70	5
15	1.66886	223.477	70	6
16	1.54141	223.477	70	7
17	1.21999	242.633	50	5
18	1.21999	242.633	50	6
19	1.21999	242.633	50	7
20	1.45304	242.633	60	5
21	1.33379	242.633	60	6
22	1.23263	242.633	60	7
23	1.68238	242.633	70	5
24	1.54431	242.633	70	6
25	1.42718	242.633	70	7
26	1.35783	255.403	50	5
27	1.35783	255.403	50	6
28	1.35783	255.403	50	7
29	1.6172	255.403	60	5
30	1.48448	255.403	60	6
31	1.37189	255.403	60	7
32	1.87245	255.403	70	5
33	1.71878	255.403	70	6
34	1.58842	255.403	70	7
35	1.4899	268.173	50	5
36	1.4899	268.173	50	6
37	1.4899	268.173	50	7
38	1.7745	268.173	60	5
39	1.62887	268.173	60	6
40	1.50533	268.173	60	7
41	2.05458	268.173	70	5
42	1.88596	268.173	70	6

6. Conclusion

The interval approach to the solution of multicriterial problems is well justified and based on reliable information. It makes possible to receive reliable results. The defect of this method is an incomplete comparability of alternatives. But this incompleteness reflects the inexactness of interval information on tradeoffs.

Bibliography

1. Berman, V.P., and G.Ye. Naumov, "Preference Relations with Interval Value Tradeoffs in Criterial Space", Automation & Remote Control, Number 3, p.p. 398-410, 1989.
2. Keeney, R.L., and H. Raiffa, Decision with Multiple Objectives, New York, Wiley, 1976.
3. Menshikova, O.R., and V.V. Podinovskii, "Construction of Preference Relation and Core in Multicriterial Problems with Non-Homogeneous Importance-Ordered Criteria", USSR Computational Mathematics and Mathematical Physics, Volume 28, Number 3, p.p. 15-22, 1988.
4. Passy, U., and Y. Levanon, "Analysis of Multiobjective Decision Problems by the Indifference Band Approach", Journal of Optimization Theory and Application, Volume 43, Number 2, p.p. 205-235, 1984.
5. Podinovskii, V.V., "Relative Importance of Criteria in Multiobjective Decision Problems", In: Multiobjective Decision Problems, Moscow, Mashinostroyeniye Publishing House, p.p. 48-92, 1978 (in Russian).
6. Stuer, R.E., "Multiple Objective Linear Programming with Interval Criterion Weights", Management Science, Volume 23, Number 3, p.p. 305-316, 1976.

ON SOME PROPERTIES OF OUTRANKING RELATIONS BASED ON A CONCORDANCE-DISCORDANCE PRINCIPLE

Denis BOUYSSOU
ESSEC
B.P. 105
F-95021 Cergy France

Abstract. The purpose of this paper is to study some properties of outranking relations based on a concordance-discordance principle. We show that, whenever the structure of the set of alternatives is sufficiently rich, imposing "nice" transitivity properties on such outranking relations always leads to a somewhat unappealing distribution of "power" among the various attributes. These results directly apply to methods, such as TACTIC, that produce a crisp asymmetric outranking relation. We explore the links between these results and classical ones in the field of Social Choice Theory and show their relevance for users of outranking methods.

I- Introduction

A classical problem in the field of MCDM is to build a preference relation on a set of multiattributed alternatives on the basis of preferences expressed on each attribute and "inter-attribute" information such as weights and/or tradeoffs. A common way (see Keeney and Raiffa (1976)) to do so is to attach a number v(x) to each alternative x and to declare that x is preferred to y if and only if v(x) > v(y). Usually, the number v(x) depends on the evaluations $x_1, x_2, ..., x_n$ of x on the n attributes and we have $v(x) = V(x_1, x_2, ..., x_n)$. When one uses such a method, the preference relation that is built has "nice" transitivity properties. However, the definition of the aggregation function V may not always be an easy task (see, *e.g.*, Roy and Bouyssou (1987)). Starting with ELECTRE I (see Roy (1968) or, for a presentation in English, Goicoechea *et al.* (1982)), a number of MCDM techniques, the so-called outranking methods, have been proposed that use an alternative way to build a preference relation based on a concordance-discordance principle (see, *e.g.*, Roy and Bertier (1973), Vansnick (1986) and the bibliography of Siskos *et al.* (1983)). In these methods, the preference relation, which is often called an outranking relation, is built through a series of pairwise comparisons. Such pairwise comparisons can be done in many ways. The idea of concordance-discordance consists in declaring that an alternative x is preferred to an alternative y if a "majority" of the attributes supports this assertion (concordance condition) and if the opposition of the other attributes is not "too strong" (non-discordance condition). In this paper we will restrict our attention to methods aiming at building a crisp (*i.e.*, nonfuzzy) and, for reasons to be explained in section 4, asymmetric preference relation[1].

In order to be more specific, suppose that we have defined on each attribute i a complete and transitive binary relation R_i allowing to compare in terms of preference the evaluations of the alternatives on this attribute[2]. When comparing two alternatives x and y, it is possible to partition the set of attributes N between attributes favoring x, attributes favoring y and "neutral" attributes, *i.e.*, $P(x, y) = \{i \in N : x_i\ P_i\ y_i\}$, $P(y, x) = \{i \in N : y_i\ P_i\ x_i\}$ and $I(x, y) = I(y, x) = \{i \in N : x_i\ I_i\ y_i\}$. Using the idea of concordance, it is declared that x is preferred to y when the "coalition" of attributes in P(x, y) (or in $P(x, y) \cup I(x, y)$) is considered more important than the "coalition" of attributes in P(y, x). For practical purposes, the importance of a coalition of attributes is usually determined in an additive way after having attached a weight p_i to each attribute. For instance, in TACTIC (Vansnick (1986)), we have, for the concordance part of the method:

$$x\ P\ y \Leftrightarrow \sum_{i \in P(x, y)} p_i > \rho \sum_{i \in P(y, x)} p_i \tag{1}$$

where ρ is a threshold greater than 1.

It is worth noting that preference relations based on the idea of concordance do not make use of the magnitude of the "differences" between evaluations and are only based on the ordinal information conveyed by the relations R_i.

This idea of concordance may however be criticized since on some attributes in P(y, x) the difference between y_i and x_i may be so large as to cast a major doubt on the validity of x P y even if these attributes are of limited importance. One simple way to circumvent this problem is to combine the idea of concordance with that of discordance. Taking discordance into account amounts to defining a set D of ordered pairs of evaluations on the various attributes such that $(y_i, x_i) \in D$ for some $i \in N$ implies Not(x P y) regardless of the comparison of the importance

---

[1] A (crisp) binary relation S on a set K is a subset of K^2. Throughout the paper we will classically write a S b instead of $(a, b) \in S$. We say that a binary relation S on a set K is (for all $a, b, c \in K$):
- complete if a S b or b S a,
- asymmetric if a S b implies Not (b S a),
- transitive if a S b and b S c imply a S c,
- negatively transitive if Not(a S b) and Not(b S c) imply Not(a S c),

It is without circuit if for all $k \geq 1$ and all $a_1, a_2, \ldots, a_k \in K$, $a_1\ S\ a_2$, $a_2\ S\ a_3$, ..., $a_{k-1}\ S\ a_k$ imply Not $a_k\ S\ a_1$.

[2] We respectively note P_i and I_i the asymmetric and symmetric parts of R_i *i.e.* [$x_i\ P_i\ y_i$ iff $x_i\ R_i\ y_i$ and Not $y_i\ R_i\ x_i$] and [$x_i\ I_i\ y_i$ iff $x_i\ R_i\ y_i$ and $y_i\ R_i\ x_i$], a similar convention holding for all binary relations used in this paper.

of the various coalitions of attributes as modelled using the idea of concordance. This set D is often defined through a binary relation $V_i \subset P_i$ that reads "is very strongly preferred to" such that $(y_i, x_i) \in D \Leftrightarrow y_i \, V_i \, x_i$. Thus taking the idea of discordance into account leads to a poorer relation that the one that would have been obtained using concordance alone. For instance, we have in TACTIC:

$$x \, P \, y \Leftrightarrow \sum_{i \in P(x, y)} p_i > \rho \sum_{i \in P(y, x)} p_i \text{ and Not}(y_i \, V_i \, x_i) \text{ for all } i \in P(y, x) \quad (2)$$

The interest of such methods and the way to assess the weights p_i and the relations V_i have been discussed elsewhere (see, *e.g.*, Roy (1971) or Roy and Vincke (1981)). It should be emphasized that, contrary to methods attaching a number to each alternative, the preference relation that is obtained in that kind of methods may not possess "nice" transitivity properties[1]. The purpose of this paper is to study under what conditions these relations may possess such properties as transitivity or absence of circuit. We present our notations and definitions in the next section. In section 3, we investigate some properties of relations that are only based on the idea of concordance. In section 4, we generalize our results so as to take the idea of discordance into account. In a final section we analyze the links between these results and more classical ones in the Theory of Social Choice and stress their relevance for users of outranking methods..

II- Definitions and Notations

Let N be a set of attributes. To each attribute $i \in N$, we associate a set X_i which will be interpreted as a set of possible levels on attribute i and a binary relation R_i on X_i. Let X be a set of alternatives such that $X \subset \prod_{i \in N} X_i$. When there will be no risk of confusion, x_i will designate the element of X_i being the ith component of $x \in X$.
Throughout the paper we suppose that the following structural conditions hold:

S1 R_i is complete and P_i is transitive,

S2 N is finite and $|N| = n \geq 2$.

Condition S1 implies a minimal consistency requirement on the preferences on each attribute. It should be noticed that S1 is compatible with a semiorder or an interval order structure on each X_i (on these notions see, *e.g.*, Roubens and Vincke (1985)). Condition S2 is hardly restrictive in the context of MCDM.

---

[1] This not to say that these relations are useless for decision-aid purposes. A number of techniques have been devised in order to rank the alternatives or to choose one of them on the basis of such relations, see, *e.g.*, Roy and Vincke (1981).

We will also use a structural condition aiming at introducing a minimal diversity among the alternatives to be compared. For k = 2, 3, ..., we consider:

D(k) for all $i \in N$, there is a set $Y_i \subset X_i$ with $|Y_i| = k$ such that for all $a_i, b_i \in Y_i$ with $a_i \neq b_i$ either $a_i\ P_i\ b_i$ or $b_i\ P_i\ a_i$, furthermore $Y = \prod_{i \in N} Y_i \subset X$.

Condition D(k) implies that on each attribute it possible to find k levels that can be distinguished in terms of strict preference and that all possible combinations of these k levels are in X. This condition may seem very restrictive. It should however be noticed that the alternatives which have to be in X if D(k) holds are very similar to the "artificial" alternatives that are used in many methods to assess inter-attribute information (see, *e.g.*, Keeney and Raiffa (1976) or Roy *et al.* (1986)). Furthermore, the set of alternatives has to be rich enough if we want such properties as transitivity or absence of circuit to be meaningful. Therefore we will use this condition throughout the paper though weaker diversity conditions could sometimes have been used at the cost of a greater complexity. As it will become apparent in the next sections, this structural condition plays a crucial part in our proofs in allowing to transfer classical "multi-profile" results into a "single profile" context (see section 5). Most of our results use D(3).
An aggregation procedure in MCDM is a rule allowing to build a preference relation on X on the basis of the relations R_i on X_i and inter-attribute information. As already mentioned, we restrict our attention here to methods producing an asymmetric relation P on X.
A usual condition relating P and the R_i is:

U for all $x, y \in X$, $[x_i\ P_i\ y_i$ for all $i \in N] \Rightarrow x\ P\ y$.

Condition U is a hardly controversial unanimity condition for strict preference. A slightly more restrictive unanimity condition is:

U* for all $x, y \in X$, $[x_i\ R_i\ y_i$ for all $i \in N$ and $x_j\ P_j\ y_j$ for some $j \in N] \Rightarrow x\ P\ y$.

As soon as all weights p_i are strictly positive it is easy to see that, when a relation P is obtained by (1), it satisfies U* and, thus, U. We are not aware of any MCDM aggregation procedure that does not satisfy U. However, as exemplified by ELECTRE I (see Roy (1968)), some procedures may fail to satisfy U*.
The following condition aims at capturing a crucial property underlying the idea of concordance:

NC for all $x, y, z, w \in X$, $[x_i\ R_i\ y_i \Leftrightarrow z_i\ R_i\ w_i$ and $y_i\ R_i\ x_i \Leftrightarrow w_i\ R_i\ z_i$ for all $i \in N] \Rightarrow$ $[x\ P\ y \Rightarrow z\ P\ w]$.

Condition NC has been introduced by Fishburn (1975 and 1976) under the name of noncompensation. It has been studied by Bouyssou and Vansnick (1986) and Bouyssou (1986). This condition implies that preference among two alternatives x and y only depends on the subsets of

N for which x_i R_i y_i and y_i R_i x_i and seems at the heart of the idea of concordance. It is not difficult to see that when a relation P is obtained by (1) it satisfies condition NC.
Condition NC does not convey any notion of monotonicity which seems a crucial part of any aggregation procedure. Combining NC with an idea of monotonicity, we obtain:

M for all $x, y, z, w \in X$, $[x_i\ P_i\ y_i \Rightarrow z_i\ P_i\ w_i$ and $x_i\ I_i\ y_i \Rightarrow z_i\ R_i\ w_i$ for all $i \in N] \Rightarrow$ $[x\ P\ y \Rightarrow z\ P\ w]$.

Condition M implies that, when x is strictly preferred to y, if the set of attributes for which there is a strict preference for x over y is enlarged then x remains strictly preferred to y regardless of what is happening on the other attributes. It is easy to see that condition M implies condition NC. When a binary relation P is obtained by (1), it is not difficult to see that it satisfies condition M.
Suppose that $\underline{P}$ is a preference relation obtained using the idea of concordance. Taking discordance into account leads to a relation $P \subset \underline{P}$, *i.e.* a relation in which a number of preferences have been deleted. Generalizing the conditions we just introduced, we obtain:

WNC for all $x, y, z, w \in X$, $[x_i\ R_i\ y_i \Leftrightarrow z_i\ R_i\ w_i$ and $y_i\ R_i\ x_i \Leftrightarrow w_i\ R_i\ z_i$ for all $i \in N] \Rightarrow$ $[x\ P\ y \Rightarrow \text{Not}(w\ P\ z)]$.

WM for all $x, y, z, w \in X$, $[x_i\ P_i\ y_i \Rightarrow z_i\ P_i\ w_i$ and $x_i\ I_i\ y_i \Rightarrow z_i\ R_i\ w_i$ for all $i \in N] \Rightarrow$ $[x\ P\ y \Rightarrow \text{Not}(w\ P\ z)]$.

It is easy to see that [WM $\Rightarrow$ WNC] and that if a binary relation $\underline{P}$ satisfies condition NC (resp. M) then all binary relations P such that $P \subset \underline{P}$ satisfy condition WNC (resp. WM)[1].

III- Properties of outranking relations based on concordance

The purpose of this section is to investigate some consequences of conditions NC and M when they are coupled with some particular properties of P. It is easy to see that a binary relation P built using relation (1) satisfies U and M (and, thus NC). Simple numerical examples inspired from Condorcet's paradox show that, in general, such a relation will not be transitive and will have circuits. Two simple ways to avoid these phenomena can be envisaged.
The first one is to chose the threshold ρ very close to one so that x P y is obtained only when all the attributes are unanimous to support this proposition. Such a solution is however extremely ineffective and leads to a relation P that is very poor.

---

[1] It is worth noting that this would not be the case if discordance were applied to a binary relation satisfying NC but not asymmetric, as this is the case for example in ELECTRE I.

The second one consists in giving a very large weight to a particular criterion so that the relation P more or less coincides with the preference relation on that criterion. This is not a very attractive solution however since it amounts to amounts to giving much power to a single criterion. Of course many other ways to obtain a relation P with "nice" transitivity properties can be envisaged. The following results show that, when the structure of X is sufficiently rich, imposing "nice" transitivity properties on P always leads to a somewhat unappealing distribution of power among the attributes. We sum up our results in:

Theorem 1. Given S1-S2,

(a) [D(3), NC, U, P is asymmetric and negatively transitive] $\Rightarrow$
there is an $i \in N$ such that for all $x, y \in X$, $x_i\ P_i\ y_i \Rightarrow x\ P\ y$.

(b) [D(3), NC, U, P is asymmetric and transitive] $\Rightarrow$
there is a unique $O \subset N$ such that for all $x, y \in X$:
$x_i\ P_i\ y_i$ for all $i \in O \Rightarrow x\ P\ y$
$x_j\ P_j\ y_j$ for some $j \in O \Rightarrow$ Not($y\ P\ x$).

(c) [D(n), M, P is without circuit] $\Rightarrow$
there is an $i \in N$ such that for all $x, y \in X$, $x_i\ P_i\ y_i \Rightarrow$ Not($y\ P\ x$).

In proving theorem 1, the following definitions will be useful. We say that $A \subset N$ is

- decisive if, for all $x, y \in X$, $x_i\ P_i\ y_i$ for all $i \in A \Rightarrow x\ P\ y$,
- almost decisive if, for all $x, y \in X$, $x_i\ P_i\ y_i$ for all $i \in A$ and $y_j\ P_j\ x_j$ for all $j \notin A \Rightarrow x\ P\ y$,
- semi-decisive if, for all $x, y \in X$, $x_i\ P_i\ y_i$ for all $i \in A \Rightarrow$ Not($y\ P\ x$),
- almost semi-decisive if, for all $x, y \in X$, $x_i\ P_i\ y_i$ for all $i \in A$ and $y_j\ P_j\ x_j$ for all $j \notin A \Rightarrow$ Not($y\ P\ x$).

We state without proof the following obvious facts as:

Lemma 1.

(a) M $\Rightarrow$ NC $\Rightarrow$ [$x\ P\ y$ (resp. Not($y\ P\ x$) for some $x, y \in X$, such that $x_i\ P_i\ y_i$ for all $i \in A$ and $y_j\ P_j\ x_j$ for all $j \notin A \Rightarrow$ A is almost decisive (resp. almost semi-decisive)]

(b) M $\Rightarrow$ [A is almost semi-decisive $\Rightarrow$ A is semi-decisive].

We also have:

Lemma 2.

For all $A \subset N$, [S1, S2, NC, U, D(3), P is asymmetric and transitive] $\Rightarrow$ [A is almost decisive (resp. almost semi decisive) $\Rightarrow$ A is decisive (resp. semi decisive)].

Proof of lemma 2.

If A = N, the conclusion follows. If not, use D(3) with $Y_i = \{a_i, b_i, c_i\}$ and $a_i\ P_i\ b_i$ and $b_i\ P_i\ c_i$ to construct the following alternatives:

	A	N\A
x	a_i	d_i
y	b_i	a_i
z	c_i	e_i

where $a_i \; P_i \; d_i$ and $a_i \; P_i \; e_i$. Such alternatives are in X by D(3). By U we get y P z.
Suppose that A is almost decisive. We thus have x P y and the transitivity of P leads to x P z, independently of the comparison of d_i and e_i on the attributes of N\A. In view NC, this proves that A is decisive.
Suppose now that A is almost semi decisive. If z P x, the transitivity of P implies y P x contradicting the fact that A is almost semi-decisive. We thus have Not(z P x), independently of the comparison of d_i and e_i on the attributes of N\A. In view NC this proves that A is semi-decisive. □

Proof of theorem 1.

Proof of part (a).
We prove that if A is decisive and |A| >1 then some proper subset of A is decisive. In order to do so let B be a proper subset of a decisive set A and use D(3) with $Y_i = \{a_i, b_i, c_i\}$ and $a_i \; P_i \; b_i$ and $b_i \; P_i \; c_i$ to construct the following alternatives:

	B	A\B	N\A
x	a_i	c_i	b_i
y	b_i	a_i	c_i
z	c_i	b_i	a_i

A being decisive, we have y P z. If x P z then NC implies that B is almost decisive and, by lemma 2, decisive (since asymmetry and negative transitivity imply transitivity). If Not(x P z) then Not(y P x) would imply by negative transitivity Not(y P z) violating the fact that A is decisive. Thus, we have y P x. Given NC and lemma 2, we conclude in this case that A\B is decisive.
By U we know that N is decisive. Repeating the previous argument leads to the conclusion that a singleton is decisive, completing the proof of part (a).

Proof of part (b).
By U, we know that N decisive. Given S2, there exists at least one decisive set of minimal cardinality. Let O be one of them. We have $x_i \; P_i \; y_i$ for all $i \in O \Rightarrow$ x P y.
If |O| = 1, then we obviously have that $x_j \; P_j \; y_j$ for some $j \in O \Rightarrow$ Not(y P x).
If not consider $\{i\} \subset O$ and use D(3) with $Y_i = \{a_i, b_i, c_i\}$ and $a_i \; P_i \; b_i$ and $b_i \; P_i \; c_i$ to construct the following alternatives:

	{i}	O\{i}	N\O
x	c_i	a_i	b_i
y	a_i	b_i	c_i
z	b_i	c_i	a_i

O being decisive, we have y P z. If x P z, then, given NC, O\{i} is almost decisive and thus decisive, by lemma 2, violating the fact that O is a decisive set of minimal cardinality . We thus have Not(x P z). But this implies Not(x P y) since x P y, y P z and Not(x P z) would contradict the transitivity of P. Given NC, Not(x P y) implies that {i} is almost semi-decisive and thus semi-decisive by lemma 2. Therefore all singletons in O are semi-decisive.
The proof of (b) is completed observing that O is necessarily unique. In fact suppose that there are two sets O and O' with O ≠ O' satisfying the conclusions of (b). Consider the following alternatives which by D(3) are in X:

	O	O'\O	N\O∪O'
x	a_i	b_i	a_i
y	b_i	a_i	a_i

We have, by construction, x P y and Not(y P x), a contradiction.

Proof of part (c).
Suppose, in contradiction with the thesis, that no singleton is semi-decisive. Given M and in view of part (b) of lemma 1, this implies that no singleton is almost semi-decisive, *i.e.* that: for all $i \in N$, $x_i \; P_i \; y_i$ and $y_j \; P_j \; x_j$ for all $j \neq i$ imply y P x.
We use D(n) to construct the following n alternatives with $Y_i = \{x_i^1, x_i^2, \ldots, x_i^n\}$ and $x_i^1 \; P_i \; x_i^2$, $x_i^2 \; P_i \; x_i^3 \ldots, x_i^{n-1} P_i \; x_i^n$,

	{1}	{2}	{3}	...	{n}
x^1	x_1^1	x_2^n	x_3^{n-1}		x_n^2
x^2	x_1^2	x_2^1	x_3^n		x_n^3
x^3	x_1^3	x_2^2	x_3^1		x_n^4
...					
x^n	x_1^n	x_2^{n-1}	x_3^{n-2}		x_n^1

We have $x^1 \; P \; x^2$, $x^2 \; P \; x^3$, ..., $x^n \; P \; x^1$, which violates the fact that P has no circuits and completes the proof of part (c). □

Theorem 1 and its proof have strong connections with classical results in Social Choice Theory that will be explored in section 5. We briefly comment here each part of theorem 1.
(a). This result says that if a concordance relation is a weak order then some attribute dictates its strict preferences to all others. Given D(3), this distribution of "power" among the various

attributes seems very undesirable. Very similar results have been proved by Fishburn (1975), Plott *et al.* (1975), Roberts (1980), Pollack (1979), Parks (1976) and Kemp and Ng (1976). As its proof suggests, this result appears as a MCDM counterpart of Arrow's theorem that takes advantage of the structure of cartesian product of the set of alternatives[1]. Fishburn (1975) proves that if U is replaced by U* then P is obtained lexicographically *i.e* there is a linear order >> on N such that for all $x, y \in X$:

$x\ P\ y \Leftrightarrow [x_i\ P_i\ y_i$ for some $i \in N$ and for every $k \in N$ such that $y_k\ P_k\ x_k$ there is a $j \in N$ such that $j >> k$ and $x_j\ P_j\ y_j]$.

When D(3) is strengthened to D(4), it is easy to see that it is possible to replace [P is asymmetric and negatively transitive] by [P is asymmetric and for all $x, y, z, w \in X$, $x\ P\ y$ and $y\ P\ z \Rightarrow x\ P\ w$ or $w\ P\ z$] or by [P is asymmetric and for all $x, y, z, w \in X$, $x\ P\ y$ and $z\ P\ w \Rightarrow x\ P\ w$ or $z\ P\ y$] without altering the conclusion. Thus asking for a semi-order or even an interval order instead of a weak order as the result of an aggregation satisfying U and NC does not change the situation.

(b). This result is a single profile counterpart of a result of Weymark (1983). It says the transitivity of P together with D(3), NC and U generates what is usually called an "oligarchy" of attributes. A good example of this situation is offered by the (strict) dominance relation defined by:

$x\ P\ y \Leftrightarrow x_i\ P_i\ y_i$ for all $i \in N$.

With this relation the oligarchy is the entire set N. Smaller oligarchies lead to a richer relation P at the cost of giving more "power" to a smaller number of attributes.

(c). This result is a single profile counterpart of a result of Blau and Deb (1977). It shows that an asymmetric concordance relation without circuit gives much power to a single attribute when the structure of X is very rich[2].

IV- Properties of outranking relations based on concordance-discordance

Suppose that $\underline{P}$ is a preference relation on X obtained using an idea of concordance. Taking discordance into account leads to a relation $P \subset \underline{P}$. In general, discordance may impoverish $\underline{P}$ in rather an uncontrollable way. For instance, the transitivity of $\underline{P}$ may be destroyed in P. On the contrary, if $\underline{P}$ has circuits, it may happen than discordance destroys all these circuits.

---

[1] It should be noted that a similar result holds if we suppose that we are working on a binary relation R that is complete and transitive when the last part of NC is changed to $[x\ R\ y \Rightarrow z\ R\ w]$, which is more in line with the usual presentation of Arrow's theorem.

[2] Such a rich structure for X can be criticized. Using D(3), a similar result could be obtained strengthening condition M to a condition of "positive responsiveness" leading to a single profile counterpart of a result of Mas-Colell and Sonnenschein (1972).

The only thing that seems reasonable to ask on P is the absence of circuit. This is the case if $\underline{P}$ has no circuit[1]. But, as we mentioned, it may also happen that P has no circuit because all the circuits in $\underline{P}$ have been destroyed by discordance. It is therefore easy to see that imposing that P has no circuit coupled with WNC or WM will not create results similar to those of section 3. One simple way, among others, to obtain similar results is to limit the extent of discordance. For instance we may impose that there is no discordance between the elements of $Y = \prod_{i \in N} Y_i$ where the sets Y_i are those used in the diversity condition D(k). This amounts to supposing that, though the elements of Y_i can be distinguished in terms of strict preference, they are sufficiently "close" to one another not to generate discordance effects. This hypothesis seems in line with the way "artificial" alternatives are introduced in X in order to assess inter-attribute information (see, *e.g.*, Roy *et al.* (1986)). Therefore we reformulate conditions WM as:

WM* for all $x, y, z, w \in X$, $[x_i\ P_i\ y_i \Rightarrow z_i\ P_i\ w_i$ and $x_i\ I_i\ y_i \Rightarrow z_i\ R_i\ w_i$ for all $i \in N] \Rightarrow$
$[x\ P\ y \Rightarrow \text{Not}(w\ P\ z)$, furthermore if $z, w \in Y$ then $z\ P\ w]$.

Though this condition may seem *ad hoc*, it has a simple interpretation. If a relation P has been obtained by applying discordance to a relation satisfying M then it satisfies WM* if there is no discordance between the elements of Y. We have the following:

Theorem 2.
Given S1-S2,
[D(n), WM*, P is without circuit] $\Rightarrow$
there is an $i \in N$ such that for all $x, y \in X$, $x_i\ P_i\ y_i \Rightarrow \text{Not}(y\ P\ x)$.

Proof of theorem 2.
It is easy to see that if P satisfies WM* on X then it satisfies M on Y. Since P has no circuit on Y, we know from theorem 1 (c) that there is an $i \in N$ such that for all $x, y \in Y$, $x_i\ P_i\ y_i \Rightarrow \text{Not}(y\ P\ x)$. Suppose now that for some $z, w \in X$, we have $z_i\ P_i\ w_i$ and $w\ P\ z$. From D(3), we know that there are $r, s \in Y$ such that $r_i\ R_i\ s_i \Leftrightarrow z_i\ R_i\ w_i$ and $s_i\ R_i\ r_i \Leftrightarrow w_i\ R_i\ z_i$ for all $i \in N$. Therefore, WM* implies that $s\ P\ r$, a contradiction. This completes the proof. □

V- Relation with Social Choice Theory.

As already argued, the results of section 3 are transpositions of classical results in the field of Social Choice Theory. In order to understand better the extent of this transposition, it is worth recalling here the classical result of Arrow.

---

[1] Note that this is not so if discordance is applied to a relation that is not asymmetric. In that case, discordance may create as well as destroy circuits in the asymmetric part of the relation. This the reason why we restrict our attention here to asymmetric relations.

A central theme in Social Choice Theory is to study how the preferences of several individuals can be aggregated in a "reasonable" way. Let X be a set of objects called "alternatives" and N a finite set, the $|N| = n$ elements of N being interpreted as "voters". We define $\mathcal{R}_X$ as the set of all binary relations on X. A Social Aggregation Procedure is a function:

$$G: \quad E \subset [\mathcal{R}_X]^n \rightarrow F \subset \mathcal{R}_X$$
$$(R_1, R_2, \ldots, R_n) \mapsto G(R_1, R_2, \ldots, R_n) = R$$

associating[1] a binary relation on X to n-uples of binary relations on X.
Let R_X be the set all all complete and transitive binary relations on X. We introduce the following conditions:

D $\quad E = [R_X]^n$ [Domain]

C $\quad F = R_X$ [Codomain]

I $\quad$ for all $(R_1, R_2, \ldots, R_n)$, $(R_1', R_2', \ldots, R_n') \in E$, and for all $x, y \in X$, [Independence]
$[x R_i y \Leftrightarrow x R_i' y$ and $y R_i x \Leftrightarrow y R_i' x$ for all $i \in N] \Rightarrow$
$[x R y \Leftrightarrow x R' y$ and $y R x \Leftrightarrow y R' x]$

UN $\quad$ for all $(R_1, R_2, \ldots, R_n)$ E, and for all $x, y \in X$, [Unanimity]
$[x P_i y$ for all $i \in N] \Rightarrow x P y$

ND $\quad$ for all $i \in N$, [Nondictatorship]
$x P_i y$ and Not$(x P y)$ for some $(R_1, R_2, \ldots, R_n) \in E$ and some $x, y \in X$

In this context we have the following:

Theorem (Arrow (1963)): When $|X| \geq 3$, there is no Social Aggregation Procedure satisfying D, C, I, UN and ND.

The relations between Arrow's result and part (a) of theorem 1 have been explored by many authors (*e.g.*, Fishburn (1975), Roberts (1980), Pollack (1979), Parks (1976) and Kemp and Ng (1976)). It will suffice to recall here that, in the framework of Arrow's theorem, all possible profiles of weak orders are in the domain of the Social Aggregation Procedure and that condition I relates the result of the aggregation between two different profiles. This multi-profile formulation is at the crux of the "impossibility" result. On the contrary, in theorem 1 only one particular profile of preference relation on each attribute is used. But D(k) requires the set X to contain alternatives for which the preferences on each attribute are conflictual. This diversity together with NC lead to the results of theorem 1. Thus, condition D(k) appears as the counterpart of condition D in the single-profile context.

---

[1] in the sequel, it is understood that $R = G(R_1, R_2, \ldots, R_n)$ and $R' = G(R_1', R_2', \ldots, R_n')$. As before, P denotes the asymmetric part of R.

Both theorems use a similar unanimity condition (U and UN) and there is an obvious correspondence between condition C in Arrow's theorem and the requirement that P be asymmetric and negatively transitive in theorem 1 (a). It is worth observing that when I is coupled with UN, C and D, it implies the following "neutrality" condition:
for all $(R_1, R_2, ..., R_n), (R_1', R_2', ..., R_n') \in E$, and for all $x, y, z, w \in X$,
$[x\, R_i\, y \Leftrightarrow z\, R_i'\, w$ and $y\, R_i\, x \Leftrightarrow w\, R_i'\, z$ for all $i \in N] \Rightarrow$
$[x\, R\, y \Leftrightarrow z\, R'\, w$ and $y\, R\, x \Leftrightarrow w\, R'\, z]$,
which is an analogue of NC with several profiles.
Parts (b) and (c) of theorem 1 have similar multi-profile correspondents in the Theory of Social Choice (see Weymark (1983) and Blau and Deb (1977)).
Consider, for instance, the concordance part of TACTIC as defined by (1). Given a set of alternatives X evaluated on a set of attributes N, once the weights p_i and the threshold ρ have been chosen, it is possible to see (1) as a way to aggregate any n-uples of weak orders defined on X. It is easy to see that this aggregation satisfies conditions D, UN and I and consequently violates either C or ND. Thus, Arrow's result applies directly to the context of MCDM and may lead the reader to question the interest of the reformulation presented in theorem 1.
Apart from bringing to the attention of the MCDM community several results that are less famous than Arrow's theorem, it seems to us that the single-profile approach is of special interest in the context of MCDM. We refer to Sen (1986), Fishburn (1987), Roberts (1980) and Pollack (1979) for a thorough analysis of the comparison between the multi-profile and the single-profile approaches in Social Choice Theory[1]. We will only emphasize here what seem to be the advantages of a single profile formulation in the context of MCDM.
It should first be observed that the multi-profile format, when applied to MCDM, does not make use of the fact that alternatives are multiattributed, which, in our opinion, is a fundamental characteristic of MCDM. Second, it is worth noting that the proof of theorem 1 is, by far, simpler than the corresponding proofs for the multi-profile case. Third, in order to apply a multi-profile result to MCDM, one has to suppose that all inter-attribute information remains unchanged when aggregating different profiles, whereas it is well-known, in practice, that the modelling of such information crucially depends on the particular problem that is under study. Finally, whereas a multi-profile result is of direct interest to the analyst wishing to create a method that "works" in all cases, it is of limited interest to the user of the method who is always confronted to a single profile, *i.e.* to a given set of alternatives with given evaluations on several attributes. Define a MCDM problem as a set of alternatives evaluated on several attributes. Suppose that you want to informally sum up multi-profile results for a user of MCDM techniques. A "multi-profile explanation" could sound like:

---

[1] In particular, Roberts (1980) studies to what extent multi-profile results have single-profile analogues.

"You have applied an outranking method based on a concordance-discordance principle with given weights and threshold to a particular problem. If the concordance relation you obtain has 'nice properties', you should check whether your choice of weights and threshold does not amount to giving much power to a single attribute. If this is not the case, it can be proved that it is possible to find at least one problem with the same number of alternatives and attributes as your problem such that, if you apply the same method with the same weights and threshold, to this new problem, the resulting concordance relation will not have 'nice properties'".

An informal explanation of single-profile results could be:

"You have applied an outranking method based on a concordance-discordance principle to a particular problem. If the concordance relation you obtain has 'nice properties', you should check whether your choice of weights and threshold does not amount to giving much power to a single attribute. If this is not the case, it can be proved that if you add some alternatives to your problem, very similar to those you have used to assess your weights, the concordance relation will loose its nice properties unless you modify the weights and threshold so as to give much power to a single attribute."

The choice of a particular presentation is, to a certain extent a matter of taste. However, if the emphasis is on the user of the method and if it is agreed that modelling inter-attribute information often implies to take into consideration "artificial" alternatives and is highly specific to a particular problem, it seems that the single-profile approach we have used has definite advantages.

References

Arrow K.J. (1963), *Social Choice and Individual Values*, Wiley, New-York.

Blau J.H. and R. Deb (1977), Social decision functions and the veto, *Econometrica*, **45**, 871-879.

Bouyssou D. (1986), Some remarks on the notion of compensation in MCDM, *European Journal of Operational Research*, **26**, 150-160.

Bouyssou D.and J.C. Vansnick (1986), Noncompensatory and generalized noncompensatory preference structures, *Theory and Decision*, **21**, 251-266.

Fishburn P.C. (1975), Axioms for lexicographic preferences, *Review of Economic Studies*, **42**, 415-419.

Fishburn P.C. (1976), Noncompensatory preferences, *Synthese*, **33**, 393-403.

Fishburn P.C. (1987), *Interprofile Conditions and Impossibility*, Harwood Academic Publishers, Chur.

Goicoechea A., D.R. Hansen and L. Duckstein (1982), *Multiobjective Decision Analysis with Engineering and Business Applications*, Wiley, New-York.

Keeney R.L. and H. Raiffa (1976), *Decisions with Multiple Objectives, Preferences and Value Tradeoffs*, Wiley, New-York.

Kemp M.C. and Y.K. Ng (1976), On the existence of social welfare functions, social orderings and social decision functions, *Economica*, **43**, 59-66.

Mas-Colell A. and H. Sonnenschein (1972), General Possibility theorems for group decisions, *Review of Economic Studies*, **39**, 185-192.

Parks R.P. (1976), An impossibility theorem for fixed preferences: a dictatorial Bergson-Samuelson welfare function, *Review of Economic Studies*, **43**, 447-450.

Plott C.R., J.T. Little and R.P. Parks (1975), Individual choice when objects have 'ordinal' properties, *Review of Economic Studies*, **32**, 403-413.

Pollack R.A. (1979), Bergson-Samuelson social welfare functions and the theory of social choice, *Quarterly Journal of Economics*, **93**, 73-90.

Roberts K.W. (1980), Social choice theory: the single-profile and multi-profile approaches, *Review of Economic Studies*, **47**, 441-450.

Roubens M. and Ph. Vincke (1985), *Preference Modelling*, Springer Verlag, Heidelberg.

Roy B. (1968), Classement et choix en présence de points de vue multiples (la méthode ELECTRE), *RIRO*, **8**, 57-75.

Roy B. (1971), Problems and methods with multiple objective functions, *Mathematical Programming*, **1**, 239-266.

Roy B. and P. Bertier (1973), La méthode ELECTRE II - Une application au media-planning, in M. Ross (Ed.), *OR'72*, 291-302, North Holland, Amsterdam.

Roy B. and D. Bouyssou (1987), *Procédures d'agrégation multicritère conduisant à un critère unique de synthèse*, Working Paper, LAMSADE, Université de Paris-Dauphine.

Roy B., M. Présent and D. Silhol (1986), A programming method for determining which Paris metro station should be renovated, *European Journal of Operational Research*, **24**, 318-334.

Roy B. and Ph. Vincke (1981), Multicriteria analysis: survey and new directions, *European Journal of Operational Research*, **8**, 207-218.

Sen A.K. (1986), Social Choice Theory, in K.J. Arrow and M.D. Intriligator (Eds.), *Handbook of Mathematical Economics Vol. III*, 1073-1181, North Holland, Amsterdam.

Siskos J., G. Wascher and H.M. Winkels (1983), *A bibliography on outranking approaches (1966-1982)*, Working Paper, LAMSADE, Université de Paris-Dauphine.

Vansnick J.C. (1986), On the problem of weights in multiple criteria decision making - The noncompensatory approach, *European Journal of Operational Research*, **24**, 288-294.

Weymark J.A. (1983), Arrow's theorem with social quasi-orderings, *Public Choice*, **42**, 235-246.

AN INTERACTIVE COMBINED BRANCH-AND-BOUND/TCHEBYCHEFF ALGORITHM FOR MULTIPLE CRITERIA OPTIMIZATION

Dr. Anthony Durso

United States Army and the RAND Corporation

Washington, D.C. 20037

A new interactive algorithm, designed to solve a general class of multiple criteria decision making problems, is discussed using branch-and-bound techniques and the augmented weighted Tchebycheff metric. The algorithm is powerful because few mathematical assumptions are necessary for it to work, it is simple to use and solution vectors are guaranteed to be nondominated.

Efficiencies, such as the identification of redundant subsets and the elimination of certain optimization runs, are included in the algorithm in order to help speed up its termination. Quasi-dominated and quasi-duplicate points are defined and their identification and elimination are built into the algorithm. These new concepts were developed to accommodate decision maker "fuzziness" and to further expedite the termination of the algorithm. The concept of secondary criteria is also introduced in an attempt to help decision makers deal with indecision.

In the later part of the paper, a practical application of the new algorithm will be discussed. The problem is a bicriteria manpower planning problem done in coordination U.S. Army Total Personnel Command.

AN INTERACTIVE COMBINED BRANCH-AND-BOUND/TCHEBYCHEFF ALGORITHM FOR MULTIPLE CRITERIA OPTIMIZATION

Dr. Anthony Durso

United States Army and the RAND Corporation

Washington, D.C. 20037

A new interactive algorithm, designed to solve a general class of multiple criteria decision making problems, is developed using branch-and-bound techniques and the augmented weighted Tchebycheff metric. The new algorithm is powerful because few mathematical assumptions are necessary for it to work, because it is simple to use, and because solution vectors are guaranteed to be nondominated.

Efficiencies, such as the identification of redundant subsets and the elimination of certain optimization runs, are included in the algorithm in order to help speed up its termination. Quasi-dominated and quasi-duplicate points are defined and their identification and elimination are built into the algorithm. These new concepts are intended to simulate decision maker "fuzziness" and to further expedite the termination of the algorithm. The concept of secondary criteria is introduced into the algorithm, in an attempt to help decision makers deal with indecision.

1. Introduction

Since the end of World War II, a great deal of work has been done to develop efficient methods to solve single objective mathematical programming problems. These problems are of the form:

$$\text{Maximize } f(x) \text{ subject to } x \in X,$$

where $X \subset R^n$ is the set of feasible points and f is a real-valued function defined on R^n.

Historically, decision makers (DMs) in the government and in the private sector have depended on the solutions of single objective (criterion) mathematical programming problems as aids in the decision making process. However, the world has become exceedingly complex, over time, and DMs have learned that important business decisions can no longer be made strictly in this fashion [Zeleny (1982)].

In practice, important decisions are made after carefully evaluating solution alternatives involving more than one criterion (multiple criteria). As a result, a need has existed for the development of multiple criteria decision making (MCDM) problem solving techniques. The MCDM problem can be written as:

$$\text{Maximize } \mathbf{f}(\mathbf{x}) \text{ subject to } \mathbf{x} \in X, \qquad (P_X)$$

where $\mathbf{f}(\mathbf{x}) = [f_1(\mathbf{x}), f_2(\mathbf{x}), \ldots, f_p(\mathbf{x})]$ is a vector-valued criterion function defined over a set X, that is assumed to be a closed and bounded subset of R^n, and where the real-valued function $f_k(\mathbf{x})$, k=1, . . . , p, gives the value of the kth criterion at point **x**. Let Z be the image of X under the mapping f, i.e.,

$$Z \equiv \{\mathbf{z} \in R^p | \mathbf{z} = \mathbf{f}(\mathbf{x}) \text{ for some } \mathbf{x} \in X\}.$$

Throughout this paper it is assumed that Z is closed, bounded, and nonempty, and that, for each criterion function $f_j(\mathbf{x})$, the DM prefers (or "likes better") a higher value to a lower value. Although the DM seeks an element of X to choose as a decision, the set Z is of more direct interest since it indicates the criteria values that may be achieved. The above problem can be written equivalently as:

$$\text{Maximize } \mathbf{z} \text{ subject to } \mathbf{z} \in Z. \qquad (P_Z)$$

Since **z** is a vector, the meaning of the expression "Maximize **z**" is not obvious. Therefore, a few basic definitions need to be introduced before proceeding.

Definition 1 Let (P_Z) be the problem as defined above and let $\beta_k = \max_{z \in Z} z_k$. Then $\beta = (\beta_1, \beta_2, \ldots, \beta_p)$ is the *ideal* of (P_Z). If $\beta \in Z$, then β is the *ideal solution* of (P_Z).

Definition 2 z' is an *inferior solution* of problem (P_Z) if $\mathbf{z}' \in Z$ and $\exists\, \mathbf{z} \in Z$ such that $z'_k \le z_k \ \forall\, k \in \{1, 2, \ldots, p\}$ and $z'_k < z_k$ for at least one index k. A solution which is not inferior is called a *nondominated solution* (Pareto optimal solution).

In light of the two previous definitions, if $\beta \in Z$, then the meaning of "Maximize z" is obvious. If $\beta \notin Z$, then the DM can restrict his or her decision to the ***set of nondominated solutions***, which shall be denoted by N:

$$N \equiv \{z \in Z | z \text{ is nondominated}\}.$$

This leads to the next definition.

Definition 3 An *optimal solution* to an MCDM problem is a nondominated solution that the DM finds no less preferable than any other feasible solution.

In order to ensure that the optimal values of the decision variables x_j, j=1, . . . , n, are available for the DM, we will be working with the following form of the MCDM problem:

$$\text{Maximize } \mathbf{z} \text{ subject to } \mathbf{f}(\mathbf{x}) = \mathbf{z},\ \mathbf{x} \in X. \qquad (P_{X,\,Z})$$

Solutions to $(P_{X,\,Z})$ are vectors $(\mathbf{x}, \mathbf{z}) \in R^{n+p}$.

2. Theoretical Foundation of the Algorithm

An MCDM problem solving algorithm should only provide the DM with nondominated solutions. If the algorithm is to be truly flexible and versatile, it must be based on a very general characterization of N, the set of nondominated solutions. Soland (1979) provides a good general characterization of N, which is presented here.

Definition 4 [Steuer (1986, p. 148)]: A function v: $R^p \rightarrow R$ is *coordinatewise increasing* iff for all $z^1, z^2 \in R^p$ such that $z^1 \geq z^2, z^1 \neq z^2$, then $v(z^1) > v(z^2)$.

Now let f(x), X, Z, and N be defined as before and let

$$G_Z \equiv \{\text{functions on } R^p \text{ that are coordinatewise increasing on } Z\}.$$

For $g \in G_Z$ and $b \in R^p$, define the scalar maximization problem $(P_{g,\,b})$:

$$\begin{aligned} &\text{Maximize } g(z) \\ &\text{subject to } z \in Z \\ &\qquad z \geq b. \qquad\qquad (P_{g,\,b}) \end{aligned}$$

Theorem 1 [Soland (1979)]. Let g be an arbitrary element of G_Z. Then $z^\circ \in N$ iff z° is optimal in $(P_{g,\,b})$ for some $b \in R^p$.

Proof:

$\Leftarrow$ Suppose that z° is optimal in $(P_{g,\,b})$ for some b. If $z^\circ \notin N$, then $\exists\, z' \in Z$ such that $z' \geq z^\circ$ and $z'_k > z^\circ_k$ for at least one index k. Thus $g(z') > g(z^\circ)$ and since z' is feasible in $(P_{g,\,b})$, z° cannot be optimal in $(P_{g,\,b})$, which is a contradiction.

$\Rightarrow$ If $z^\circ \in N$, consider the problem $(P_{g,\,b})$ with $b = z^\circ$, which implies z° is feasible. If z° is not optimal in $(P_{g,\,b})$, there must exist an $z' \in Z$ such that $z' \geq b = z^\circ$ and $g(z') > g(z^\circ)$. Since g is coordinatewise increasing, $z'_k > z^\circ_k$ for some index k. Thus $z^\circ \notin N$, which is a contradiction. □

This general characterization is a useful one. For any arbitrary coordinatewise increasing function, g, if the MCDM problem has an optimal solution, it is nondominated. Conversely, with g still arbitrary, for every nondominated solution z°, there exists at least one b such that z° is optimal. The flexibility and versatility needed to develop an MCDM interactive algorithm is present in this general characterization of N.

Steuer (1986) has also been able to develop a general characterization of N, using Tchebycheff metrics. His mathematical development of this characterization is fairly extensive and it is accomplished in three parts: one for the discrete-finite case, one for the polyhedral case and one for the nonlinear and infinite discrete cases. Unfortunately, this characterization does not include mixed integer mathematical programming problems. However, by capitalizing on the usefulness of Theorem 1, a more general characterization of N can be developed, one that includes mixed integer programming problems.

Consider the mathematical programming problems

$$\begin{aligned} &\text{Maximize } f_i(\mathbf{z}) \\ &\text{s.t.} \quad \mathbf{z} \in Z, \\ &\qquad\quad \mathbf{z} \geq b \end{aligned} \tag{$P_{f_i,b}$}$$

for $i=1, \ldots, p$, where $f_i(\mathbf{z}) = z_i$ and $b \in R^p$. Since the constraint set of $(P_{f_i,b})$ is closed and bounded, and assuming this set to be nonempty, each $(P_{f_i,b})$, $i=1, \ldots, p$, has an optimal solution, say $\hat{\mathbf{z}}^i$. However, since the possibility of more than one optimal solution exists for a given i, there is no guarantee that $\hat{\mathbf{z}}^i$ is nondominated.

Now consider the mathematical programming problems

$$\begin{aligned} &\text{Maximize } h_i(\mathbf{z}) \\ &\text{s. t.} \quad \mathbf{z} \in Z \\ &\qquad\quad \mathbf{z} \geq b \end{aligned} \tag{$P_{h_i,b,\rho}$}$$

for $i=1, \ldots, p$, where

$$h_i(z) = [1-\rho(p-1)]z_i + \rho \sum_{\substack{j=1 \\ j \neq i}}^{p} z_j,$$

ρ is a very small positive number, e.g. 10^{-4}, and, as before, $b \in R^p$. Since the constraint set of $(P_{h_i,b,\rho})$ is closed and bounded, and assuming this set to be nonempty, each $(P_{h_i,b,\rho})$ has an optimal solution; call it $\overline{\mathbf{z}}^i$. Since ρ is positive, each $h_i \in G_Z$. Therefore, by Theorem 1, $\overline{\mathbf{z}}^i \in N$, $i=1, \ldots, p$. In comparing $(P_{f_i,b})$ and $(P_{h_i,b,\rho})$, we see that the slight modification in replacing $f_i(z)$ by $h_i(z)$ serves to insure that the solution obtained is a nondominated solution of (P_Z).

Definition 5 Consider the problems $(P_{h_i,b,\rho})$. Let $\overline{\mathbf{z}}^i$ be an optimal solution to $(P_{h_i,b,\rho})$ and let $\beta_i^* \equiv \overline{z}_i^i$, $i=1, \ldots, p$. We say that $\beta^* = (\beta_1^*, \ldots, \beta_p^*)$ is an *ideal* of $(P_{h_i,b,\rho})$ and a *quasi-ideal* of $(P_{f_i,b})$. If $\beta^* \in Z$, then β^* is an *ideal solution* of $(P_{h_i,b,\rho})$ and a *quasi-ideal solution* of $(P_{f_i,b})$.

Definition 6 Let β^* be an ideal of $(P_{h_i,b,\rho})$ and let $\lambda \in \overline{\Lambda}$, where

$$\overline{\Lambda} = \{\lambda \in R^p | \lambda_i \geq 0, \sum_{i=1}^{p} \lambda_i = 1\}.$$

Then

$$||\beta^* - \mathbf{z}||_\infty^\lambda \equiv \max \{\lambda_i|\beta_i^* - z_i|, i=1, \ldots, p\}$$

is a member of the family of weighted Tchebycheff metrics for measuring the distance between $\mathbf{z} \in R^p$ and $\beta^* \in R^p$.

Definition 7 Let β^* be an ideal of $(P_{h_i,b,\rho})$, $\lambda \in \overline{\Lambda}$, and $\rho > 0$. Then

$$|||\beta^* - \mathbf{z}|||_\infty^\lambda \equiv ||\beta^* - \mathbf{z}||_\infty^\lambda + \rho \sum_{i=1}^{p} |\beta_i^* - z_i|$$

is a member of the family of augmented weighted Tchebycheff metrics for measuring the distance between $\mathbf{z}$ and β^*.

In order to find a point $\mathbf{z} \in Z$, and satisfying $\mathbf{z} \geq b$, that is closest to β^* according to the weighted Tchebycheff metric, we have the weighted Tchebycheff program

$$\begin{aligned}
&\text{Minimize } \alpha \\
&\text{s. t.} \quad \mathbf{z} \geq \mathbf{b} \\
&\qquad \alpha \geq \lambda_i(\beta_i^* - z_i),\ i=1, \ldots, p \\
&\qquad f(\mathbf{x}) = \mathbf{z} \\
&\qquad \mathbf{x} \in X \\
&\qquad \alpha \geq 0. \qquad\qquad (P_{\alpha,b})
\end{aligned}$$

In order to find a point $\mathbf{z} \in Z$, and satisfying $\mathbf{z} \geq b$, that is closest to β^* according to the augmented weighted Tchebycheff metric, we have the augmented weighted Tchebycheff program

$$\begin{aligned}
&\text{Minimize } \alpha + \rho \sum_{i=1}^{p} (\beta_i^* - z_i) \\
&\text{s.t.} \quad \mathbf{z} \geq b \\
&\qquad \alpha \geq \lambda_i\ (\beta_i^* - z_i),\ i=1, \ldots, p \\
&\qquad f(\mathbf{x}) = \mathbf{z} \\
&\qquad \mathbf{x} \in X \\
&\qquad \alpha \geq 0. \qquad\qquad (P_{\alpha,\rho,b})
\end{aligned}$$

The formulations of $(P_{\alpha,b})$ and $(P_{\alpha,\rho,b})$ differ from those given in Steuer (1986), because of the inclusion of constraints placing lower bounds on b, the inclusion of the constraint $\alpha \geq 0$, and the replacement of Steuer's ideal β^{**} by the quasi-ideal β^*. If β' is the ideal of $(P_{f_i,b})$, $i=1, \ldots, p$, then according to Steuer, $\beta^{**} = \beta' + \epsilon$, where ϵ_i, $i=1, \ldots, p$, is a small positive number. If β^{**} is substituted for β^* in the definitions of the weighted and augmented weighted Tchebycheff metrics, then $\beta_i^{**} - z_i$ is strictly positive for all $i=1, \ldots, p$, the absolute value sign can be omitted, and the problems $(P_{\alpha,b})$ and $(P_{\alpha,\rho,b})$ can be directly written, as given. A solution to either $(P_{\alpha,b})$ or $(P_{\alpha,\rho,b})$ is a vector of the form $(\mathbf{x}, \mathbf{z}, \alpha) \in R^{n+p+1}$, where $\mathbf{z}$ is a closest criterion vector to β^* and $\mathbf{x}$ is its inverse image in decision space.

The proof of nondominance of the augmented weighted Tchebycheff program is originally attributed to Steuer (1986). However, as already indicated, modifications have been made in an attempt to provide a more general characterization of N, one that can include continuous, integer and mixed integer mathematical programming problems.

Theorem 2 Let $M \equiv \{z \in Z \mid (x, z, \alpha)$ solves $(P_{\alpha,b})$ for some $x \in X$ and $\alpha \in R\}$. If $M \neq \phi$, there exists a $\mathring{z} \in M$ such that $\mathring{z} \in N$.

Proof. Let $\tilde{z} \in M$ with $(\tilde{x}, \tilde{z}, \tilde{\alpha})$ a solution to $(P_{\alpha,b})$. If $\tilde{z} \notin N$, consider $(P_{g,b})$ with $g \in G_Z$ arbitrary and $b = \tilde{z}$. The constraint set of this problem is not empty since $\tilde{z}$ is a feasible solution. Let $\mathring{z}$ solve $(P_{g,\tilde{z}})$, and let $\mathring{x} \in X$ such that $f(\mathring{x}) = \mathring{z}$. By Theorem 1, $\mathring{z} \in N$. Since $\mathring{z} \geq \tilde{z}$, $(\mathring{x}, \mathring{z}, \tilde{\alpha})$ is feasible for $(P_{\alpha,b})$, and hence a solution to $(P_{\alpha,b})$. Thus $\mathring{z} \in M \cap N$. □

It is important to point out that, by Theorem 2, a nondominated solution to $(P_{\alpha,b})$ exists and can be computed. However, in cases where more than one optimal solution to $(P_{\alpha,b})$ exists, Theorem 2 loses its attractiveness because of complications arising from the separation of nondominated solutions from dominated solutions. The next theorem shows that the augmented weighted Tchebycheff program yields a general characterization of the nondominated set N.

Theorem 3 Let $\lambda \in \overline{\Lambda}$ and $\rho > 0$ be arbitrary. Then $\overline{z} \in N$ iff $(\overline{z}, \overline{x}, \overline{\alpha})$ is optimal in $(P_{\alpha,\rho,b})$ for some $b \in R^p$, some $\overline{\alpha} \in R$, and some $\overline{x} \in X$ such that $\overline{z} = f(\overline{x})$.

Proof.

$\Leftarrow$ Suppose $(\overline{z}, \overline{x}, \overline{\alpha})$ is optimal in $(P_{\alpha,\rho,b})$. If $\overline{z} \notin N$, then $\exists\ z' \in Z$ and $x' \in X$ such that $z' = f(x')$, $z' \geq \overline{z}$ and $z'_k > \overline{z}_k$ for at least one index k. Then $\beta^*_i - z'_i \leq \beta^*_i - \overline{z}_i$ $\forall i$, so that $(z', x', \overline{\alpha})$ is feasible for $(P_{\alpha,\rho,b})$. Also,

$$\overline{\alpha} + \rho \sum_{i=1}^{p} (\beta^*_i - z'_i) < \overline{\alpha} + \rho \sum_{i=1}^{p} (\beta^*_i - \overline{z}_i).$$

Thus $(z', x', \overline{\alpha})$ is a better solution to $(P_{\alpha,\rho,b})$ than $(\overline{z}, \overline{x}, \overline{\alpha})$, contradicting the assumption that $(\overline{z}, \overline{x}, \overline{\alpha})$ is optimal in $(P_{\alpha,\rho,b})$. Hence $\overline{z} \in N$.

$\Rightarrow$ If $\overline{z} \in N$, consider $(P_{\alpha,\rho,b})$ with $b = \overline{z}$.

Then $(\overline{z}, \overline{x}, \overline{\alpha})$ is feasible for $(P_{\alpha,\rho,b})$, where

$$\overline{\alpha} \equiv \max\Big\{0, \max\big\{\lambda_i(\beta^*_i - \overline{z}_i), i=1, \ldots p\big\}\Big\}.$$

If $(\overline{z}, \overline{x}, \overline{\alpha})$ is not optimal in $(P_{\alpha,\rho,b})$, there must exist (z', x', α') feasible for $(P_{\alpha,\rho,b})$ such that

$$\alpha' + \rho \sum_{i=1}^{p} (\beta^*_i - z'_i) < \overline{\alpha} + \rho \sum_{i=1}^{p} (\beta^*_i - \overline{z}_i).$$

Now $\alpha' < \overline{\alpha} \Rightarrow \alpha'$ does not satisfy the constraints

$$\alpha \geq \lambda_i (\beta^*_i - \overline{z}_i), i=1, \ldots, p,$$

so it must be that

$$\sum_{i=1}^{p} (\beta_i^* - z_i') < \sum_{i=1}^{p} (\beta_i^* - \overline{z}_i),$$

which implies

$$\sum_{i=1}^{p} z_i' > \sum_{i=1}^{p} \overline{z}_i .$$

Since $z' \geq b = \overline{z}$, we conclude that $z_k' > \overline{z}_k$ for some index k, contradicting the assumption that $\overline{z} \in N$. We conclude that $(\overline{z}, \overline{x}, \overline{\alpha})$ is indeed optimal in $(P_{\alpha,\rho,b})$.

□

Given Theorem 3, it has been shown that all nondominated criterion vectors are computable from $(P_{\alpha,\rho,b})$ and that all criterion vectors returned from $(P_{\alpha,\rho,b})$ are nondominated. The general characterization of N, using the augmented weighted Tchebycheff metric, has now been established.

3. General Description

The algorithm, to be discussed in this paper, is a modification of the branch-and-bound algorithm of Marcotte and Soland (1986). As such, its theoretical foundation comes from Theorem 1 [Soland (1979)]. In addition, nondominated solutions, generated from the augmented weighted Tchebycheff program, are considered. These points are included in the algorithm in an attempt to accelerate the convergence of problems where the preferred solution may be located in the relative interior of the nondominated set N.

Let n_j be the node to be examined. Node n_1 corresponds to the original set Z, and each node n_j for $j \geq 2$ corresponds to a proper subset Z_j of Z. Each Z_j is specified by a vector $b^j = (b_1^j, \ldots, b_p^j)$ of lower bounds placed on the various criterion values. The definition of Z_j is thus

$$Z_j \equiv \left\{z \in Z | z \geq b^j\right\}.$$

For notational convenience, some of the lower bounds b_k^j are allowed to be $-\infty$. Since Z is bounded, this causes no difficulty. In fact, node n_1 is defined by $b^1 = (-\infty, -\infty, \ldots, -\infty)$, so that $Z_1 = Z$. For $j \geq 1$, N_j is specified to be the set of nondominated points that belong to Z_j, i.e.,

$$N_j \equiv N \cap Z_j.$$

Let $\beta^{j*} = (\beta_1^{j*}, \ldots, \beta_p^{j*})$ be the ideal that corresponds to the problems $\left(P_{h_i,b^j,\rho}\right)$ and the quasi-ideal that corresponds to the problems $\left(P_{f_i,b^j}\right)$, $i=1, \ldots, p$. Since ρ is fixed, β^{j*} is determined by b^j, as is Z_j. Thus, there is a one-to-one correspondence between Z_j and β^{j*}; for notational convenience, therefore, we shall refer to β^{j*} as the quasi-ideal of Z_j. β_k^{j*} serves as the upper bound for each z_k, $z \in Z_j$.

The branching process begins by finding a nondominated point $z^j \in N_j$. If β^j is replaced by β^{j*}, by solving the problems $(P_{h_i,b,\rho})$, $i=1, \ldots, p$, p nondominated points are available for examination. By examining these points, the DM is able to begin learning about the set of nondominated feasible solutions. However, since these p points are calculated in order to determine β^{j*}, the DM only gets to learn something about the extremes of N_j. If the DM could also see a nondominated solution that is "more centrally" located in N_j, in relation to β^{j*}, then the DM's knowledge of $N_j \subset Z_j$ is more comprehensive. This "centralized" nondominated point can be provided by solving an equally weighted augmented Tchebycheff program. The resulting nondominated point, along with the previously calculated p nondominated points, can be shown to the DM. Branching begins once the DM chooses the most preferred of the p+1 points.

Before going on to the next step in the branching process, we define the set

$$I_j \equiv \{k \mid 1 \leq k \leq p, \beta_k^{j*} - z_k^j \geq \delta_k\}.$$

The δ_k are small positive quantities that represent the "fuzziness" associated with the DM's preferences during the decision making process. That is to say, if we compare $z^1, z^2 \in R^p$ and $z_k^1 - z_k^2 \leq \delta_k \, \forall k$, then the DM does not prefer z^1 to z^2. If δ is allowed to change at node n_j, then $\delta(j) = (\delta_1(j), \ldots, \delta_p(j))$. In this case, the $\delta_k(j)$ are small positive quantities that represent the "fuzziness" associated with the DM's preferences during the decision making process at node n_j. Throughout this paper, δ and $\delta(j)$ can be used interchangeably. In practice, the choice of whether to use δ or $\delta(j)$ belongs to the DM.

If I_j is not empty, for every nondominated point z of Z_j, other than z^j, there exists at least one index $k \in I_j$ such that either $z_k \geq z_k^j + \delta_k$ or $z_k < z_k^j + \delta_k$. As a result, we can write

$$N_j \subset \{z^j\} \cup \left[\bigcup_{k \in I_j} (Z_j \cap \{z \mid z_k \geq z_k^j + \delta_k\}) \right] \cup \left[\bigcup_{k \in I_j} (Z_j \cap \{z \mid z_k < z_k^j + \delta_k\}) \right] \subset Z_j .$$

Given the definition of I_j, if the DM's preferred solution belongs to Z_j, which implies it belongs to N_j, then it belongs to the set

$$\{z^j\} \cup \left[\bigcup_{k \in I_j} (Z_j \cap \{z \mid z_k \geq z_k^j + \delta_k\}) \right].$$

Branching is actually accomplished when new subsets of the form, $Z_j \cap \{z \mid z_k \geq z_k^j + \delta_k\}$, are created for each $k \in I_j$. Depending on the number of elements in the set I_j, up to p new nodes are created each time branching takes place. Because of the way the branching is done, each node n_j of the branch-and-bound tree corresponds to a set Z_j of the form $Z \cap \{z \mid z \geq b^j\}$ for some vector b^j, where $b_k^j = -\infty$ if no lower bound has been imposed on the kth criterion. As a result, the maximization of an arbitrary coordinatewise increasing function $g \in G_Z$ over Z_j results in a nondominated solution z^j by Theorem 1. As already shown, the solution of an augmented weighted Tchebycheff program over Z_j is also a nondominated solution. In the case where I_j is empty, there are no other $z \in N_j$ that are preferred to z^j. Therefore, z^j is a preferred solution for subset Z_j, or for node n_j.

At each branching, the quasi-ideals of the new nodes are calculated. The DM is then asked to choose the most preferred quasi-ideal. Marcotte and Soland (1986) require the DM to totally order the (quasi)-ideals at a particular branching. However, it is also possible to ask the DM to select just the preferred quasi-ideal. As a result, the order in which the intermediate nodes are examined may differ, which can possibly require the DM to make fewer comparisons. Regardless of the manner in which the DM selects the preferred quasi-ideal at a given branching, he or she must subsequently decide if the preferred quasi-ideal is also preferred to the incumbent solution. The incumbent solution is the nondominated point, among those already considered, that is preferred by the DM.

By choosing the preferred quasi-ideal over the incumbent, the DM has basically indicated that he or she believes a "better" nondominated solution might be found in the subset currently being examined. If the DM chooses the preferred quasi-ideal over the incumbent solution, then the node representing the subset under consideration is placed on the *master list*. All the nodes at a given branching, other than the preferred node, are placed on another list known as the *partial list*. Both lists are "queues" of intermediate nodes, i. e., nodes waiting to be branched. The last node added to the master list is the first to be branched. The nodes on the partial list are added to the master list only after the corresponding preferred node has been examined and removed from the master list. The order in which intermediate nodes are placed on the partial list is directly related to the manner in which the DM selects the corresponding preferred node. If the DM does not choose the preferred ideal over the incumbent, then the corresponding node is not added to the master list. The node is "pruned" or "fathomed." Additionally, since the nodes on the partial list are not more preferable than the corresponding preferred node, they are also excluded from further consideration. The algorithm terminates when the master list is empty. This happens after all the nodes that were created are either examined or fathomed. However, the algorithm can be stopped at any time by the DM. This can happen whenever the DM is satisfied with the incumbent solution and chooses to proceed no further.
Proof of finite convergence of the algorithm follows.

Proof:

Suppose we ask the DM to specify certain values $\delta_k > 0$, $k=1, \ldots, p$, and we define I_j as before:

$$I_j = \{k | 1 \leq k \leq p, \beta_k^{*j} - z_k^j \geq \delta_k\}.$$

Let the lower bounds γ_k^ℓ at node n_ℓ correspond to the set

$$Z_\ell = Z_j \cap \{z | z_{k_1} \geq z_{k_1}^j + \delta_{k_1}\}.$$

These lower bounds are defined recursively. The vector of lower bounds for n_1 is $\gamma^1 = (-\infty, \ldots, -\infty)$ and if n_ℓ is a successor of n_j, the γ_k^ℓ are defined as follows:

$$\gamma_{k_1}^\ell = z_{k_1}^j + \delta_{k_1}, \ \gamma_k^\ell = \gamma_k^j \text{ if } k \neq k_1$$

Then

$$\begin{aligned}\gamma_{k_1}^\ell - \gamma_{k_1}^j &= z_{k_1}^j + \delta_{k_1} - \gamma_{k_1}^j \\ &= z_{k_1}^j - \gamma_{k_1}^j + \delta_{k_1}.\end{aligned}$$

Then

$$\begin{aligned} z_{k_1}^j &\geq \gamma_{k_1}^j \\ \Rightarrow z_{k_1}^j &- \gamma_{k_1}^j \geq 0. \end{aligned}$$

$$\therefore \ \gamma_{k_1}^\ell - \gamma_{k_1}^j \geq \delta_{k_1}.$$

If we follow a path from the root of the tree, either a finite lower bound is imposed on a criterion that did not have one before (in the path being considered) or the lower bound for the k_1th criterion is increased by at least δ_{k_1}. It is therefore clear that every path beginning at the root of the tree is of finite length since the MCDM problem is bounded. Since every tree with infinitely many nodes has a path of infinite length, we conclude that the branch-and-bound tree is finite, and hence the algorithm is finite. This is the desired result.

4. Algorithmic Efficiency and Flexibility

In order to help speed up the termination of the algorithm, a number of efficiencies were built into it. The first is a series of node suppression tests, designed to identify nodes whose corresponding subsets are entirely contained in other subsets. These redundant subsets are generated in the branching process, but they are excluded from examination. The concept was initially presented by Marcotte and Soland (1986), but the number of tests has been increased considerably. Mathematical proofs of the node suppression tests are rather long, so they have been omitted for the sake of brevity.

It has been observed that under certain conditions, some nodes have the same quasi-ideals as their immediate predecessors. This observation allows for a significant reduction in the number of optimization problems that need to be solved, thereby decreasing the algorithm's run time.

Special checks for dominance are made. If it is determined that one ideal vector dominates another, or an ideal vector dominates a nondominated point, then the DM is required to make fewer comparisons. A new concept called *quasi-dominance* is introduced, which incorporates "fuzziness" through the use of the δ_k's.

Definition 8 If $z_k^2 - z_k^1 \leq \delta_k \; \forall k=1, \ldots, p$, and $z_k^2 - z_k^1 < \delta_k$ for at least one index k, then we say that z^2 is *quasi-dominated* by z^1.

Special checks for identifying duplicate nondominated points are also made. Another new concept called *quasi-duplicate nondominated points* is also introduced.

Definition 9 If $|z_k^2 - z_k^1| \leq \delta_k \; \forall k=1, \ldots, p$, then we say that z^2 is a *quasi-duplicate* of z^1.

It should be pointed out that if z^2 is quasi-dominated by z^1 and the inverse also holds true, i.e., z^1 is quasi-dominated by z^2, then z^1 and z^2 are *mutually quasi-dominated.* If z^1 and z^2 are mutually quasi-dominated, then they are quasi-duplicates.

If two or more nondominated vectors are compared, and one or more are determined to be quasi-dominated or to be quasi-duplicates, then these points are never shown to the DM. Clearly, by conducting these tests, fewer nondominated points will need to be compared by the DM. If a quasi-ideal vector is compared to another vector, e.g., the incumbent vector or another quasi-ideal vector, and it is determined to be quasi-dominated or to be a quasi-duplicate, then this node can be fathomed. When a node is fathomed, it is not branched, descendant nodes are not created, and resultingly, the number of comparisons made by the DM are decreased.

Throughout the MCDM literature, there is a great deal of discussion about DM indifference or indecision. However, very little has been done about dealing with it. It is felt that in order for an algorithm to be versatile, it should account for indecision. Indecision is handled, when comparing nondominated points, by introducing the concept of a *secondary criterion.* Schilling et al (1982) allude to the existence of such criteria.

Definition 10 A *secondary criterion* is a decision variable, or a function of decision variables, that is not considered to be significant enough to be classified as a primary criterion, but whose value gives the decision maker a clearer understanding of the decision at hand.

The classification of some criteria as secondary criteria can be important, because it is desirable to keep the number of primary criteria as small as possible. Since the branch-and-bound tree grows by p nodes each time we branch, the amount of time needed to solve the problem can become too great. Additionally, it is prudent to minimize the amount of information a DM must digest and process each time a decision is made. As a result, if a DM is unable to choose the preferred vector from among the vectors that he or she is shown, the value of a secondary criterion is displayed as the $(p+1)$st component of each vector. If the DM is still indecisive, after the introduction of the first secondary criterion, the remaining secondary criteria are displayed one-by-one until a decision is made. If the DM cannot make a decision after all secondary criteria are displayed, then the Tchebycheff point is chosen as the preferred nondominated point. Since the Tchebycheff point has a "measure of central tendency" in relation to β^{j*}, its choice as the preferred nondominated point is considered to be a reasonable compromise.

If the DM is indecisive when choosing among p "ideals", since secondary criteria values do not exist, they cannot be used as "tie breakers." In this case, the first ideal encountered is automatically selected. It is felt that this choice is justified, because we may only be changing the order in which the nodes are eventually examined. The possibility of having a slightly higher run time is preferred to a "stalemate".

5. Test Results

The algorithm was tested on seven problems taken from the literature. In each case a value function was used to simulate a decision maker's choices. As a result, known optimal solutions were available for comparative purposes, i.e., it was possible to determine whether or not the algorithm provided the correct solution. Problems with 2-5 criteria were chosen, whose value functions had varied mathematical properties, i.e., problems with linear, nonlinear, and nondifferentiable value functions. In addition, some problems were formulated and solved as linear MCDM problems, but were later reformulated and solved as integer and mixed integer MCDM problems.

As was previously discussed in this paper, currently existing MCDM algorithms were designed to solve specific types of problems or problems with restrictive mathematical properties. The algorithm discussed in this paper, was designed to solve a general class of MCDM problems with few mathematical assumptions. Developers of MCDM algorithms also use different measures of effectivess to determine the efficiency of an algorithm. As a result, no attempt was made to compare the new algorithm with any that currently exists. Instead, as many as six different variations of each version (continuous, integer, mixed integer) of each test problem was compared with up to three different variations of δ. The six variations of the algorithm are listed below.

1. Branching from the preferred nondominated point; totally ordered partial lists.
2. Branching from the Tchebycheff point; totally ordered partial lists.
3. Branching from the preferred nondominated point; partially ordered partial lists.
4. Branching from the Tchebycheff point; partially ordered partial lists.
5. Branching from the preferred nondominated point; unordered partial lists.
6. Branching from the Tchebycheff point; unordered partial lists.

Comparisons of the different variations of the algorithm were based upon the number of iterations needed for problems to terminate; binary comparisons and actions, which are defined below.

Definition 11 A *binary comparison* is a comparison of two vectors $z^1, z^2 \in Z$ that yields one of the following three results:

1. z^1 is preferred to z^2, written $z^1 \succ z^2$,
2. z^2 is preferred to z^1, written $z^2 \succ z^1$,
3. there is indifference between z^1 and z^2, written $z^1 \sim z^2$.

Definition 12 A *ranking* is a comparison of two or more vectors, $z^i \in Z$, $i=1, \ldots, r$ (where $r \leq p$), that yields an ordering of the r vectors, say $z^1, \ldots, z^r$, such that each vector in the ordering is at least as preferable as the following one, i.e., $z^1 \succsim z^2 \succsim \ldots \succsim z^r$.

It should be noted that there is an asymmetric relationship between a binary comparison and a ranking, i.e., every binary comparison is a ranking, but not every ranking is a binary comparison.

Definition 13 A *decision* is a comparison of two or more vectors $z^i \in Z$, $i=1, \ldots, r$ (where $r \leq p+1$), that yields one of the vectors, say z^k, that is at least as preferable as the others, i.e., $z^k \succsim z^i$, $i=1, \ldots, r$.

Here again, it can be noted that every binary comparison is a decision, but not every decision is a binary comparison.

Definition 14 An *action* is either a decision or a ranking.

In order to identify the variation of the algorithm that performed most efficiently from a qualitative perspective, a comparison was made of the average number of iterations, actions, and binary comparisons that each required to solve the test problems. The average number of iterations, actions, and binary comparisons are listed in Table 1.

	Variation					
	1	2	3	4	5	6
Iterations	11.54	9.57	12.60	9.29	13.37	9.77
Actions	40.17	37.00	30.86	29.66	32.00	29.37
Binary Comparisons	66.84	60.94	47.74	43.80	49.58	44.62

Table 1

In terms of iterations, variations 2, 4, and 6 are about equally effective. In terms of actions and binary comparisons, variations 3, 4, 5, and 6 are about equally effective, with variations 4 and 6 providing slightly better results.

Given these observations, variations 1 and 2 were discarded and variations 3, 4, 5, and 6 were retained in the FORTRAN program written to implement the algorithm.

6. A Practical Application

In order to test the usefulness and acceptability of the new algorithm on a practical problem, it was used to solve a manpower planning problem for the United States Total Army Personnel Command located in Alexandria, Virginia.

The problem was modelled as a general multi-period network with linear side constraints, which meant that the constraint set was linear. The problem ultimately contained about 200 constraints and 300 decision variables. The decision maker, Colonel William L. Hart decided upon using two primary criteria, one to minimized and one to be maximized. He also chose to use four secondary criteria.

In the first iteration, since the decision maker opted to use two primary criteria in his interactive session, he was shown three nondominated points to choose from. Prior to making his selection, however, he decided to examine all four secondary criteria. After a few minutes of deliberation, he made his choice and decided to stop the problem. The results were eventually briefed to the Chief of Staff of the Army, who approved them in concept. The Army estimated that annual cost savings of between 2.7 and 5 million dollars would be realized by implementing the manpower policy changes suggested by the solution to the MCDM problem.

7. Conclusions

The algorithm discussed in this paper was designed to solve a general class of MCDM problems with minimal mathematical restrictions and with the needs of the decision maker in mind. During the testing of the algorithm on problems taken from the literature, it was shown that it is capable of solving continuous, integer and mixed integer linear mathematical programs with relative ease. The algorithm has a sense of realism built into it, because it considers DM fuzziness and it helps the DM deal with indecision through the introduction of secondary criteria. The feedback received from the Army, who tested the algorithm on a real problem, was also most encouraging. Users found the algorithm easy to work with; they were pleased to be able to make on the spot modifications to a problem and to be provided with updated, accurate solutions in real time.

8. Acknowledgments

I am most grateful to Professor Richard M. Soland of the George Washington University, who served as my academic advisor and doctoral dissertation director. His understanding, encouragement and much needed technical direction have been of great value to me.

9. Bibliography

Durso, A., "An Interactive Combined Branch-and-Bound/Tchebycheff Algorithm for Multiple Criteria Optimization," Doctoral Dissertation, The George Washington University, Washington, D.C. (1988).

Hart, W. L., Private Communication (1988).

Marcotte, O. and R. M. Soland, "An Interactive Branch-and-Bound Algorithm for Multiple Criteria Optimization," ***Management Science***, 32 (January 1986), 61-75.

Schilling, D. A., A. McGarity and C. Revelle, "Hidden Attributes and the Display of Information in Multiobjective Analysis," ***Management Science***, 28 (March 1982), 236-242.

Soland, R. M. "Multicriteria Optimization: A General Characterization of Efficient Solutions," ***Decision Sciences***, 10 (January 1979), 26-38.

Steuer, R. E., **Multiple Criteria Optimization: Theory, Computation, and Application**, Wiley, New York, 1986.

Zeleny, M., **Multiple Criteria Decision Making**, McGraw-Hill, New York, 1982.

FACTS AND FICTIONS ABOUT THE ANALYTIC HIERARCHY PROCESS

ERNEST H. FORMAN
PROFESSOR OF MANAGEMENT SCIENCE
GEORGE WASHINGTON UNIVERSITY
WASHINGTON, D.C. 20052

ABSTRACT

Despite the many books and journal articles that have appeared about the Analytic Hierarchy Process (AHP), some important misconceptions about AHP remain. This paper discusses issues which underlie these misconceptions, including the cause and significance of "rank reversals", situations allowing or preventing rank reversals, the constraint of a 9 point scale, the roles of redundancy, intransitivities, and inconsistencies, the accommodation of objectivity and uncertainty, the similarities of AHP and MAUT, and opportunities to combine MCDM methodologies in real world decisions.

The Analytic Hierarchy Process developed in the early 1970's by Thomas Saaty [10] has gained wide popularity and acceptance throughout the world. Despite the many books and journal articles that have appeared about AHP, some misconceptions remain about the similarities and differences between AHP and other MCDM methodologies as well as about some aspects of AHP itself. During discussions with other researchers and practitioners at the December 1989 MCDM meeting in Bangkok, I became aware of the need for the MCDM community to elaborate the basic differences between those MCDM techniques that are based on optimizations involving continuous decision variables and those techniques that evaluate a discrete number of alternatives; differences not only in methodology, but in intent, the former generally being better suited for the design phase of decision making, the latter for choice.

The MCDM meetings have helped me broaden my views and to develop an appreciation of the opportunities for combining MCDM techniques to address decisions, for example, using AHP to evaluate tradeoffs among efficient solutions in a multi criteria optimization problem, or using priorities developed with AHP as objective function coefficients in a 0-1 integer programming resource allocation problem, or as objective function coefficients in goal programming problem.

Before beginning to understand how to integrate the various MCDM methodologies in analyzing real world problems, we need to examine the purposes, strengths and weaknesses of the specific MCDM methodologies. This paper will discuss issues relating to the purpose, strengths and weaknesses of AHP. The issues will be presented in the form of "facts" and "fictions" in order to address some of the misconceptions that I have read about, or heard when AHP is discussed by other MCDM practitioners. The categorizations as "facts" or "fictions" are, of course, a matter of the author's judgment. In some cases, the brief discussion that accompanies a "fact" or "fiction" may be an adequate explanation for the reader, while in others, it may just be the beginning to a more in depth inquiry or debate.

Contents

FACTS and FICTIONS

FACT: AHP uses pairwise comparisons.

Discussion: Pairwise comparisons can be made either verbally, numerically, and now, graphically as well. The ability of AHP to derive ratio scale priorities from approximate pairwise verbal comparisons relative to *any* attribute, quantitative or qualitative, is one of AHP's major attractions to decision makers.

FICTION: AHP cannot accommodate objectivity.

Discussion: This fiction probably stems from the observation that pairwise comparisons *can* be made in a verbal mode about qualitative factors. However, pairwise comparisons *also* can and often are based on hard data. If one wishes to assume a linear utility and has hard, ratio scale data, then pairwise comparisons will be perfectly consistent and equivalent to using absolute numbers. The AHP has been implemented to accommodate both absolute judgments or a non-redundant set of pairwise judgments based on hard data to arrive at the same priorities.

FACT: AHP need not satisfy the "independence of irrelevant alternative" assumption of MAUT.

Discussion: This is true and is an important difference between AHP and MAUT. A consequence of not satisfying the independence of irrelevant alternative assumption of MAUT is to allow "rank reversals" or what can also be called "rank adjustments". Any alternative evaluated in the AHP relative mode is a fortiori relevant because all other alternatives are evaluated in terms of it. An "irrelevant" alternative is one that is dominated, and according to MAUT practitioners should be eliminated from consideration. However, others argue that a dominated alternative is not necessarily irrelevant to a decision. For example, Huber, Payne, and Puto [7] state that " the very presence of [a] dominated alternative results in quite different choice probabilities among the remaining alternatives than in the pristine state, where such items are never considered".

FICTION: Rank reversal with AHP is caused by the Eigenvector calculations.

Discussion: Rank reversal does *not* occur because of the eigenvector calculations, because of the nine point scale, or because of inconsistencies in judgments, and may, depending on the situation, be a *desirable* phenomenon. *Any* technique that decomposes and synthesizes in a relative fashion by normalizing priorities to maintain a ratio scale[1], will allow rank reversals regardless of whether it uses

[1]Note that this normalization is also performed in MAUT analysis at all but the lowest or alternative level of the analysis. Also note that if such normalizations were not performed (continued...)

pairwise comparisons, eigenvector calculations, or demands perfect consistency. This occurs because of an *abundance* or *dilution* (or what is also called a *substitution* effect), as is discussed in Saaty [14], Dyer [2], Forman [3], and Huber [8]. In addition, the importance of criteria may change and what may have been the most important criterion may, because of common occurrence in the alternatives, become less important in evaluating the alternatives. Rank reversal under AHP is never capricious or arbitrary.

FICTION: A technique allowing rank reversals is flawed.

Discussion: On the contrary, rank reversals are often desirable. Consider the evaluation and ranking of employees in a small office with half a dozen or so employees. John, who is reasonably good at analysis and good at relating to customers is the only employee who is proficient in the use of personal computers. As in many offices today, the ability to use PC's well is fairly important and because of this, John is evaluated as the most valuable of the employees, with Susan ranking as second most valuable. Subsequently, two more employees are hired, each of whom can use PC's fairly well, but not as well as John. A new evaluation is performed and Susan is evaluated as the most "valuable" with John now second. This is reasonable and occurs because the allocation of value for ability with PC's has been diluted, with John still getting the largest share, but not as much as before (see Forman [3]). The ability of an evaluation methodology such as AHP to adjust the rank based on such dilution is a *benefit*, not a deficiency. The need for a methodology to allow for such adjustments has been long recognized. For example, in an article "Market Boundaries and Product Choice: Illustrating Attraction and Substitution Effects", Huber and Puto [8] state that:

> "Choice researchers have commonly used two general approaches to account for the way proximity of a new item affects choice. These approaches differ primarily in the way item similarity, as derived from the dimensional structure of the alternatives is assumed to affect the choice process. The first proposition (proportionality) assumes that the new item takes share from existing items in proportion to their original shares (i.e., no similarity effect)."

This approach would *preclude* rank reversal. Huber and Puto continue to say:

> "The second proposition (substitutability) assumes that the new item takes share disproportionately from more similar items -- i.e. the closer the added item is to existing items in the set, the more it "hurts" them (a negative similarity effect)."

---

[1](...continued)
at all but the lowest level, interval level priorities would be produced at these levels and it would be necessary to multiply interval level priorities with other interval level priorities. This would be unacceptable because the product of two interval level scales is meaningless.

Huber, Payne and Puto [7] note that:

> "... the similarity hypothesis asserts that a new alternative takes disproportionate share from those with which it is most similar. Researchers have shown that the similarity effect is operant for individual or aggregate choice probabilities."

Substitutability requires that rank reversals be permitted. Decision makers must decide which of these two approaches (proportionality or similarity) to use, based on the effect, proportionality or substitution, they think is most in play. Huber and Puto argue that

> "a substitution effect will be more salient where multi-attribute decision making occurs. It should, therefore, be most apparent in major purchases (where attribute-based processing is more cost effective) and in product classes for which a limited number of attributes emerge that permit easy comparisons across alternatives."

Since MCDM techniques like AHP are now facilitating comparisons across alternatives for more than just a limited number of attributes, the substitution effect should become even more common.

FICTION: If it is desirable to preclude rank reversals then AHP should not be used.
Discussion: Rank reversal's CANNOT and DO NOT occur when AHP is used in a "ratings", or what Saaty calls an "absolute" mode. In order to use the absolute mode one needs expert knowledge from previous experience so that normative standards can be developed and applied to the alternatives. If one feels that allowing a rank reversal is inappropriate for a particular decision, then one can use the absolute mode. Doing this makes AHP look almost identical to multi-attribute value theory as discussed below.

FACT: Normalization is required if the ratio scale property is to be maintained when combining values over several dimensions.
Discussion: We cannot add numbers from different ratio scales and get meaningful results, but we can if the numbers belong to the same ratio scale. Normalization puts the numbers under different criteria on the same ratio scale, so that when we multiply by the weight of the corresponding criteria and add, the result also belongs to the same ratio scale. On occasion, people have wanted to normalize to different constants on each criterion so that an alternative scoring the best on that criterion gets the full weight of the criterion. But, according to the above, the result would not belong to a ratio scale.

FACT: AHP allows inconsistency / intransitivities.
Discussion: The theory of AHP does <u>not</u> demand perfect consistency, but provides a measure of the inconsistency in each set of judgments. This measure is an important by-product of the process of deriving priorities based on pairwise comparisons. It is natural for people to want to be consistent because consistency is a prerequisite to logical thinking. However, our knowledge of the real world is hardly ever perfectly consistent, and we can learn new things only by allowing for some inconsistency with what we already know.

The most common cause of inconsistency is a clerical error. When entering one or more judgments into a computer, the wrong value, or perhaps the inverse of what was intended is entered. Clerical errors can be very detrimental and often go undetected in many computer analyses. Because of the AHP inconsistency measure, one can easily find and correct such errors.

A second cause of inconsistency is understanding, which may be caused by a lack of information. If one has little or no information about the factors being compared, then judgments will appear to be random, and a high inconsistency ratio will result indicating a need to gather more information.

Another cause of inconsistency is lack of concentration during the judgment process. This can happen if the those making judgments become fatigued at which point it is time to take a break.

Still another cause of a high inconsistency ratio is that some occurrences in the real world *are* inconsistent. To illustrate this point, it is not too uncommon in sports for Team A to defeat Team B, after which Team B defeats Team C, after which Team C defeats Team A![2]

A final cause of inconsistency is "inadequate" model structure. Ideally, one would structure a complex decision in a hierarchical fashion such that factors at any level are comparable, within an order of magnitude or so, of other factors at that level. Practical considerations might preclude such a structuring and it is still possible to get meaningful results. Suppose for example, we compared several items that differed by as much as two orders of magnitude.

FICTION: An AHP analysis with low inconsistency is a good analysis.

Discussion: It is important that the a low inconsistency not become the goal of the decision making process. A low inconsistency is necessary but not sufficient for a good decision. It is more important to be accurate than consistent. It is possible to be perfectly consistent but consistently wrong.

---

[2]Although judgments may be intransitive, the resulting priorities may be considered as transitive by the decision maker. If, for example, the intensity of judgment about Team A over Team B were the same as the intensity for Team B over Team C and that was the same as the intensity for Team C over Team A, then the resulting AHP priorities would show each of the teams to be equal to one another. However, if team A were judged to be VERY STRONG over team B, while team B was judged to be STRONG over team C, which is judged to be MODERATE over team A, the resulting AHP priorities would produce a measure of strength with Team A the strongest, Team B the second strongest, and team C the weakest. The resulting *priorities* are thought of as being transitive, even though the judgments that produced them are not.

FICTION: AHP is constrained because of its 9 point scale.

Discussion: The 9 point scale is not required by the basic AHP axioms [13] but rather is based on empirical studies [10]. It also fits in nicely with the AHP principle of structuring complexity so that factors are compared to those within about an order of magnitude. However, because, one factor might be 9 times greater that of a second, which in turn might be 5 times greater than a third, and because the resulting priorities are based on second, third, and higher order dominances, AHP can produce priorities far beyond an order of magnitude. A higher than usual inconsistency ratio may result because of the extreme judgments necessary. If one recognizes this as the cause, (rather than a clerical error for example), one can accept the inconsistency ratio even though it is greater than 10%

FICTION: Inconsistencies and intransitivities are bad.

Discussion: Inconsistencies and intransitivities exist everywhere. Any theory that attempts to treat the real world *must* accept them. In fact, inconsistencies and intransitivities are commonly accepted. For example, a scientist wishing to measure the length of an object accurately may calculate the average of a large number of measurements. The set of measurements will obviously contain inconsistencies (unless there is no error of measurement and they are all exactly the same.) Furthermore, if the length of three different objects of approximately the same length are being measured, it is also likely that the measurements would contain some intransitivities as well. Despite these inconsistencies and intransitivities, the average of the measurements will provide results more accurate than any individual measurement.

Proponents of MAUT who may criticize AHP because it is contrary to the MAUT intransitivity axiom should note what Luce and Raiffa said about the MAUT transitivity axiom on page 25 of their book *Theory of Games* [9]:

> "No matter how intransitivities exist, we must recognize that they exist, and we can take only little comfort in the thought that they are an anathema to most of what constitutes theory in the behavioral sciences today."
> "We may say that we are only concerned with behavior which is transitive, adding hopefully that we believe this need not always be a vacuous study. Or we may contend that the transitive description is often a 'close' approximation to reality. Or we may limit our interest to 'normative' or 'idealized' behavior in the hope that such studies will have a metatheoretic impact on more realistic studies. In order to get on, we shall be flexible and accept all of these as possible defenses, and to them add the traditional mathematician's hedge: transitive relations are far more mathematically tractable than intransitive ones."

FACT: The ability of AHP to accommodate inconsistencies and intransitivities is a strength, not a weakness.
Discussion: The previous discussion addressed the need to accommodate inconsistencies and intransitivities in any practical decision methodology. AHP not only accommodates inconsistencies and intransitivities (that may arise within a set of redundant judgments) but is able to use redundancy to derive value functions and utilities that are more accurate than could be derived without redundancy.

FICTION: AHP cannot deal with uncertainties.
Discussion: Scenarios are included in AHP models in order to accommodate uncertainties. Pairwise comparisons can be used to derive a subjective probability distribution for the relative likelihoods of the scenarios. Alternatively, if available, probabilities from "objective" probability distributions can be incorporated.

FACT: AHP and MAUT (MAVT) are more alike then they are different.
Discussion: If we compare the AHP absolute or ratings approach with MAUT, the only significant difference is the way the value function is derived. AHP uses pairwise comparisons while MAUT uses lotteries. For example, in deriving the value function for the reliability of cars each car might be evaluated on the following scale:

Much better than average repair record
Better than average repair record
Average repair record
Worse than average repair record
Much worse than average repair record

The value function with AHP is derived through pairwise comparisons about the relative preference of cars rated with one of these intensities as opposed to another. A typical question would be:

How much more preferable to you is a car with a "better than average repair record" than a car with an "average repair record"?

Pairwise questions such as the above can be answered not only numerically or graphically, but they can be also be answered with approximate words from the AHP verbal scale: EQUAL, MODERATE, STRONG, VERY STRONG, and EXTREME. Decision makers usually have less difficulty committing to such approximate judgments and therefore find the verbal mode very easy to use. AHP does not merely make it easy to solicit judgments, but derives "accurate" priorities from these judgments. The accuracy of the priorities derived from such judgments with AHP has been validated numerous times in situations involving objective factors such as the brightness of light or the size of geometric shapes [10][5].

In contrast, the MAUT approach that most often appears in the literature[3], first assigns numerical values to the intensities, and indicates a best and worst case as follows:

Best Case:	10.0	Much better than average repair record
	7.5	Better than average repair record
	5.0	Average repair record
	2.5	Worse than average repair record
Worst Case:	0.0	Much worse than average repair record

Then a series of hypothetical questions in the form of a fictitious lottery are asked in order to derive the value function. A typical question would be:

If the "sure thing" were 7.5, then what would you do: take the sure thing or take a gamble with a 50/50 chance of getting either 9 or 7, or are you indifferent?

In contrast to the relative pairwise comparisons with AHP, it is generally acknowledged that the lottery questions are relatively difficult to answer.

Other differences between AHP and MAUT involve the way an analysis is structured when there are numerous attributes, and how weights are derived for the attributes. These differences are becoming less apparent as MAUT practitioners have begun to use the AHP approach for these aspects of the problem. Thus, *if* the absolute or ratings approach of AHP is compared to MAUT, the only significant difference is the questioning used to derive the shape of the value function.

FACT: AHP can be used to produce identical results as Expected Value theory.

Discussion: Saaty [12] has shown that AHP can be used to produce identical results as Expected Value Theory. However, AHP is not intended to duplicate expected value theory where there is essentially only one dimension. When parts of a decision involve quantitative values which can be combined into one scale (e.g. discounted cash flows), they should be so combined with such tools as electronic spreadsheets before being considered in an AHP formulation.

FICTION: The criteria weights in AHP can be automatically calculated or adjusted based on the alternative values.

Discussion: As AHP has gained popularity, there have been numerous suggestions and contributions from researchers throughout the world. While many have been, and are being incorporated into the theory and practice of AHP, some of them remain controversial. One such controversial suggestion is that criteria weights should be derived or altered

---

[3]Because they fell this approach is often too difficult, MAUT practitioners such as Watson and Buede [16] recommend other approaches including SMART, the percent distribution approach, and the numeraire and balance-beam approaches. Watson and Buede believe that latter two approaches are the more reliable.

based on the alternative values in order to avoid rank reversal [1][2][15]. While Saaty was first to point out how criteria weights *could* be adjusted [12], he recognized that this was not advisable in problems involving multiple criteria [11]. In most AHP analyses, criteria weights should be derived from judgments about the relative importance of the criteria, which in turn depend on the relative values of the alternatives. However, it is nonsense to assume that the dependance of the criteria importance on the alternative values can be relegated to a formula (see Forman [4]) because an increase in the values of alternatives on one dimension may increase the importance of that dimension in one circumstance but decrease the importance of that dimension in others. The resolution of what happens must be left to human judgment and cannot be answered by a formula.

FACT: A "great deal" of time may be required for the pairwise comparisons in a typical AHP analysis.
Discussion: This is perhaps the major "weakness" of AHP. If a moderately complex problem is being analyzed and there is a need to do it in a few minutes, then AHP may be too time consuming, as will almost any rational, compensatory decision analysis approach.

FICTION: "Too much" time is required for making the many pairwise comparisons with AHP.
Discussion: The time is spent in analyzing a decision with AHP is usually much less than the time spent for data collection and data analysis, and often of more importance to the decision being made. Consideration should be given to the benefits of the time spend making pairwise, redundant judgments. First of all, the pairwise comparisons results in a well structured discussion of the relative importance and preferences of factors that improves both communication and insight among members of a group involved in making a decision. Secondly, the redundancy in the pairwise comparisons permits the derivation of ratio scale priorities from approximate verbal judgments. The redundancy obviously means that more than a minimal number of judgments are required. The "tradeoff" as I see it, is between the accuracy of the derived priorities from what can be approximate judgments, and the time required to make the judgments. Harker [6] has developed a procedure for calculating AHP priorities with "missing" judgments. This procedure has been incorporated in AHP to allow one the choice of making fewer judgments, but, with a corresponding loss of accuracy. Saaty believes, and I agree, that for the majority of important complex problems analyzed with AHP, decision makers are not averse to taking the time required to make all of the redundant judgments, provided that the time is divided into several sessions of reasonable length, which are held in comfortable surroundings that are conducive to open and informative discussions.

The "facts" and "fictions" presented above are intended to encourage additional discussion and understanding among members of the MCDM community. A substantial body of literature about AHP is now available. The following references should provide a starting point for those wishing to read more about issues such as those discussed in this paper. A recent textbook, *The Analytic Hierarchy Process - Applications and Studies*, edited by Golden, Wasil and Harker and published by Springer-Verlag, is recommended for those who want to read about some AHP applications.

REFERENCES

[1] Belton, V. and T. Gear, "On a Shortcoming of Saaty's Method of Analytic Hierarchies", *Omega* Vol. 11 No. 3 pp. 226-230, 1982.

[2] Dyer, J. S., "Remarks on The Analytic Hierarchy Process", *Management Science*, Vol. 36, No. 3, pp. 249-258, 1990.

[3] Forman, Ernest H., "Relative vs. Absolute Worth", *International Journal of Mathematical Modeling*, Vol. 9, Number 3-5, pp 195-202, 1987.

[3] Forman, E. H., "AHP is Intended for More than Expected Value Calculations", *Decision Sciences*, Volume 21, Number 3, pp 670-672, 1990.

[4] Forman, E. H. "Deriving Ratio Level Measures from Verbal Judgments", *George Washington University Working Paper*, 1990.

[5] Harker, P. T., "Incomplete Pairwise Comparisons in The Analytic Hierarchy Process", *Mathematical Modelling*, Vol. 9, pp 837-848, 1987.

[6] Huber, J., Payne, J. W., and C. Puto, "Adding Asymmetrically Dominated Alternatives: Violations of Regularity and the Similarity Hypothesis", *Journal of Consumer Research*. Vol. 9, June, pp. 90-98, 1982.

[7] Huber, J. and C. Puto, "Market Boundaries and Product Choice: Illustrating Attraction and Substitution Effects", *Journal of Consumer Research* Vol. 10, June, pp. 31-44, 1983.

[8] Luce, R.D. and H. Raiffa, *Games and Decisions*. John Wiley and Sons, Inc., New York, 1957.

[9] Saaty, T. L., *The Analytic Hierarchy Process*, McGraw-Hill, New York, 1990.

[10] Saaty, T. L. and L. G. Vargas, "The legitimacy of Rank Reversal", *Omega* Vol 12 pp. 514-516, 1984.

[11] Saaty, T. L., "A Note on AHP and Expected Value Theory", *Socio Economic Planning Science*, Vol. 20, No. 6. pp 397-398, 1986.

[12] Saaty, T. L., "Axiomatic Foundations of the Analytic Hierarchy Process", *Management Science*, Vol. 32, pp. 841-855, 1986.

[13] Saaty, T. L., "An Exposition of the AHP in Reply to the Paper Remarks on the Analytic Hierarchy Process", *Management Science*, Vol. 36, No. 3, pp. 259-268, 1990.

[14] Schoner, B. S. and C. W. Wedley, "Ambiguous Criteria Weights in AHP: Consequences and Solutions", *Decision Sciences*, Vol. 20, Summer. pp. 462-475, 1989.

[16] Watson, S. R. and D. M. Buede, *Decision Synthesis -- The Principles and Practice of Decision Analysis*, Cambridge University Press, pp 200-203, 1987.

DECISION SUPPORT SYSTEM FOR MANAGEMENT OF PATIENT NUTRITION: AN INTERACTIVE AHP/GOAL PROGRAMMING APPROACH

Josef Jablonský

Prague School of Economics
Department of Econometrics
130 67 Prague 3, Czechoslovakia

Abstract

The paper offers a decision support system for prescribing of total parenteral nutrition for critically ill patients at intensive care units. The basic problem is to determine accurately intake of food for each patient. To solve this problem the system uses interactive goal programming approach in conjunction with the analytic hierarchy process (AHP). To prevent over- or underprescribing of nutritional components the objective function of the goal programming model minimizes the weighted sum of positive and negative deviations from given requirements. These deviations are weighted in proportion to their importance which depends on the patient´s state. To determine these weights the AHP model is utilized. The system is realized on PC´s which enables interactive solution of the problem. The numerical results of the model are discussed in the last part of the paper. At this time, the usage of the system is being verified in the clinical practice.

1. Introduction

In this paper we present a simple support system which could help to propose the intravenous feeding composition for patients cured at intensive care units. These patients are usually critically ill and require individual infusion prescribing and calculations. It is mostly impossible to apply standard procedures to get a satisfactory proposal of their nutrition.

The physician must take into consideration many factors (patient´s condition, limitations depending on the type of his disease, the composition of available stock solutions for intravenous feeding etc.) during the process of construction of feeding proposal. This process could be divided into three relatively independent steps.

1. Specification of requirements of all nutritional components (water, energy, amino acids, minerals, vitamins, trace elements etc.) for the patient.
   This problem is not subject of this study. We suppose that the requirements of all nutritional components are known - e.g. as a result of physician´s consultation with the nutritional team, a computer-aided nutritional decision or expert system. The expert system for specification of nutritional components as a component of the large decision support system for intensive care units was created in the Department of the Clinical Biochemistry of the District Hospital in Kladno, Czechoslovakia and is being widely utilized in clinical practice.
2. Realization of the requirements by means of available stock solutions.
   This problem is too complex (with regard to the relatively large number of nutritional components and stock solutions) to be solved in requested accuracy with physician´s judgment only. The system proposed in this study makes possible to construct the final structure of patient´s feeding but physicians can influence all calculations by the interactive communication during the process of solution. Thus, the system makes possible to prevent over- and underprescribing of nutritional components and to save physician´s time and in this way it can reduce volumes of discarded fluid, spare material for administering solutions and make intravenous feeding cost effective.
3. Time specification of intravenous feeding.
   In the second step, the nutritional daily proposal is

obtained. But it is very important to make its time specification. There are lots of conditions which must be respected in this process (e.g. prescribing of each solution takes some minimum time, many solutions must not be combined). The most of these conditions are not included in the present version of the model (and probably will not be included because of their variety) so that the proposal received in the previous step may not represent feasible solution of the problem. In this case, it is necessary to modify the primary proposal.

The aim of this study was to explore the possibility of using multiple objective programming methodology in effective prescribing of intravenous nutrition. To solve this problem the system utilizes goal programming techniques in conjunction with the analytic hierarchy process (AHP).

2. Problem definition

The available stock solutions for intravenous feeding can be divided into three basic groups. Their classification is evident from the following survey:

- solutions with energy -
  - glucose - in different concentrations (5 to 40%), (G1-G4)
  - amino acids - in different concentrations either with saccharides or without them, (A1-A6)
  - fat - one kind of solution, (F1)
- without energy - in physiological concentrations, (P1-P5)
- without energy - concentrated. (C1-C12)

The nutritional components considered in the model can be classified in the following way:

- water (H_2O),
- energy - total energy (TE),
  - energy derived from fat (Fat),
- nitrogen (N),
- trace elements (Na,K,P,Ca,Cl,Mg),
- vitamins (not implemented in the present version),
- other components (not implemented).

In this paper, we will consider 28 stock solutions (each of them contains at least two of the nutritional components) and 10 components which were indicated in the previous surveys. However, since the system is constructed as a open system, it is possible in a simple way to change (add or delete) the number of solutions or components. This allows to make the current list of solutions suitable for patient's state and his type of vein and for solutions which are currently available. Table 1 presents the complete list of solutions (with minimum quantity for prescribing) and components, minimum and maximum daily requirements of components, a set of real requirements for a patient and the composition of all 28 solutions (per unit of minimum volume of each solution).

Let us suppose that all patient's nutritional requirements are known (e.g. as a product of an expert system). Quantity of each component is still controlled by the physician according to patient's actual condition and some modifications can be done. Then, the individual requirements are supposed to be "true" and the model tries to minimize deviations from them.

There are a lot of other conditions which influence the final nutritional mix. Most of them depend on the patient's instantaneous state - in many cases the physician must prohibit usage of certain solutions, can prefer a group of solutions to another etc. It is impossible to respect so many various conditions in the model which must be simple to give results very promptly. The model includes only one condition of this kind - "non-protein energy related to 1g of nitrogen" must get near to the according value which is given by the ratio

$$\frac{TE - 105.N}{N},$$

where TE is total energy requirement in kJ/day and N is nitrogen requirement in g/day. This condition must come true with very high priority.

Minimum volumes of solutions for prescribing are defined in Table 1. The decision maker prefers such proposal which respects these minimum volumes in order to prevent discarding of large volumes of expensive solutions. Thus, the variables in the model should be defined as the integer variables which would express the number of minimum volumes of available solutions. This condition is very important with respect to the cost effectiveness and it must be especially respected for the first two groups of solutions which are prescribed in large volumes (these solutions are available in bottles 400ml and the minimum volume is usually just one bottle or half a bottle).

Table 1
Definition of solutions and nutritional components

Comp. / Sol.(ml)	W	TE	F	N	Na	K	Cl	Ca	P	Mg
G1 (200)	200	168								
G2 (200)	200	334								
G3 (200)	200	668								
G4 (50)	50	334								
F1 (100)	100	909	825							
A1 (200)	200	469		1.2	7					
A2 (200)	176	350		1.7			2.8			
A3 (200)	192	128		1.22	8	5	4.8			0.5
A4 (200)	184	256		2.44	8	7.2	9.6			0.5
A5 (200)	172	471		1.22	8	5	4.8			0.5
A6 (200)	164	599		2.44	8	7.2	9.6			
P1 (200)	200				30.8		30.8			
P2 (200)	200				26	1.1	22.4	0.2		0.2
P3 (200)	200				29.4	0.8	31	0.46		
P4 (200)	200				26	0.8	22	0.36		
P5 (200)	100				1.2	1.2				
C1 (5)	5				8.5		8.5			
C2 (5)	5					5	5			
C3 (5)	5				2.5					
C4 (5)	5				5					
C5 (5)	5					5				
C6 (5)	5						5			
C7 (5)	5					5			5	
C8 (5)	5				2.5				1.5	
C9 (5)	5						5	2.5		
C10 (5)	5							1.25		
C11 (5)	5						5			2.5
C12 (5)	5									2
Unit	ml	kJ	kJ	g	mmol		mmol		mmol	
Minimum	500	1000	0	0	10	0	10	0	0	0
Maximum	6000	16000	5000	25	800	500	800	50	100	50
Real	3600	9400	0	11	220	100	200	5	15	0

3. AHP/goal programming formulation

The problem of the prescribing of parenteral nutrition was described in brief from the biochemical point of view in previous sections. To facilitate the process of prescribing the integer goal programming model is proposed. The goal programming approach was elected because it makes possible in a simple way to perform interactive solution of the problem (see Korhonen [3]) which is very important in this case.

The model minimizes the weighted sum of deviations from "true" requirements specified in the first step. Therefore, the objective function of the model measures the total penalty corresponding to the proposed nutritional mixture. In the primary version of the model published in Jablonsky [2] two basic kinds of penalties and three-sided penalty functions have been used. The decision maker could influence many parameters to obtain satisfactory solution. However, the experience demonstrated that decision makers were usually unable to utilize all possibilities given by the model. This experience is taken into consideration in its new version and that is why the model is more simple and accessible for decision makers.

The model can be formulated as follows:

minimize

$$z = s_1 \sum_{i=1}^{m} \frac{w_i}{b_i} (d_{i1}^- + d_{i1}^+) + s_2 \sum_{i=1}^{m} \frac{w_i}{b_i} (d_{i2}^- + d_{i2}^+) \qquad \text{Eq.(1)}$$

subject to

$$\sum_{j=1}^{n} a_{ij}x_j + (d_{i1}^- + d_{i2}^-) - (d_{i1}^+ + d_{i2}^+) = b_i \ , \ i=1,2,\ldots,m, \qquad \text{Eq.(2)}$$

$$p_{m+1}^{min} \frac{b_{TE}-105b_N}{b_N} \leqq \frac{\sum_{j=1}^{n} (a_{TE,j}-105a_{N,j})x_j}{\sum_{j=1}^{n} a_{N,j}x_j} \leqq p_{m+1}^{max} \frac{b_{TE}-105b_N}{b_N}, \qquad \text{Eq.(3)}$$

$$g_j \leqq x_j \leqq h_j \,, \qquad x_j - \text{integer} \,, \qquad j=1,2,\ldots,n, \qquad \text{Eq.}(4)$$

$$\begin{aligned}
0 &\leqq d_{i1}^- + d_{i2}^- \leqq b_i - p_i^{min}.b_i \,, \\
0 &\leqq d_{i1}^+ + d_{i2}^+ \leqq p_i^{max}.b_i - b_i \,, \\
0 &\leqq d_{i1}^- \leqq b_i - p_i^1 b_i \,, \qquad i=1,2,\ldots,m, \qquad \text{Eq.}(5) \\
0 &\leqq d_{i1}^+ \leqq p_i^2 b_i - b_i \,, \\
& d_{i1}^- \,, d_{i2}^- \geqq 0 \,,
\end{aligned}$$

where

- m is the number of nutritional components,
- n is the number of available stock solutions,
- a_{ij} is the quantity of the ith component in the unit of volume of the jth solution ($a_{TE,j}$ and $a_{N,j}$ correspond to the total energy and nitrogen requirement),
- x_j are integer variables which are interpreted as the number of minimum volumes of solutions,
- b_i are daily requirements of components (b_{TE} and b_N total energy and nitrogen requirements),
- g_j, h_j are lower and upper bounds of variables,
- d_{i1}^-, d_{i2}^- (d_{i1}^+, d_{i2}^+) are negative (positive) deviational variables - their positive values indicate underprescribing (overprescribing) of the ith component,
- $p_i^{min} \leqq p_i^1 \leqq 1 \leqq p_i^2 \leqq p_i^{max}$ are coefficients which specify relative scale (in two levels) for over- and underprescribing of the ith component,
- w_i are weights coefficients which measure the relative importance of requirements,
- s_1, s_2 are coefficients which penalize in two levels over- and underprescribing.

Constraints Eq.(2) are balance constraints which correspond to m-nutritional requirements. The constraint Eq.(3) express the "non-protein energy related to 1g nitrogen" condition. The range of satisfactory values of this condition is given by p_{m+1}^{min} and p_{m+1}^{max} coefficients. The objective function (1) measures the total penalty for feasible solutions of the problem (positive and negative deviations in both levels are weighted by w_i coefficients - both levels are differentiated by penalties s_1 and s_2).

Two positive and negative deviational variables are used in the model. The first ones (d_{i1}^-, d_{i1}^+) specify under- or overprescribing of the first level and the second ones (d_{i2}^-, d_{i2}^+) of the second level. These levels are given by decision maker by means of p_i^{min}, p_i^1, p_i^2 and p_i^{max} coefficients. They can be expressed as follows:

- level 1 (penalty s_1) -
  - overprescribing - $d_{i1}^- = d_{i2}^- = d_{i2}^+ = 0,\ d_{i1}^+ > 0,$
  - underprescribing - $d_{i1}^+ = d_{i2}^- = d_{i2}^+ = 0,\ d_{i1}^- > 0,$
- level 2 (penalty s_2) -
  - overprescribing - $d_{i1}^- = d_{i2}^- = 0,\ d_{i1}^+ = p_i^2 b_i - b_i,\ d_{i2}^+ > 0,$
  - underprescribing - $d_{i1}^+ = d_{i2}^+ = 0,\ d_{i1}^- = b_i - p_i^1 b_i,\ d_{i2}^- > 0.$

Except bounds for levels 1 and 2 the decision maker must determine weights w_i which are extremelly important and should reflect his knowledge and experience. They can be easily received by means of the AHP model presented in Figure 1.

The AHP model can be extended according to the number of nutritional components. Some results given by the model will be discussed in the last section of the paper.

4. Interactive solution of the problem

The model described in the previous section is a traditional linear integer programming problem. There are lots of professional software packages on personal computers intended for solution of linear optimizing problems (with integer conditions). But, with regard to the fact that the solution of any problem should not take more than 2 minutes of computing time on PC and

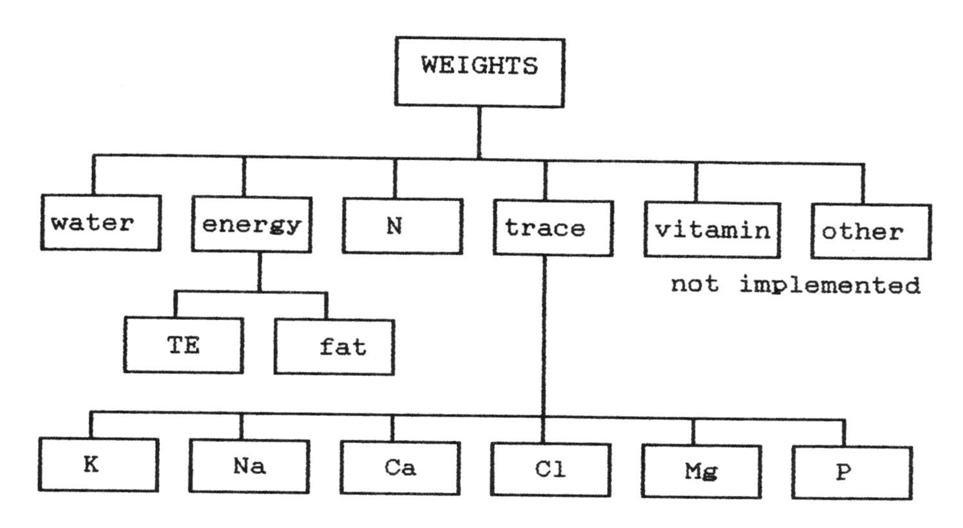

Figure 1
Specification of weights by means of the AHP model

with regard to the number of integer variables, it is impossible to use any of these packages because even the best of them do not give results in the desired time and hence, they disable interactive solution. Thus, it was necessary to find a way how to get a satisfactory integer solution in a short time. The very simple approach was elected - the integer solution is obtained as the round optimal solution of the problem without integer conditions. Therefore, it is not the optimal integer solution but only its estimation.

The first experiments and numerical solution of the problem had been performed with the XA (EXtended Application) professional linear programming software package (Sunset Software Technology, 1988) which is one of the best systems for solution of linear programming problems. The results received by the model were very interesting and the system was accepted among physicians well. Due to these facts, it was decided to develop an original software package. This commercial software product will be distributed as the part of the large decision support system for intensive care units.

The computational system must enable decision maker (usually a physician) to influence the final nutritional proposal during the solution in the interactive way. In the process of system installation the decision maker must specify the following data:

- the set of nutritional components which should be taken into consideration,
- the set of available solutions and their composition with regard to the set of specifying components,
- weights coefficients which measure the relative importance of requirements (e.g. by means of the AHP model),
- coefficients which specify relative scale of components and determine levels for over- and under-prescribing,
- penalties for two levels of over- and under-prescribing.

These data are not usually subjects for interactive modification because they can be common for most of patients (this fact, however, does not mean that it is impossible to change them).

The starting point of the process of finding nutritional proposal for the concrete patient is the specification of remaining data:

- requirements of nutritional components,
- lower and upper bounds for volumes of solutions (in this way physician can prohibit usage of some solutions and influence usage of the others),
- minimum volumes of solutions which can be prescribed.

Then, they are modified (except requirements) during the interactive solution.

A visual interactive approach was proposed in order to facilitate to decision maker solution of the problem. It was necessary to respect two basic conditions in this approach - decision maker should have possibility to influence not only goal values (deviations from requirements) but also he must be able to change the set of available solutions and to determine what solutions will be used in the nutritional mixture.

In the first step of a solution procedure the optimal solution of the problem Eq.(1)-Eq.(5) is presented to the decision maker. If this problem does not have a feasible solution it is necessary to modify input data (e.g. p_i^{min}, p_i^{max}, or some requirements) - the system can propose what data should be modified to obtain a feasible solution. If it is not a satisfactory solution the decision maker can experiment with different values of variables (volumes of solutions) or he can bound or fix deviational variables - he directly influences how the requirements will be fulfilled. In this way he generates alternate solutions until the satisfactory solution will be found.

An original software was developed to implement this approach on personal computers. The software is based on visual interaction with decision maker and is constructed as a very user-friendly oriented system. The visual interaction is realized by means of graphical representation of requirements (deviations from "true" requirements are expressed as bar graphs varied inside given ranges). The visual interaction is a little bit similar to the approach used by P.Korhonen in his Pareto Race (see Korhonen [3]). The program is written in Turbo Pascal.

5. Numerical illustration

In this last section of the paper we would like to demonstrate some results given by the model.

First, it is necessary to determine importance coefficients by means of the AHP model (shown in Figure 1). Comparative judgments of elements in both levels of the model (comparisons have been performed by expert biochemist) and weights derived from them are shown in the following tables:

Table 2
Pairwise comparison matrices

	W	E	N	T	w_i
Water	1	2	3	1/2	0.251
Energy	1/2	1	1	1/6	0.103
Nitro.	1/3	1	1	1/6	0.092
Trace	2	6	6	1	0.554
					1.000

En.	TE	Fat	w_i
TE	1	3	0.077
Fat	1/3	1	0.026
			0.103

Trace	Na	K	Cl	Ca	P	Mg	w_i
Na	1	1	3	3	8	4	0.176
K	1	1	4	4	8	5	0.200
Cl	1/3	1/4	1	1	3	2	0.062
Ca	1/3	1/4	1	1	2	1	0.051
P	1/8	1/8	1/3	1/2	1	1/2	0.023
Mg	1/4	1/5	1/2	1	2	1	0.042
							0.554

According to this set of weights and to penalty coefficients (s_2 is five times greater than s_1 in this case) the computational experiments are performed. Results of prescribing for the concrete set of requirements are presented in Table 3 (the requirements are taken from Table 1). The table contains:

- the set of real requirements,
- absolute quantities of components proposed by the model (the final proposal is here constructed in three steps),
- deviations from requirements in each step (in [%]),
- quantities of components received according to the proposal constructed by the physician.

Table 3
Numerical results of the model

Requir.	Water	TE	N	Na	K	Cl	Ca	P	Mg
	3600	9400	11	220	100	200	5	15	0
Model									
1 Abs.	3614	9506	10.7	216	100	197	5	15	1.5
1 [%]	0.4	1.1	-2.5	-1.8	0	-1.5	0	0	-
2 Abs.	3830	9881	12.4	217	101	200	5	15	1.5
2 [%]	6.4	5.1	12.9	-1.4	1 0	0	0	0	-
3 Abs.	3449	9196	10.7	217	96	197	5	15	1.5
3 [%]	-4.2	-2.2	-2.5	-1.4	-2.0	-1.5	0	0	-
Phys. Abs.	3510	6673	12.2	240	84	289	5	3	3

Table 4 presents the feeding composition for all three steps of interactive solution and for the physician´s proposal.

The solution obtained in the first step is a quite satisfactory proposal with regard to the deviations from requirements. However, composition of solutions in this proposal is not acceptable for decision maker because it contains solution G4 (solution rich in energy which is prescribed only seldom). Thus, usage of this solution is excluded in the second step. It is evident that values of deviations are mostly worse in the second step than in the first one. Moreover, the decision maker can not be satisfied with this proposal

Table 4
Composition of nutritional mixture

Sol.	Step 1	Step 2	Step 3	Phys.	Sol.	Step 1	Step 2	Step 3	Phys.
G2	-	200	-	-	P5	300	400	400	-
G3	2000	2000	2000	1200	C1	90	90	90	-
G4	200	-	-	-	C2	-	-	-	45
A2	400	600	400	-	C4	35	35	35	-
A3	-	-	-	400	C5	60	60	55	-
A4	600	-	-	-	C7	15	15	15	15
A6	-	600	600	800	C8	-	-	-	15
P1	-	-	-	1200	C9	10	10	10	10

with regard to the fact that three solutions (G2,A2, A6) are used in half a bottle quantity. So, usage of G2 solution is still excluded and usage of A2 solution is limited to 400ml. According to these limitations the final proposal is calculated. This proposal is acceptable with regard to deviations from requirements as well as to the composition of used solutions (five bottles of G5, one bottle of A2 and P5, and one and half a bottle of A6 - volumes of concentrated C solutions are not so important with regard to minimum volumes which can be used).

It is obvious that by the model in all steps obtained results are much better with regard to given requirements than the physician´s proposal. This fact has been confirmed by many experiments which were realized by the model.

Acknowledgements

The author would like to express thanks to dr.A. Jabor, the head of the Department of the Clinical Biochemistry of the District Hospital in Kladno, who made possible to complete the work of this study.

References

[1] Jablonsky,J., Fiala,P. and Manas,M.: "Multiple Criteria Optimization" (in Czech), SPN, Prague, 1986.

[2] Jablonsky,J. and Jabor,A.: "Prescribing of Parenteral Nutrition via Goal Programming", In: Tabuca-

non,M.T. and Chankong,V. (Eds.): Multiple Criteria Decision Making: Applications in Industry and Service, Asian Institute of Technology, pp.765-778, Bangkok, 1989.

[3] Korhonen,P. and Laakso,J.: "A Visual Interactive Method for Solving the Multiple Criteria Problem", European Journal of Operational Research, Vol.24, No.2., pp.277-287, 1987.

[4] Rehman,T. and Romero,C.: "Goal Programming with Penalty Functions and Livestock Ratio Formulation", Agricultural Systems, Vol.23, pp.117-132, 1987.

[5] Saaty,T.L.: "Rank Generation, Preservation, and Reversal in the Analytic Hierarchy Decision Process", Decision Sciences, Vol.18, No.2, pp.157--177, 1987.

[6] Steuer,R.E.:"Multiple Criteria Optimization:Theory, Computation and Application", J.Wiley, New York, 1986.

SEQUENTIAL DECISION MAKING AND RESTRUCTURABLE MODELLING

GREGORY E. KERSTEN*, WOJTEK MICHALOWSKI* AND STAN SZPAKOWICZ**

* DECISION ANALYSIS LABORATORY
SCHOOL OF BUSINESS, CARLETON UNIVERSITY
OTTAWA, ONTARIO, CANADA K1S 5B6

** DEPARTMENT OF COMPUTER SCIENCE, UNIVERSITY OF OTTAWA
OTTAWA, ONTARIO, CANADA K1N 6N5

ABSTRACT

THE PAPER DEALS WITH MODELLING AND SUPPORT OF A PROCESS OF SEQUENTIAL DECISION MAKING. THIS IS A PROCESS WHERE DECISIONS FOLLOW ONE FROM ANOTHER, AND THERE IS A STRONG INTERDEPENDENCE BETWEEN AN AGENT AND THE DECISION ENVIRONMENT. RULE-BASED FORMALISM IS USED TO DEVELOP REPRESENTATIONS OF A DECISION PROBLEM, AND TO MODEL TRANSITIONS BETWEEN ITS STATES. RESTRUCTURABLE MODELLING USES RULES TO STRUCTURE THE PROBLEM REPRESENTATIONS, AND TO COMBINE QUANTITATIVE MODELS WITH SYMBOLIC REASONING.

The research presented in this paper was partially supported by grants from the Natural Sciences and Engineering Research Council of Canada. The authors would like to thank Dr. Ian Lee for helpful comments and suggestions.

1. Introduction

In this paper we outline an approach to the modelling of sequential decision making. The approach is based on our research on negotiation modelling (Kersten *et al.*, 1988; Kersten *et al.*, 1991), and on experiments with NEGOPLAN, an expert system shell for negotiation support (Matwin *et al.*, 1989; Michalowski *et al.*, 1991).

We consider an agent who makes decisions in a complex and dynamic decision environment. This environment is usually non-uniform. Some entities (individuals, organizations) close to the agent react to an agent's actions, and influence its decisions. We refer to those as the *relevant others* (for example, the opponents in negotiations, neighbors, competitors, partners). There also are numerous other beings for which the agent is an unknown entity, which do not react to its particular actions, but whose actions may strongly influence the agent's behavior. We refer to those, collectively, as the *environment*.

The environment describes the circumstances which are beyond the control of the agent or the relevant others, and which constitute the context of decision making. The examples of the environment are market forces, government agencies, juridical system, *etc*. To sum up, the entities considered are: (1) the agent; (2) the relevant others; and (3) the environment. We may need to understand the relevant others, to develop unique models for determining their behavior in order to represent their reactions to the agent's actions. We often need statistical methods to determine the states of the environment.

At any moment, a decision problem is at a particular state, and it can be transformed into other states. Consequently, we view time as a tree whose branching points are states in which the decisions are made and implemented (Horwich, 1988).

The traditionally accepted decision-theoretic paradigms single out three theoretical issues in decision-making: (1) development of a representation of a decision problem; (2) determination of a procedure which transforms the representation into decision alternatives; and (3) choice of an alternative (Merkhofer, 1987). These issues have influenced much of the research on decision making and support. For sequential decision making fourth issue emerges: determination of procedures which create and modify problem representations. Such procedures allow for *restructurable modelling*, that is, modelling of systems with an evolving structure (Kersten *at al*, 1991; Michalowski *et al.*, 1991).

The modelling and analysis of decision problems in a sequential setting is typical of the behavior of 'intelligent' support systems or agents. The level of support increases if the system can make complex decisions independently in an attempt to determine their present and future consequences. Our work aims at expanding the design requirements of the Phoenix system (Cohen *et al.* 1989) and it is an example of how the concepts of intelligent decision making and rationality can be embodied in a computer system.

2. Basic Concepts

The agent has a conceptual view of the class of problems which contains the initial decision problem. This general view is captured as a *generic representation* $\mathcal{GR}$. A generic representation is created using the agent's knowledge $\mathbb{AK}$ about the specific class of decision problems (for example, buying a car, managing a project, making a capital investment). The $\mathcal{GR}$ is specified using additional information about the state of the world, and all its components. This specification at time "t" is captured as a *specific representation* $\mathcal{SR}_t$, which is used to determine a decision.

An initial specific representation denoted as $\mathcal{SR}_0$ may be different than $\mathcal{GR}$. This is because the agent may consider some elements of the $\mathcal{GR}$ as irrelevant, or because it does not have information about them. The subsequent $\mathcal{SR}_t$ $(t > 0)$ may expand beyond $\mathcal{GR}$, if the agent acquires new information.
The agent uses its knowledge $\mathbb{AK}$ to develop representations of the relevant others $\mathcal{OR}$ and the environment $\mathcal{ER}$. Both representations reflect the agent's perception. The triad $(\mathcal{SR}_t, \mathcal{OR}, \mathcal{ER})$ describes the state of the world at time $t \geq 0$. We restrict our considerations to the situation when $\forall\, t \geq 0$, $\mathcal{GR} \cap \mathcal{SR}_t \neq \emptyset$ and $\mathcal{SR}_t \cap \mathcal{SR}_{t+1} \neq \emptyset$. The first condition indicates that a specific representation inherits some elements from a generic one. The second condition states that two subsequent specific representations have common elements.

The following assumptions underlie creation of the representations, their relationships, and description of transitions between them:

1. The *separability* assumption: the $\mathcal{SR}_t$, $\mathcal{OR}$ and $\mathcal{ER}$ can be developed independently.
2. The *synthesizability* assumption: availability of $\mathbb{AK}$ subsumes all of $\mathcal{SR}_t$, $\mathcal{OR}$, $\mathcal{ER}$, and the possible modifications of the $\mathcal{SR}_t$. Although future representations are not known *a priori*, their elements, possible relationships between them, and their interpretations are known.
3. The *solvability* assumption: algorithms and procedures which the agent may need to solve its problem or to modify it are known *a priori.*
4. The *hierarchical decomposition* assumption: to represent a decision problem, it is sufficient to form a $\mathcal{SR}_t$, such that it is decomposed into a *principal* (top-level) problem and a number of subproblems to produce a hierarchy.

3. Problem Representation

3.1. Preliminaries

Sequential decision making can be viewed as a process of achieving certain goals (for example, the goal of *buying a car*). Usually these goals are complex semantic constructs which can be described in simpler terms. The most general goal is referred to as the *principal* goal, while the simpler terms are its *subgoals*. For instance, the principal goal of negotiating the purchase of a car may be described as buying a *comfortable car with different options* which, in turn, can be further explained in detail. These terms, in the simplest form, are *statements* which define the agent's view during the analysis of the problem. In that sense, sequential decision making is a process of making and validating judgements about the correctness of some of these statements. The relationship between the principal goal and other subgoals is seen as a hierarchy. High-level goals are expressed in terms of more detailed ones. The simplest goals - the statements about the decision problem - are referred to as *facts*. The facts are evaluated, whereupon they become the *metafacts*. Information exchange in sequential decision making is focused on the interpretation of facts, in the sense that only facts can be the building blocks of decisions. The hierarchical decomposition of goals into statements is an implication which is represented by the *rules*: The operations on these statements (needed for rules' analysis) conducted within a predicate calculus, are described by Kersten *et al.* (1991).

Validation of the statements and exchange of information about them is governed by another set of principles represented by *metarules* (Matwin *et al.*, 1989). The metarules may require some additional information for its evaluation. This information is provided by the incorporation of a *window*. A window refers to the tests and calculations performed on the arguments of metafacts. These operations are executed by external procedures. A window is not part of the $\mathcal{SR}_t$, nor does it describe a particular statement. It may refer to the conditions which restrict the valuation of the statements, or it may contain a call to a procedure which belongs to the representation of the environment. Incorporating windows into the problem representation allows us to enrich qualitative reasoning with quantitative considerations.

3.2. Generic Representation

A generic representation $\mathcal{GR}$ consists of rules which are created using agent's knowledge $\mathbb{AK}$, and windows which contain calls to the computational procedures. An example of a $\mathcal{GR}$ is:

```
car <- options & offered_price(X) & comfort.
options <- cruise_control & trip_computer.
options <- tinted_windows and power_locks.
comfort <- horsepower(Y) and road_handling.
comfort <- ac & road_handling.
```

In order to reason about the statements in $\mathfrak{GR}$, the variables must be instantiated. This happens when, for example, a specific car, i.e. VW Passat costing \$19,000 (`X = 19`) becomes the subject of negotiation.

The agent's knowledge about the dynamics of a decision process is conceptualized as the representation of the relevant others $\mathfrak{OR}$, and of the environment $\mathfrak{ER}$. The $\mathfrak{OR}$ consists of metarules, and the $\mathfrak{ER}$ consists of the variables and procedures which simulate states and reactions of the environment. For example, a car buyer may expect that a dealer will propose an extended warranty free of charge, if the car's delivery time need not be immediate. This situation is represented by the following metarule:

```
buyer: offered_price (19) ::= true &
buyer: delivery_time( T ) ::= true &
{ T > 30 }
==>
dealer: extended_warranty ::= true.
```

The example below shows a metarule which is used, inter alia, to read certain parameters of the environment:

```
buyer: cycle( looking_for_a_car ) ::= true &
buyer: compact ::= true &
{ get_info( price, X ),  get_info( lease, R ) }
==>
buyer: price( X ) ::= true &
buyer: lease( R ) ::= true.
```

3.3. Specific Representation and Solutions

Initial specific representation $\mathfrak{SR}_0$ for a car purchase problem described by the $\mathfrak{GR}$ presented in section 3.2, is given below:

```
car <- options & offered_price(19) & comfort.
options <- cruise_control & trip_computer.
options <- tinted_windows and power_locks.
comfort <- horsepower(180) and road_handling.
comfort <- ac & road_handling.
```

A feasible $\mathcal{AR}_t$ has a solution which is obtained by such valuations of the facts which will infer the value *true* of the principal goal (in this case *car*).

Decision theory recognizes that a sequential decision problem to be solved may have a natural redundancy. Redundancy in the $\mathcal{AR}_t$ comes in the form of disjunctions, modelled with *competitive rules* which have the same consequent but different antecedents (Kersten and Szpakowicz (1990). In order to create a solution, only one of the competitive rules may be chosen. Selecting one combination of disjunctions converts $\mathcal{AR}_t$ into a *reduced* representation $\mathcal{RR}_t$.

We say that $\mathbf{v}_t$ is a *solution* of $\mathcal{AR}_t$ and $\mathbf{a}_t^v$ is a set of facts in this solution *iff* for any $\mathcal{RR}_t \subset \mathcal{AR}_t$ the value *true* of the principal goal can be inferred from the values assigned to the facts $\mathbf{a}_t^v$.

Kersten and Szpakowicz (1990) describe the procedure which transforms $\mathcal{AR}_t$ into a $\mathbf{v}_t$. The procedure is non-discriminatory in that it does not favor any particular way of choosing rules for a $\mathcal{RR}_t$, and it does not reflect the possible preferences of the agent.

Two extensions of this procedure have been proposed: binding of facts, and weighting. The first extension is implemented as the set-option (Matwin *et al.*, 1989), and it is used when the agent wants to pre-set the values of some facts $\mathbf{a}_t^v$. Binding may introduce inconsistency in the $\mathcal{AR}_t$, so its feasibility must be checked. The second extension is the use of weights assigned to facts (Kersten and Szpakowicz, 1990) in order to determine the most preferred $\mathbf{v}_t$.

Since $\mathbf{a}_t^v$ need not contain all facts included in $\mathcal{AR}_t$, it is possible to partition the facts in the $\mathcal{AR}_t$ into those which are found in $\mathbf{v}_t$, and the remainder. The solution $\mathbf{v}_t$ (facts $\mathbf{a}_t^v$) cannot be altered, while one can assign any value to other facts (denoted $\mathbf{a}_t^r = \mathbf{a}_t \setminus \mathbf{a}_t^v$) without rendering the principal goal *false*. Such a partition enables one to distinguish the set of *inflexible* facts $\mathbf{a}_t^v$, and the set of *flexible* facts $\mathbf{a}_t^r$.

4. Restructurable Modelling

Sequencing decisions in time involves the transformation of $\mathbf{v}_t$ into $\mathbf{v}_{t+1}$. This may indicate that the agent needs to perform activities, such as the modification of the $\mathcal{SR}_t$, generation of a new $\mathbf{v}_{t+1}$, or a change of valuation of the facts. Two general classes of activities can be distinguished:

(i) for a given $\mathcal{SR}_t$ formation of $\mathcal{RR}_{t+1}$ and determination of $\mathbf{v}_{t+1}$. Such operation leads to changes in the flexible and inflexible facts: $\mathcal{SR}_{t+1} = \mathcal{SR}_t$, $\mathcal{RR}_{t+1} \neq \mathcal{RR}_t, \mathbf{v}_{t+1} \neq \mathbf{v}_t$.

(ii) New $\mathcal{SR}_t$ needs to be created, thus $\mathcal{SR}_t \rightarrow \mathcal{SR}_{t+1}$, and $\mathcal{SR}_{t+1} \neq \mathcal{SR}_t$, $\mathcal{RR}_{t+1} \neq \mathcal{RR}_t, \mathbf{v}_{t+1} \neq \mathbf{v}_t$.

An attempt should be made to adjust the decision without changing $\mathcal{SR}_t$. If this is not possible, then generation of $\mathbf{v}_{t+1}$ involves a restructuralization of the sequential decision problem representation.

4.1. Restructuralization

A restructuralization of $\mathcal{SR}_t$ occurs when $\mathbf{v}_{t+1}$ cannot be obtained through adjustment (change of type (i)), that is, when the current representation $\mathcal{SR}_t$ cannot process the recently acquired information. The act of restructuralization is described by a *restructuring metarule* which governs the creation of $\mathcal{SR}_{t+1}$ (Kersten *et al.*, 1991). The new goal introduced by this metarule may be a complex one, and it may consist of several subgoals. For example, the restructuring metarule given below redefines a buyer's car purchase problem by eliminating all the statements related to different options, and by changing the price offered for a car:

```
dealer: price( 19 ) ::= false &
dealer: power_locks ::= false &

dealer: ac ::= false
==>
modify (   car <-
               offered_price(17) & comfort ).
```

A transformation of $\mathcal{SR}_t$ into $\mathcal{SR}_{t+1}$ implies changes in the agent's perception of the goals and the relationships among them. Assume, for example, that `car` was defined in $\mathcal{SR}_t$ as:

```
car <- offered_price(X) & comfort.
```

The restructuring metarule of the form

```
buyer: deal_made( X ) ::= true
==>
modify ( car <- sign_deal_at( X ) ).
```

changes this goal's definition to one of the form

```
car <- sign_deal_at( X ).
```

with a *specific* value of variable `X` (price). If the trigger contains `X = 17`, the rule actually produced will be

```
car <- sign_deal_at( 17 ).
```

Restructuring the decision problem by transforming it to $\mathcal{SR}_{t+1}$ should be the last remedy available to the agent. It is always better to adjust a decision for a given $\mathcal{SR}_t$ than to restructure it in order to find another decision.

The sequential decision process can be described by a sequence of states s_1, ..., s_t, ... The termination of the process is modelled by termination metarules. These metarules capture the conditions necessary to occur for a process to be halted. The solution that immediately precedes the activation of a termination metarule is a final decision, and its metafacts describe the statements acceptable to both sides. A deadlock is signalled by the inability to consider a termination metarule or a restructuring metarule after other methods of decision adjustment have been exhausted. In such a case, there must exist statements which are perceived differently by both sides, and the problem remains unsolved.

5. Conclusions

The restructurable modelling of sequential decision making allows one to take a non-procedural view of routine and unusual information processing tasks undertaken by the agents. One can model complex situations from the viewpoint of one agent, represent reactions to changes perceived by the agent, determine necessary adjustments, and obtain new representations. The modelling framework presented in this paper is a step towards creation of a participatory human/computer interactive system where a computer system responds to introduced changes and identifies the solutions that take into account the present and anticipated requirements (DeSanctis and Gallupe, 1987).

The restructurable modelling is flexible enough to accommodate different, non-standard behaviors of the agent. This flexibility may be further increased by the introduction of a choice mechanism representing the agent's preferences. This mechanism might provide meta-control of the process by selecting the components of the $\mathcal{GR}$ to be used in the creation of $\mathcal{SR}_t$ (Kickert and van Gigch, 1979). Also, this mechanism might be used to select a preferred $\mathbf{v}_t$. Meta-control should govern the process of selecting the metarules by referring to some meta-knowledge about the anticipated responses of the relevant others and the environment. The existence of a meta-control should open the possibility of developing and controlling the whole process of sequential decision making, and to generate different decision making strategies (van Gigch, 1987).

A distinction between flexible and inflexible facts results from the technique used to solve $\mathcal{SR}_t$. Flexible facts have an impact on the decision flexibility - the system's capacity to adapt to the challenges posed by the environment or the relevant others (Kickert, 1985). Decision flexibility should be incorporated into a choice mechanism as it is a meta-criterion directed towards future events and inputs (Merkhofer and Saade, 1978). Integrating the preferences and cost-benefit analysis with decision flexibility gives the ability to generate different decision strategies oriented towards present and future viewpoints.

The concept of intelligent decision making, as well as selected aspects of the design of closed-loop engineering systems, may be applied to the creation of autonomous support systems. Future support systems will, if necessary, be able to act independently of the human decision maker. To do so, they will be required to mimic mental information processing, including shifts in perception and cognition. Within the classical paradigm of decision support systems technologies (the model base, data base and interface) or expert system technologies (knowledge base and inference engine), this requirement is impossible to satisfy due to the static representation of a decision problem to be supported. However, such a possibility is being created by the development of restructurable modelling.

References

Cohen, P.R., M.L. Greenberg, D.M. Hart and A.E. Howe (1989) "Trial by Fire: Understanding the Design Requirements for Agents in Complex Environments", *AI Magazine*, **10**(3), 34-48.

DeSanctis, G. and R.B. Gallupe (1987), "A Foundation for the Study of Group Decision", *Management Science*, **33**(5), 589-610.

Horwich, P. (1988) *Asymmetries in Time. Problems in the Philosophy of Science*, Cambridge: MIT Press.

Kersten, G.E., W. Michalowski, S. Szpakowicz and Z. Koperczak (1991) Restructurable Representations of Negotiation, *Management Science*, (in print).

Kersten, G.E., W. Michalowski, S. Matwin and S. Szpakowicz (1988) "Rule-based Modelling of Negotiation Strategies", *Theory and Decision*, **25**, 225-257.

Kersten, G.E. and S. Szpakowicz (1990) "Rule-based Formalism and Preference Representation: An Extension of NEGOPLAN", *European J. of Operational Research*, **45**(2-3), 309-323.

Kickert, W.J.M. (1985) "The Magic Word Flexibility", *Int. Studies of Management and Organizations*, **14**(4), 6-31.

Kickert, W.J.M. and J.P. van Gigch (1979) "A Metasystem Approach to Organizational Decision-Making", *Management Science*, **12**, 1217-1231.

Matwin, S., S. Szpakowicz, Z. Koperczak, G.E. Kersten and W. Michalowski (1989) "NEGOPLAN: An Expert System Shell for Negotiation Support", *IEEE Expert*, **4**(4), 50-62.

Merkhofer, M.W. (1987) *Decision Science and Social Risk Management*, Dordrecht: Reidel.

Merkhofer, M.W. and W.M. Saade (1978) "Decision Flexibility in a Learning Environment", Menlo Park: SRI International.

Michalowski, W., Z. Koperczak, G.E. Kersten and S. Szpakowicz (1991) "Disaster Management with NEGOPLAN", *Expert Systems with Applications*, **2**, 107-120.

van Gigch, J.P. (1987) *Decision Making about Decision Making. Metamodels and Metasystems*, Cambridge: Abacus Press.

COMPUTER-BASED EXPLANATION OF MULTIATTRIBUTE DECISIONS

David A. Klein, Martin Weber, Edward H. Shortliffe

IBM Thomas J. Watson Research Center, Yorktown Heights, NY, USA
Institut für Betriebswirtschaftslehre, Universität Kiel, Kiel, Germany
Stanford University School of Medicine, Stanford, CA, USA

Abstract

Experience with expert systems indicates that automated explanation of the rationale for computer-based advice is central to acceptance by decision makers. In this paper, we present strategies for computer-based explanation of measurable multiattribute value functions, and we demonstrate these strategies with implemented systems in marketing, process control, and medicine. Our approach illustrates the feasibility of explaining value-theoretic models automatically in the qualitative style of expert systems.

1. Introduction

Experience with expert systems indicates that automated explanation of the rationale for computer-based advice is central to acceptance by decision makers [28]. More than a decade ago, researchers in artificial intelligence (AI) noted that consultative systems "are unlikely to be accepted by users if they function as 'black boxes' that simply print their final answers" [23], and investigators have since performed empirical studies to verify the utility of automated explanation facilities [6, 25]. Explanation has become a focal topic of AI research, with experiments in a variety of application areas, from medicine [24] to financial planning [15].

A number of authors have discussed the need for automated explanation in the context of computer-based decision-analytic models [3, 5, 7, 17, 19], which historically have been avoided by expert-systems designers for lack of an explanation methodology [1, 20]. By developing computer-based methods for generating qualitative explanations of decision-analytic models, we potentially can promote the acceptance of such models and broaden the scope of their employment among nontechnical decision makers.

In this paper, we present computer-based strategies for explaining measurable multiattribute value functions [4], and we describe the role of these strategies in *Interpretive Value Analysis (IVA)* [10], our broader framework for modeling tradeoff-intensive decisions in computer systems. We demonstrate the explanation strategies with implemented systems in marketing, process control, and medicine. Our approach illustrates the feasibility of explaining value-theoretic models automatically in the qualitative style of expert systems.

The paper is organized as follows. Section 2 provides a brief review of Multiattribute Value Theory. In Section 3, we describe requirements for explaining value-theoretic choices. Section 4 outlines the structure of IVA to provide a context for describing the explanation strategies reported in Section 5. Section 6 provides examples of IVA-based explanation modules that build on the strategies of Section 5. In Section 7 we note limitations of IVA's explanations, describe related research, and speculate on the extension of our approach to additional normative decision models. Section 8 provides a summary of the paper and reviews our conclusions.

2. Review: Multiattribute Value Theory

Multiattribute Value Theory addresses the problem of modeling **multiattribute choices under certainty,** in which multiple, potentially *mutually competitive* objectives underlie choices among alternatives, and the outcomes of choices are known with virtual certainty. The problem of modeling multiattribute choices under certainty can be stated as follows.[1] Let **a** designate a feasible **alternative,** and let A denote the set of all such alternatives. With each $\mathbf{a} \in A$, we associate n indices of value that reflect our **objectives,** describing the degree to which our objectives are satisfied in the context of **attribute values** $a_1, \ldots, a_n$ of alternatives. We define a **multiattribute value function** that maps each $\mathbf{a} = (a_1, \ldots, a_n)$ into a scalar index of value, and we select alternatives that maximize this function. The appropriate form for the value function depends on the relationship among the decision maker's objectives; the form employed in most practical applications [27] is the **additive multiattribute value function:**

$$v(a) = v(a_1, \ldots, a_n) = \sum_{i=1}^{n} w_i v_i(a_i) \qquad \text{Eq. (1)}$$

1. Each $a \in A$ is represented by a vector of **attribute values** $(a_1, \ldots, a_n)$.
2. v_i is the **component value function** for attribute i, with $v_i(\text{worst } a_i) = 0$, $v_i(\text{best } a_i) = 1$, and $0 \leq v_i(a_i) \leq 1$ for all a_i. The component value functions express the relative desirability of various levels of their respective attributes.
3. w_i is the **weight** for attribute i, $0 < w_i < 1$ and $\sum w_i = 1$. The weights indicate the relative importance of each attribute as it changes from its best to its worst value.

The appropriate measurement scale for value depends on the tastes of the analyst and the client. For example, the theory of measurable multiattribute value functions [4] is based on difference measurement, which lends meaning to statements such as "alternative **a1** is *much better* than alternative **a2,**" whereas classical utility theory [26] supports only ordinal statements, such as "**a1** is *better* than **a2.**"

The construction of a value function can be facilitated by the employment of a **value tree,** which represents explicitly the decomposition of the decision maker's overall objective (the root of the tree) into a hierarchically structured set of more detailed objectives. Figure 1 depicts a simple value tree from a system that evaluates process-control actions [12].

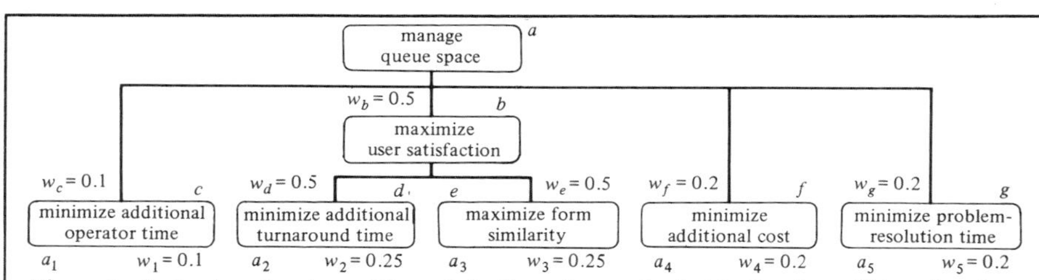

Figure 1: A simple value tree. The satisfaction of a nonleaf objective (e.g., b) is a function of the satisfaction of its children (d and e) and of these children's local weights (w_d and w_e). Leaf objectives are associated with attributes in the value function (e.g., d is associated with a_2); the satisfaction of a leaf objective is represented by its associated component value function, as we described. The weight for an attribute can be calculated by taking the product of the local weight of the associated leaf objective and the weights of its ancestors. w_2, for example, is given by $w_d w_b = (0.5)(0.5) = 0.25$.

[1] Portions of this description are adapted from [9].

The multiattribute value function associated with the tree of Figure 1 is

$$v(a_1, a_2, a_3, a_4, a_5) = 0.1v_1(a_1) + 0.25v_2(a_2) + 0.25v_3(a_3) + 0.2v_4(a_4) + 0.2v_5(a_5) \quad \text{Eq. (2)}$$

A number of authors provide guidelines for value-tree construction [9, 27].

3. Explanation of value-theoretic choices

The explanation of multiattribute value functions involves more than simply embedding such functions in textual or graphical displays. Consider, for example, the rote presentation of Eq. (2) in a text explanation:

> *Your best option is ACTION.1 because your value function is* $v(a_1, a_2, a_3, a_4, a_5) = 0.1v_1(a_1) + 0.25v_2(a_2) + 0.25v_3(a_3) + 0.2v_4(a_4) + 0.2v_5(a_5)$ *where* x_1 = *additional operator time in minutes,* x_2 = *additional turnaround time in minutes, ...,* x_5 = *problem-resolution time in minutes, and ACTION.1 maximizes v over all available alternatives.*

Our analysis of explanations for decisions that we collected from both decision analysts and nonanalysts suggests particular limitations of such rote explanations. For example, this explanation fails to compare alternatives explicitly, providing little insight into the tradeoffs that recommend the chosen alternative over close contenders. In addition, the rote explanation refers only to the most detailed objectives in the value tree, whereas our subjects mentioned objectives at multiple levels of abstraction, sometimes substituting an abstract objective for its children. A related deviation from our empirical observations, the explanation includes *all* the attributes that underlie the decision, whereas frequently only a subset of these attributes distinguishes the chosen alternative from its closest contenders. The explanation also mentions inappropriately the attributes by position in the value function's parameter list, whereas our subjects tended to cite compelling reasons for decisions before relatively uncompelling ones. Another shortcoming of this explanation is that it mentions only detailed quantitative values for parameters, whereas our subjects supplemented these descriptions with qualitative terms, such as *very important.*

Generally speaking, current normative decision aids burden decision makers in the spirit of our rote explanation with the task of qualitatively interpreting value-function results (although these aids surely present more elegant displays). IVA represents an attempt to encode in computer programs elements of the qualitative interpretation process demonstrated by our subjects.

4. Interpretive Value Analysis

4.1 Overview

Interpretive Value Analysis (IVA) [10] is a framework for explaining and iteratively refining measurable multiattribute value functions, comprising the following components. Our interviews with subjects suggest an **interpretation** for the value function that provides (1) a vocabulary of approximately 100 **interpretation concepts** for talking about decisions and (2) a set of computational structures that support the generation of explanations. We associate interpretation concepts with formal value-theoretic interpretations, as well as intuitive ones. The concepts serve as the primitives of IVA's **explanation strategies,** which are designed to provide the decision maker with sufficient insight into a value function's operation either (1) to become convinced that the chosen alternative is preferred or (2) to identify model parameters for correction. IVA's **refinement strategies** [13] facilitate such corrections, building on the explanation strategies.

This paper focuses on IVA's explanation component. Because this component builds on the interpretation, we provide in the following section examples of interpretation concepts that do not correspond to standard value-theoretic terms. In Section 5, we describe a sample computational structure from IVA's interpretation component that facilitates the generation of explanations.

4.2 Foundation for explanation: Sample interpretation concepts

A frequently referenced concept in IVA's explanation strategies is COMPELLINGNESS,[2] which captures the strength of a factor (objective) supporting a preference of alternative **a1** to alternative **a2**, as in the statement, "Cost is a notably compelling reason to prefer a used car to a new car." We define COMPELLINGNESS as follows.

Definition 1: COMPELLINGNESS(o **a1** **a2** $refo$) = $w(o\ refo)[v_o(\mathbf{a1}) - v_o(\mathbf{a2})]$
where

- o is an objective
- $\mathbf{a1} = (a1_1, ..., a1_n)$ and $\mathbf{a2} = (a2_1, ..., a2_n)$ are alternatives
- $refo$ is an ancestor of o in the value tree
- $w(o\ refo)$ is the **composite weight** of objective o in the context of the subtree with root $refo$, which is defined recursively as follows:

  IF o = $refo$,
  THEN return 1
  ELSE IF PARENT(o) = $refo$,
  THEN return w_o, the **primitive weight** assigned to o in the value tree
  ELSE return $[w_o][w(\text{PARENT}(o)\ refo)]$
- v_o either is component-value-function evaluation or is recursive multiattribute evaluation over CHILDREN(o), for leaf and nonleaf objectives respectively. ■

We can interpret the COMPELLINGNESS of o as a measure of o's strength in determining the overall value difference of **a1** and **a2**, other things being equal. More formally, when we consider the COMPELLINGNESS of a leaf objective o in the context of the root, we imagine two alternatives $\boldsymbol{\alpha}$ and $\boldsymbol{\beta}$ that differ only with respect to their oth attribute, with $\alpha_o = a1_o$, $\beta_o = a2_o$, and $\alpha_k = \beta_k$ for $k \neq o$. Thus,

$$v(\boldsymbol{\alpha}) - v(\boldsymbol{\beta}) = w(o\ root)[v_o(\alpha_o) - v_o(\beta_o)] = \text{COMPELLINGNESS}(o\ \mathbf{a1}\ \mathbf{a2}\ root).$$

We interpret the COMPELLINGNESS of nonleaf objectives analogously, with the nonidentical attributes of $\boldsymbol{\alpha}$ and $\boldsymbol{\beta}$ corresponding to the leaves of the value subtree with root o and with v_o as recursive multiattribute evaluation. When $refo \neq root$, we consider only the leaves of the value subtree with root $refo$ in defining $\boldsymbol{\alpha}$ and $\boldsymbol{\beta}$.

Although COMPELLINGNESS can be interpreted in this fashion as a multiattribute value difference under special conditions, this interpretation is inconsistent with measurable value theory in a superficial sense: The theory is based on the axioms of Krantz et al. [16], which assume a positive difference structure, whereas COMPELLINGNESS is defined over an algebraic difference structure (e.g., $w(o\ refo)[v_o(a1_o) - v_o(a2_o)] < 0$ whenever $a2_o$ is preferred to $a1_o$). But this discrepancy is not of pragmatic concern, because there exist axiom systems that are based on an algebraic difference structure (e.g., [22]), and the redevelopment of a measurable value theory based on such a structure would be straightforward [21]. The analysis of such discrepancies underscores the motivation for IVA's interpretation component: By scrutinizing the semantic integrity of IVA's 100 vocabulary terms with respect to a common theory, we guarantee their semantic consistency.

---

[2] We display interpretation concepts in HYPHENATED-SMALL-CAPITAL-TYPEFACE.

To generate qualitative statements in explanations such as, "Cost is a notably compelling reason to prefer a used car to a new car," we define an abstraction of COMPELLINGNESS:

Definition 2: NOTABLY-COMPELLING?(o $opop$ **a1** **a2** $refo$) $\equiv$
$|$COMPELLINGNESS(o **a1** **a2** $refo$)$| > \mu_X + k\sigma_X$
where

- o, **a1**, **a2**, and $refo$ are defined as in Definition 1
- $opop$ is an **objective population** such that $p \in opop \rightarrow p' \notin opop$ for every $p' \in$ ANCESTOR(p)
- $o \in opop$
- $x \in X = \{|$COMPELLINGNESS(p **a1** **a2** $refo$)$|, p \in opop\}$
- μ_X is the mean of X
- σ_X is the standard deviation of X
- $k > 0$ is a user-defined constant. ■

Informally, a NOTABLY-COMPELLING? objective o in a population is an outlier, and k determines the degree to which the magnitude of the COMPELLINGNESS of o must be an outlier to be considered NOTABLY-COMPELLING? with respect to that population. The constant k can be interpreted as a measure of conservatism, in that a sufficiently larger k will yield fewer NOTABLY-COMPELLING? objectives than will a smaller k.[3] Although other definitions are possible, this definition of NOTABLY-COMPELLING? has behavioral appeal. For example, for an arbitrary value of k, if the absolute values of COMPELLINGNESS are clustered arbitrarily closely, then *none* are NOTABLY-COMPELLING?. As intuition demands, it is impossible for *all* the objectives in a population to be NOTABLY-COMPELLING?. We show in [10] that NOTABLY-COMPELLING? is meaningful under difference measurement: For $a > 0$, the Boolean value of $x > \mu_X + k\sigma_X$ is unchanged when we replace each $x \in X$ with $ax + b$.

As we shall describe in the following section, interpretation concepts such as NOTABLY-COMPELLING? are employed both to classify objectives in explanations and to determine the structure and content of explanations.

5. Explanation strategies

We provide in IVA explanation strategies that allow decision makers to pose detailed queries about decisions, to generate summary comparisons of alternatives, to observe the computation of decisions in a step-by-step fashion, and to generate abstract descriptions of how decisions are computed. We demonstrate selected strategies in the following sections, employing examples generated by **VIRTUS**,[4] a software tool for building IVA-based systems that has been tested in the context of problems from marketing, process control, and medicine.

---

[3] Because the quantitative thresholds in definitions of interpretation concepts that express qualitative abstractions such as NOTABLY-COMPELLING? are user-specific, our implementation permits decision makers to adjust parameters such as k directly.

[4] VIRTUS is Latin for *value*, as in the value of an alternative in a choice among alternatives. In the examples, the decision maker's inputs are displayed in **bold-faced type**, VIRTUS's responses appear in *italics*, our annotations appear in {normal typeface enclosed in braces}, and alternatives are set in CAPITAL letters. Because refined text generation is not a focus of our work, we have made cosmetic corrections, such as capitalization and pluralization.

5.1 Generation of responses to detailed queries

IVA's interpretation concepts provide a space of queries that can be made available to the decision maker directly. VIRTUS, for example, includes an **interpretation-concept invocation** facility that permits decision makers to view interpretation-concept results by specifying interpretation concepts and their operands (alternatives and objectives). The following example is taken from a VIRTUS application for evaluating expert-systems shells [14]:

> **What is the relative quality of SHELL.B and SHELL.C with regard to development environment support?**
>
> *SHELL.B provides somewhat better development environment support than SHELL.C.*

In this simple example, the decision maker requests the result returned by ABSTRACT-RELATIVE-QUALITY, an interpretation concept that maps the quantitative result [v(SHELL.B) − v(SHELL.C)] with respect to objective *maximize development environment support* to the qualitative value *somewhat better*. The decision maker can invoke such interpretation concepts arbitrarily to conduct an interactive dialog with VIRTUS:

> **Which alternative is best with regard to overall value?**
>
> *SHELL.B is the best with regard to overall value.*
>
> **What is the relative quality of SHELL.B and SHELL.C with regard to overall value?**
>
> *SHELL.B provides substantially better overall value than SHELL.C.*
>
> **What are notably compelling reasons in the choice between SHELL.B and SHELL.C with regard to overall value?**
>
> *Price provides the most compelling reason for the choice.*
>
> **What is the relative quality of SHELL.B and SHELL.C with regard to development environment support?**
>
> *SHELL.B provides somewhat better development environment support than SHELL.C.*
>
> **What are the reasons for and against SHELL.B vs. SHELL.C with regard to development environment support?**
>
> *Quality of documentation and quality of front end are reasons to prefer SHELL.B over SHELL.C with regard to development environment support. Reliability, interactive development facilities, and syntactic familiarity to data processing programmers are reasons to prefer SHELL.C over SHELL.B.*
>
> **What factors are notably important in the determination of development environment support?**
>
> *Quality of documentation is notably important in the determination of development environment support.*
>
> {And so on}

Although useful for exploring the details of a decision in an unrestricted fashion, interpretation-concept invocation burdens the decision maker with coordinating VIRTUS's responses to achieve an understanding of a decision. The following explanation strategy imposes organization on such responses.

5.2 Generation of summary comparisons

IVA's **value-tree pruning and presentation strategies** generate summary-level comparisons of particular pairs of alternatives. Generating such a comparison involves pruning objectives from an explanation by (1) limiting the depth and breadth of the value tree in accordance with the alternative pair of interest and (2) generating an explanation that reflects the remains of the tree.

Vertical-pruning strategies determine a desirable level of abstraction for talking about objectives (in the context of comparing a particular pair of alternatives) by pruning detailed objectives that can be summarized by higher-level objectives; these strategies limit the depth of the value tree by eliminating uninteresting subtrees. The following algorithm, for example, accepts as input a value tree of arbitrary depth with root *root* and two alternatives **a1** and **a2**, and produces as output a population of interesting objectives, if any exist, and *root* otherwise.

1. Let o = PARENT(deepest leaf)
2. If CARDINAL-TRADEOFFS?(**a1 a2** CHILDREN(o)), then mark o deleted (and retain CHILDREN(o));
otherwise, mark CHILDREN(o) deleted (and retain o).
3. If $o = root$, then return the remains of the tree;
otherwise, go to 1.

The interpretation concept CARDINAL-TRADEOFFS? is employed in the algorithm to identify interesting subtrees, returning *true* only when **a1** is strictly better than **a2** with respect to at least one objective and is strictly worse with respect to at least one objective. On each iteration, the algorithm prunes either the current node in the tree (if the node's children are interesting) or the current node's children (if the children are uninteresting). For example, whenever **a1** and **a2** are CARDINALLY-EQUIVALENT? with respect to all the objectives in a subtree (one instance of CARDINAL-TRADEOFFS? = *false*), the algorithm prunes these objectives from the explanation and retains their parent, because the effects of these objectives can be summarized qualitatively by their parent. Figure 2 shows a vertically-pruned value tree from a VIRTUS application [11] that evaluates **randomized clinical trials (RCTs)** – studies that compare the relative effectiveness of medical treatments by randomly assigning alternate therapies to subjects and observing the effects.

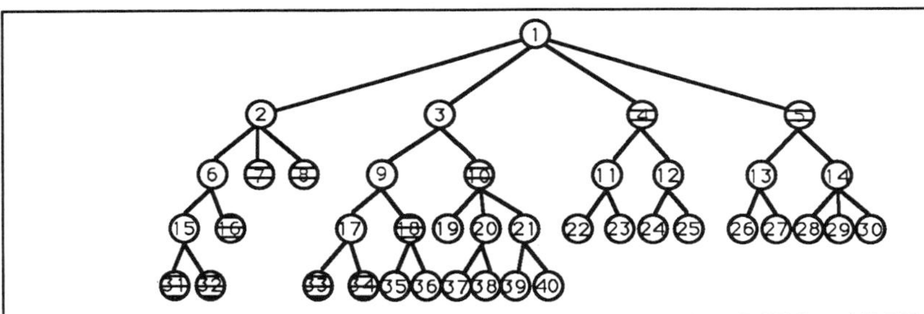

Figure 2: Vertical pruning. Comparing two particular alternatives RCT.1 and RCT.2, vertical pruning retains the following objectives (shaded), which vary in abstraction: *adjustments due to subgroups* (node 31), *subgroup analysis plan* (32), *stopping appropriateness* (16), *P-value* (7), *statistical techniques* (8), *distance of assigner from patient* (33), *quality of actual treatment assignment* (34), *blinded assignment design* (18), *equivalence at study outset* (10), *equivalence of care* (4), and *endpoint assessment* (5). Other objectives are pruned from the associated explanation.

This vertically pruned tree provides the basis for the following explanation:

Equivalence of care, equivalence at study outset, endpoint assessment, distance of assigner from patient, and adjustments due to subgroups are reasons to prefer RCT.1 over RCT.2 with regard to credibility. Quality of actual treatment assignment, blinded assignment design, stopping appropriateness, and subgroup analysis plan are reasons to prefer RCT.2 over RCT.1. P-value and statistical techniques do not at all impact the choice between RCT.1 and RCT.2.

Although more terse than an exhaustive display of value-tree leaves, this explanation is somewhat verbose, because limiting only the depth of the value tree still allows for arbitrary breadth.

The breadth of the value tree can be reduced by IVA's **horizontal-pruning strategies,** which vary in computational complexity and effectiveness. One inexpensive strategy, for example, simply selects the n most compelling objectives arguing for and against the preference of one alternative over another. This strategy is somewhat arbitrary, however, because we lack intuitive justification for omitting objective $n+1$ in the COMPELLINGNESS ranking whenever that objective differs only infinitesimally in COMPELLINGNESS from the nth. A more effective (but expensive) IVA strategy selects *clusters* of objectives for inclusion in explanations based on NOTABLY-COMPELLING?.

Combining vertical pruning with horizontal pruning yields a powerful capability for generating summary comparisons: Such explanations capture the most compelling reasons for preferring a particular alternative to another at an appropriate level of abstraction. For example, applying horizontal pruning to the vertically pruned tree of Figure 2 and embellishing with a statement of ABSTRACT-RELATIVE-QUALITY, VIRTUS generates the following summary:

> *RCT.1 is somewhat better than RCT.2. Compelling reasons to prefer RCT.2, such as quality of actual treatment assignment, are outweighed by considerations of equivalence of care and equivalence at study outset, along with less compelling reasons that recommend RCT.1.*

Whenever a decision maker finds a summary-level comparison of two alternatives unconvincing, however, he may require a more detailed comparison. In the following section, we present strategies that demonstrate qualitatively the detailed relationships between the decision maker's parameter values and the value function's results.

5.3 Generation of detailed comparisons

IVA's **difference-function-traversal strategies** generate detailed qualitative comparisons of alternatives. These comparisons demonstrate the relationships between a value function's results and parameter values by explaining the step-by-step computation of $[v(\mathbf{a1}) - v(\mathbf{a2})]$, the overall RELATIVE-QUALITY of two particular alternatives **a1** and **a2**. Difference-function-traversal strategies reflect a decompositional approach to explanation that involves (1) representing explicitly the precedence of operations that combine parameters, (2) providing a qualitative explanation for each such operation, and (3) designing control procedures that concatenate these explanations according to the operation-precedence representation. An analogous approach has been employed to support explanation in systems that model the behavior of physical devices [2].

Under IVA, generating a step-by-step comparison of a specified pair of alternatives involves stepping through a **difference function** $\boldsymbol{h}$, which we developed to reflect elements of our subjects' explanations. More specifically, a difference function captures interactions among value-function parameters for a pair of alternatives ($\mathbf{a1}$, $\mathbf{a2}$) and reflects the hierarchical structure of the value tree: The difference function corresponding to the tree of Figure 1, for example, is

$$\begin{aligned} h(\mathbf{a1},\mathbf{a2}) = {} & [w_c(v_1(a1_1) - v_1(a2_1))] \\ & + w_b[w_d(v_2(a1_2) - v_2(a2_2)) + w_e(v_3(a1_3) - v_3(a2_3))] \\ & + [w_f(v_4(a1_4) - v_4(a2_4))] \\ & + [w_g(v_5(a1_5) - v_5(a2_5))] \end{aligned} \qquad \text{Eq. (3)}$$

The step-by-step execution of Eq. (3) is represented explicitly in the **topology** shown in Figure 3.

We associate each arithmetic operation in the topology with an **operation explainer** that generates a qualitative explanation of the relationship between its operands and its result. An addition explainer, for example, provides a qualitative summary of the relationship between its input COMPELLINGNESS values for objectives that argue for and against a choice between two particular alternatives and its output RELATIVE-QUALITY of these alternatives. The following explanation is generated by the addition explainer in box *a* of Figure 3:

DASD provides infinitesimally better overall queue space management effectiveness than EXPENSIVE.PRINTING. While overall user satisfaction provides a compelling reason to prefer EXPENSIVE.PRINTING, this is outweighed by considerations of additional cost, along with other less compelling reasons, that provide motivation for preferring DASD.

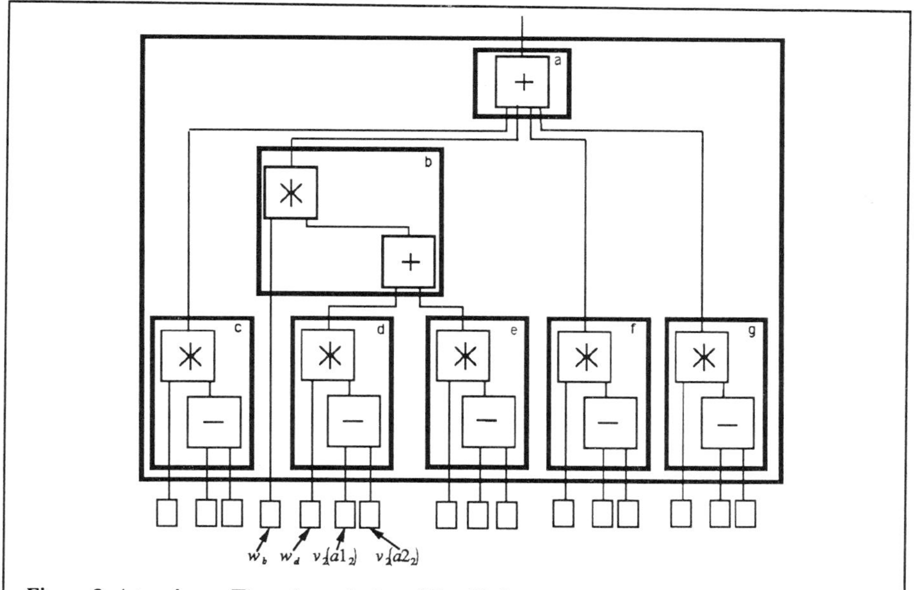

Figure 3: A topology. The subtopologies of Eq. (3) framed by boxes correspond to objectives in the value tree of Figure 1, as labeled. For uniformity, parameters are represented as nullary operations that return (user-specified) constants.

Control strategies generate explanations by stepping through a topology and concatenating the explanations generated by operation explainers. **Backward chaining** over a topology provides an account of how the final value-function result is derived from intermediate results, and of how intermediate results are, in turn, derived from value-function parameters. The following explanation, for example, is produced by backward chaining over the multiplication explainer and its connected subtraction explainer in box *f* of Figure 3:

Additional cost is a compelling factor favoring DASD over EXPENSIVE.PRINTING. While additional cost is not notably important in determining overall queue space management effectiveness, DASD provides sufficiently different additional cost from EXPENSIVE.PRINTING, relative to other factors, to make additional cost a notably compelling factor in this particular decision.

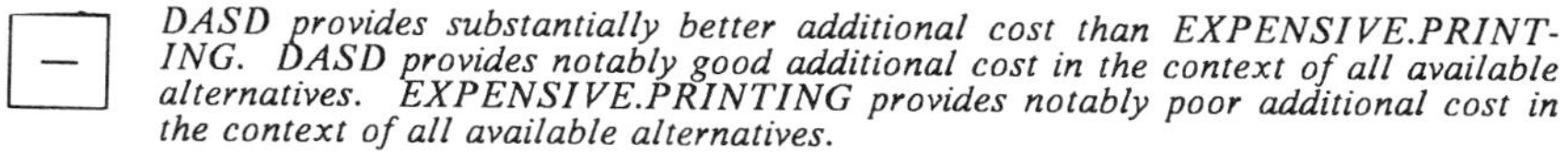

DASD provides substantially better additional cost than EXPENSIVE.PRINTING. DASD provides notably good additional cost in the context of all available alternatives. EXPENSIVE.PRINTING provides notably poor additional cost in the context of all available alternatives.

In this example, the subtraction explainer clarifies the explanation produced by the multiplication explainer, providing additional detail regarding the meaning of *sufficiently different*.

In a similar fashion, **forward chaining** provides an account of how a particular parameter participates in the determination of an intermediate result, which in turn ultimately affects the final result. **Interactive chaining** permits the decision maker to chain forward and backward over a topology in an incremental, interactive fashion. Klein [10] provides a more detailed exposition of strategies for difference-function traversal.

6. Construction of custom explanation modules

The modularity of IVA's design permits flexibility in building explanation modules for particular applications. The implementation of such modules might vary, for example, with the designs of operation explainers, with the particular horizontal-pruning strategies that are employed, and more generally, with the classes of strategies that are included.

Software designers can integrate IVA's explanation strategies to implement common explanation commands in expert systems, such as WHY. For example, a designer can provide an explanation module that generates a particular response to WHY (e.g., the summary comparison of Section 5) or a more sophisticated module that permits the decision maker to select among multiple interpretations of WHY, as in the following dialog:

DASD is the best with regard to overall queue space management effectiveness.

> **WHY**

By WHY, which of the following interpretations do you mean?

1. What is notably good about DASD?
2. How does DASD compare with the next best alternative, EXPENSIVE.PRINTING?
3. What are DASD's close contenders?
4. How does DASD compare with its close contenders?
5. List the alternatives that are not as good as DASD.
6. ...

> 1

DASD provides notably good additional cost and problem resolution time.

Questions such as (1), (3), and (5) can be answered by interpretation concepts directly, whereas responses to questions such as (2) and (4) might employ value-tree pruning and presentation or difference-function traversal.

IVA's explanation strategies also can be integrated to generate qualitative reports about decisions. VIRTUS, for example, generates reports that are organized as follows:

1. Problem Statement
2. Solution Summary
3. The Decision-Making Process
4. Brief Analysis of the Decision
5. Brief Analysis of Close Contenders
6. Detailed Analysis of Close Contenders

These reports draw on a variety of IVA's interpretation concepts and explanation strategies. The "Solution Summary," for example, includes a reference to the concept SIMILAR-ALTERNATIVES to identify the best alternative's close contenders; these contenders are compared with the best alternative by value-tree pruning and presentation in "Brief Analysis of the Decision." The alternatives are compared again by a backward-chaining, difference-function-traversal strategy that prunes uninteresting operation-explainer outputs in "Brief Analysis of Close Contenders," and once again by exhaustive, backward-chaining difference-function traversal in "Detailed Analysis of Close Contenders." In [10], we provide a complete example of such a report.[5]

[5] Langlotz [17] describes the generation of analogous reports in the context of explaining decisions based on decision trees.

7. Discussion

By developing computer-based methods for generating qualitative explanations of decision-analytic models, we potentially can promote the acceptance of such models and broaden the scope of their employment among nontechnical decision makers. Decision makers involved in each of our demonstration applications provided positive informal assessments of IVA's explanations, although we can identify limitations in our methodology. Decision makers may find it difficult to understand IVA's detailed comparisons in the context of large value trees, for example, and may be confused or inconvenienced by the menu-driven facilities described in Section 6. More generally, the particulars of IVA's explanations depend strongly on the axioms of difference measurement and on the independence assumptions that underlie the additive form.

We can speculate, however, that our approach to developing IVA is applicable to a broader class of normative decision models. In applying our approach to a different model, the system designer would determine from empirical data (as we did in our discussions with decision analysts and nonanalysts) an interpretation for the model that corresponds more closely to intuition than does its standard analytical representation. The development of the interpretation would involve generating a vocabulary of interpretation concepts and verifying its consistency with the original model. The interpretation might also include a reformulation of the model, as we provided in IVA's difference function. The designer would then construct a set of explanation strategies for organizing and presenting interpretation concepts. The simplest, least structured strategy involves making the concepts available to the decision maker directly, as we implemented in IVA's interpretation-concept-invocation strategy. The designer can synthesize more sophisticated strategies that impose organization on interpretation-concept results and also employ the concepts as evaluation functions that determine the structure and content of explanations (e.g., NOTABLY-COMPELLING? in horizontal pruning). Although a designer can implement explanation strategies that are specific to a particular model (e.g., value-tree pruning and presentation under IVA), the notion of computation traversal (as we implemented in difference-function traversal) has potentially broader application to other functional forms. In particular, the designer must develop a set of qualitative explanations for the values returned by the component operations of the model (or its altered form) and a set of control strategies that concatenate these explanations according to the precedence of operations.

IVA can be viewed as part of a growing body of research at the intersection of decision analysis and artificial intelligence. For example, a number of researchers have explored uses of and extensions to decision-analytic models as knowledge representations in expert systems [8, 17, 18, 29]. Recent research also has focused on the problem of generating explanations for probabilistic inferences [3, 5] and for recommendations based on decision trees [17].

8. Summary and conclusions

Experience with expert systems indicates that automated explanation of the rationale for computer-based advice is central to acceptance by decision makers. By developing computer-based methods for generating qualitative explanations of decision-analytic models, we potentially can promote the acceptance of such models and broaden the scope of their employment among nontechnical decision makers.

In this paper, we presented computer-based strategies for explaining measurable multiattribute value functions, and we described the role of these strategies in IVA, our broader framework for modeling tradeoff-intensive decisions in computer systems. In particular, we described techniques for supporting detailed queries about decisions, summary comparisons of alternatives, and detailed comparisons of alternatives. We demonstrated these explanation strategies in the context of IVA-based systems in marketing, process control, and medicine. Our approach illustrates the feasibility of explaining value-theoretic models automatically in the qualitative style of expert systems.

Acknowledgments

Leslie Perreault provided helpful comments on earlier drafts of this paper. Eric Clemons, Tim Finin, and Curt Langlotz provided many useful discussions. This work was supported in part by the IBM Corporation, Grants LM05208 and LM07033 from the National Library of Medicine, the IBM Graduate Fellowship Program, and the NASA Graduate Student Researcher's Program.

References

[1] Clancey, W., Details of the revised therapy algorithm, in: B. Buchanan and E. H. Shortliffe, eds., *Rule-Based Expert Systems* (Addison-Wesley, Reading, MA, 1984) 133-146.

[2] de Kleer, J., and J. S. Brown, A qualitative physics based on confluences, *Artificial Intelligence* **24** (1984) 7-83.

[3] Druzdzel, M., and M. Henrion, Qualitative propagation and scenario-based approaches to explanation of probabilistic reasoning, in: *Proceedings Sixth Conference on Uncertainty and Artificial Intelligence*, Cambridge, MA (1990) 10-20.

[4] Dyer, J., and R. Sarin, Measurable multiattribute value functions, *Oper. Res.* **27** (1979) 810-822.

[5] Elsaesser, C., Explanation of Bayesian conditioning for decision support, Research Report MP-89W00037, The MITRE Corporation, McLean, VA (1990).

[6] Erdman, H., The impact of an explanation capability for a computer consultation system, *Meth. Info. Med.* **24** (1985) 181-191.

[7] Holtzman, S., *Intelligent Decision Systems* (Addison-Wesley, Reading, MA, 1989).

[8] Horvitz, E., Reasoning under varying and uncertain resource constraints, in: *Proceedings AAAI-88*, St. Paul, MN (1988) 111-116.

[9] Keeney, R., and H. Raiffa, *Decisions with Multiple Objectives: Preferences and Value Tradeoffs* (Wiley, New York, NY, 1976).

[10] Klein, D. A., Interpretive value analysis, PhD thesis, available as RC #15278, IBM Thomas J. Watson Research Center, Yorktown Heights, NY (1989).

[11] Klein, D. A., H. Lehmann, and E. H. Shortliffe, A value-theoretic expert system for evaluating randomized clinical trials, in: *Proceedings Fourteenth Annual Symposium on Computer Applications in Medical Care*, Washington, DC (1990) 810-814.

[12] Klein, D. A., and E. H. Shortliffe, Integrating artificial intelligence and decision theory in heuristic process control systems, in: *Proceedings Tenth International Workshop on Expert Systems and Their Applications*, Avignon, France (1990) 165-177.

[13] Klein, D. A., and E. H. Shortliffe, Interactive diagnosis and repair of decision-theoretic models, in: *Proceedings Seventh IEEE Conference on Artificial Intelligence Applications*, Miami, FL (1991) 289-293.

[14] Klein, D. A., Integrating artificial intelligence and decision theory to forecast new products, in: *Computational Intelligence III: Proceedings of the International Conference "Computational Intelligence '90"* (Elsevier, Amsterdam, The Netherlands, in press).

[15] Kosey, E., and B. Wise, Self-explanatory financial planning models, in: *Proceedings AAAI-84*, Austin, TX (1984) 176-181.

[16] Krantz, D., R. Luce, P. Suppes, and A. Tversky, *Foundations of Measurement* (Academic Press, New York, NY, 1971).

[17] Langlotz, C. P., A decision-theoretic approach to heuristic planning, PhD thesis, Stanford University, Stanford, CA (1989).

[18] Pearl, J., Fusion, propagation, and structuring in belief networks, *Artificial Intelligence* **29** (1986) 241-288.

[19] Reggia, J., and B. Perricone, Answer justification in medical decision support systems based on Bayesian classification, *Comp. Bio. Med.* **15:4** (1985) 161-167.

[20] Rennels, G., E. H. Shortliffe, and P. Miller, Choice and explanation in medical management: A multiattribute model of artificial intelligence approaches, *Medical Decision Making* **1** (1987) 22-31.

[21] Sarin, R., personal communication, 1989.

[22] Scott, D., and P. Suppes, Foundational aspects of theories of measurement, *J. Symbolic Logic* **23** (1958) 113-128.

[23] Shortliffe, E. H., R. Davis, S. Axline, B. Buchanan, C. Green, and S. Cohen, Computer-based consultations in clinical therapeutics: Explanation and rule acquisition capabilities of the MYCIN system, *Comp. Biomed. Res.* **8** (1975) 303-320.

[24] Swartout, W., Producing explanations and justifications of expert consultation programs, PhD thesis, Massachusetts Institute of Technology, Cambridge, MA (1981).

[25] Teach, R., and E. H. Shortliffe, An analysis of physicians' attitudes, *Comp. Biomed. Res.* **14** (1981) 542-558.

[26] von Neumann, J., and O. Morgenstern, *Theory of Games and Economic Behavior* (Princeton University Press, Princeton, NJ, 1947).

[27] von Winterfeldt, D., and W. Edwards, *Decision Analysis and Behavioral Research* (Cambridge University Press, New York, NY, 1986).

[28] Waterman, D., *A Guide to Expert Systems* (Addison-Wesley, Reading, MA, 1986).

[29] Wellman, M., Dominance and subsumption in constraint-posting planning, in: *Proceedings IJCAI-87*, Milan, Italy (1987) 884-890.

SUPPORTING THE DECISION MAKER TO FIND THE MOST PREFERRED SOLUTION FOR A MOLP-PROBLEM

Pekka Korhonen
professor

Merja Halme
research assistant

Helsinki School of Economics and Business Administration
Runeberginkatu 14-16, 00100 Helsinki
Finland

ABSTRACT

The paper deals with the problem of assisting the decision maker to find and indentify the "most preferred" or "best" solution for a multiple objective linear programming model using an interactive approach. We consider the conditions on which the final solution can be regarded as the most preferred one. We also propose an approach that supports the decision maker in finding the most preferred solution provided a locally best solution is also globally best. An additional supporting procedure is proposed in the case when the decision maker's value function can be assumed to be pseudoconcave at the moment of final choice.

Keywords: Multiple Criteria, Interface, Decision Support, Visual, Interactive

1. INTRODUCTION

In multiple criteria decision making (MCDM) any nondominated solution can be regarded as a rational choice, if there is no information available about the decision maker's (DM) preferences. The multiple objective procedures thus require the intervention of the decision maker.

A popular way to involve the DM in the solution process is to use an interactive approach. Several dozen procedures have been developed for solving optimization problems having multiple criteria. For an excellent review of interactive multiple criteria procedures, see Steuer (1986).

In interactive procedures at each iteration, a solution, or set of solutions, is generated to the DM for examination. As a result the DM inputs information in the form of tradeoffs, pairwise comparisons, aspiration levels, etc. His/her responses are used to generate a presumably improved solution. The ultimate goal is to find the DM's "most preferred" solution. Which solution is finally chosen as the most preferred one depends on the assumptions made about the DM's behavior and the way in which these assumptions are implemented. It is essential that these assumptions are related to the DM's actual behavior, which we call the **behavioral realism** of a method.

This paper deals with the problem of finding the most preferred solution for a multiple objective linear programming (MOLP) problem using an interactive approach. First, we consider the concept *most preferred* and its interpretation. Next, we propose an approach which provides the DM with a possibility to find the most preferred solution provided the local solution is also globally best. More support is provided in the case when the DM's value function can be assumed to be pseudoconcave at the **moment of final choice**. By that we mean the moment when the DM does not him/herself find any improved solutions. In conclusion, we discuss the advantages and disadvantages of our approach, and the relaxation of the asssumptions concerning the value function.

2. THE MOST PREFERRED SOLUTION

In this paper, we consider only the most preferred solution of a MCDM-model. We do not try to discuss the relation between the most preferred solution for the model and the best solution of the DM's real world decision problem.

We are interested in how good the final solution is at the moment of final choice. All interactive methods stop in a finite number of steps, actually in a few steps. Korhonen et al. (1990) have suggested an explanation for that. Therefore the key question is: Does the method give strong enough evidence to that the final choice is really the DM's most preferred solution?

We next give a definition for the most preferred solution of a rationally behaving DM dealing with a multiple criteria decision making model:

Definition. **The most preferred** (or best) **solution** for a multiple criteria decision making model is the nondominated solution, preferred to all other solutions by the decision maker at the moment of final choice.

It is not realistic, however, to assume that in general any MCDM-method could enable the DM to compare all possible solutions to the final solution at the moment of final choice. Therefore, MCDM-methods make various assumptions on the DM's value function. The classical methods of Geoffrion et al. (1972) and Zionts and Wallenius (1976) assume that the DM presents his/her preferences according to a stable value function of certain form. In the Geoffrion et al. method (1972), the assumption is not restrictive, because the consistency of the responses from the previous iterations is not checked. In the method by Zionts and Wallenius (1976) instead, the assumption on the form plays a central role. To meet the requirements of behavioral realism, the authors propose some heuristic techniques for treating inconsistent responses.

We propose the following requirements for a "most preferred" solution. No assumptions are made on the DM's preference structure.

> "The solution can be regarded as the most preferred solution for a MCDM-model, if the DM is **convinced** that it is preferred to all other nondominated solutions, and he has a **realistic perception** of these solutions."

By **convinced** we mean that the DM believes the current solution really is the best one, which he/she can find from among a set of (nondominated) solutions. The concept **realistic perception** means that there exists no other potential (nondominated) solution that would make the DM "surprised".

Both of these aspects are important in evaluating MCDM-methods. If a method is not able to convince the DM that the final solution is the best he/she can find, the solution may not even be acceptable. On the other hand, if a method does not give a realistic perception over all nondominated solutions, he/she may erroneously believe that the final choice is the best one.

To conclude this chapter we shortly list how the requirements of being convinced and realistic perception may fail in various types of interactive approaches:

- If the DM is explicitly forced to follow some rules specified by the system, he/she may have a feeling that he "has not seen" everything. That means the DM is not convinced.

- If the DM is free to specify what he/she would like to achieve, but the system does not offer the kind of solutions he/she is striving after, even if they are achievable. Then the DM may be convinced, but has an unrealistic perception of the other solutions

- If the DM is completely free to move in the set of nondominated solutions without any restrictions, on one hand he/she will realize the abundance of solutions and may feel unable to see everything, and on the other hand he/she may not be willing to examine solutions far from the current one. In the first case, he is not convinced, and in the second case the system did not support/encourage him/her to obtain a realistic perception on other alternatives.

3. THE USE OF THE VALUE FUNCTION

MCDM-methods may be classified on the basis of the assumptions made concerning the DM's value function, which represents the DM's preference structure.

A Stable Value Function Exists and It is Known

The methods based on the above assumption is mainly developed under the title multiattribute utility theory (MAUT) (see for survey, e.g. Fishburn (1989)). The research area focuses on the structure of multicriteria or multiattribute alternatives, usually when risks and uncertainties have a significant role in the definition and assessment of alternatives. Based on the information of the DM's preferences, a value (utility) function is constructed to order all alternatives without any further intervention of the DM. Often MAUT is treated separately from MCDM.

A (Stable) Value Function Exists, but It is Only Implicitly Known

Two well known classical interactive methods (Geoffrion et al. (1972) and Zionts and Wallenius (1976)) for solving MOLP-problems are based on that assumption. An assumption of the functional form (concave or linear) of the implicit value function is made. Partial information is used by the system to generate potential solutions for the DM's evaluation. The conditions for the most preferred solution are based on the characteristics of the value function.

A Dynamic Value Function is Implicitly Known

The Korhonen and Laakso (1986a,b) method for MOLP-problems needs no assumpion concerning the value function until at the moment of final choice when it is assumed to be pseudoconcave. The dynamic version (Korhonen and Wallenius (1988)) of the method is implicitly based on the assumption that at the moment of final choice the DM's local optimum is also the global one.

No Value Function is Known

There are several MOLP-methods, which do not assume the existence of a value function. Various ideas have been used for supporting the DM to find the most preferred solution:

1) A representative set of solutions with certain properties (e.g. extreme point solutions) are generated for the DM's evaluation (e.g. Evans and Steuer (1973))

2) Heuristic assumptions concerning the behavior of the DM are used to restrict the DM's choices (e.g. Benayoun, et al. (1971), Steuer (1977))

3) The DM is assisted to (freely) explore the set of nondominated solutions until he/she is satisfied. (e.g. Wierzbicki (1980))

Discussion

The observed behavior of the DM generally is not consistent with any (stable) value function (see, e.g. Korhonen et al. (1990)). However, by that argument we cannot deny the existence of the value function. Inconsistent behavior may be caused by many reasons: the DM is careless, unable to concentrate, unable to deal with several alternatives simultaneously, etc.

It is often difficult to say, whether the inconsistencies are caused by behavioral characteristics or by the assumed, excessively simple form of the value function. For practical reasons, it is necessary to assume a simple form for the value function: linear, concave, pseudoconcave, quasiconcave etc. The actual form may be - if the function exists - far from these ones.

The value function provides tools to be used in controlling the DM's search and in checking the optimality conditions at the moment of final choice. For the latter use alone, it is sufficient to assume a form for the value function only at the moment of final choice. During the search process, no assumptions are needed whatsoever concerning the stability or form of the value function. If a MCDM-method does not make assumptions on a value function, it is difficult - if not impossible - to evaluate how good the final solution is.

The value function traditionally used in modelling MCDM-problems is "path-independent": It does not depend on information the DM has provided during the process. A more realistic value function could be found testing it against and revising it according to the DM's stated preferences. Prospect theory proposes a value function, which is based on the use of a reference alternative (see, Kahneman and Tversky (1979)). However, its utilization is difficult in interactive MCDM-methods.

4. AN APPROACH

We consider an MCDM-problem at the moment of final choice. We only assume that at that very moment the DM's choices are based on an implicit value function. Our purpose is to develop a conceptual framework and an approach for considering other potential solutions in comparison with the final solution.

Consider the case with assumptions concerning only the existence of the value function at the moment of final choice. Then we can state that the DM has found a locally most preferred solution provided he/she has used a free search type method like Pareto Race (Korhonen and Wallenius (1988)). Pareto Race provides an opportunity to freely explore the neighborhood of the current best solution. Then if the DM's efforts have not produced any improved solutions, that implies the final solution is preferred to all the others in its neighborhood. However, there is no easy way to judge if it is necessary to continue the search somewhere else on the efficient frontier.

Although Pareto Race allows the DM to freely search the neighborhood of the current solution, a thorough search is difficult to carry out, especially in the case of several objectives. To prove that the final solution is the most preferred one, we have to assume that the DM is capable of exploring the neighborhood of the current solution in such a way that he/she achieves a comphrehensive knowledge of the tradeoffs in the area. That requirement may be hard to meet in realistic size problems.

In Pareto Race the DM can impose bounds for objectives and thus deal with fewer varying objective values simultaneously, which can considerably ease the search. Still, it may not be easy.

If we can assume pseudoconcavity of the value function at the moment of final choice, we may support the DM in the following way: Generate a representative set of tradeoffs for the DM's evaluation. If none of these produces an improved solution, he/she has found his/her best solution. How the representative set of tradeoffs can be generated is considered e.g. in Halme (1990), Halme and Korhonen (1989) and Zionts and Wallenius (1980). Let's consider the following simple example:

$$\max \quad x_i,\ i=1,2,3,$$

s.t.

$$\begin{aligned}
3x_1 + \quad 2x_2 + \quad 3x_3 &\le 18,\\
x_1 + \quad 2x_2 + \quad x_3 &\le 10,\\
9x_1 + \quad 20x_2 + \quad 7x_3 &\le 96,\\
7x_1 + \quad 20x_2 + \quad 9x_3 &\le 96,\\
x_i &\ge 0.
\end{aligned}$$

Figure 1 shows a graphical representation of the problem in the criterion space, which in this case is the same as the decision variable space.

Consider an efficient solution $\mathbf{x} = (2.25, 2.25, 2.25)$ (point A in Fig. 1). For this solution, we can easily generate three nondominated tradeoffs: $\mathbf{d}_1 = (3.75, -2.25, -2.25)$, $\mathbf{d}_2 = (-0.25, 0.75, -0.25)$, and $\mathbf{d}_3 = (-2.25, -2.25, 3.75)$, which are scaled such that $\mathbf{x} + \lambda\mathbf{d}_i$, $i = 1,2,3$, is feasible for $\lambda \in [0,1]$. In Figure 1, the tradeoffs $\mathbf{d}_i$, $i=1,2,3$, are divided by two.

It is slightly more difficult to generate a representative set of nondominated tradeoffs for points B and C in Fig. 1.

The tradeoffs can be offered to the DM using the interface depicted in Figure 2, which illustrates the tradeoffs for point A. The DM can freely choose which vector of tradeoffs he/she is willing to consider. He/she chooses the corresponding "card" on the left. The elements of the tradeoff vector are shown on the right in such a way that the values of the objectives improving are shown as bars to the right and the values which get worse as bars to the left. The DM can state his/her preference ("Yes", "No", "I do not know"). The cards in the deck will get colors accordingly. If the pseudoconcavity assumption holds and the DM does not like any offered vector of tradeoffs, the optimal solution is found.

Leafing in the card deck can be eased by e.g. defining an objective (objectives) the DM insists on improving.

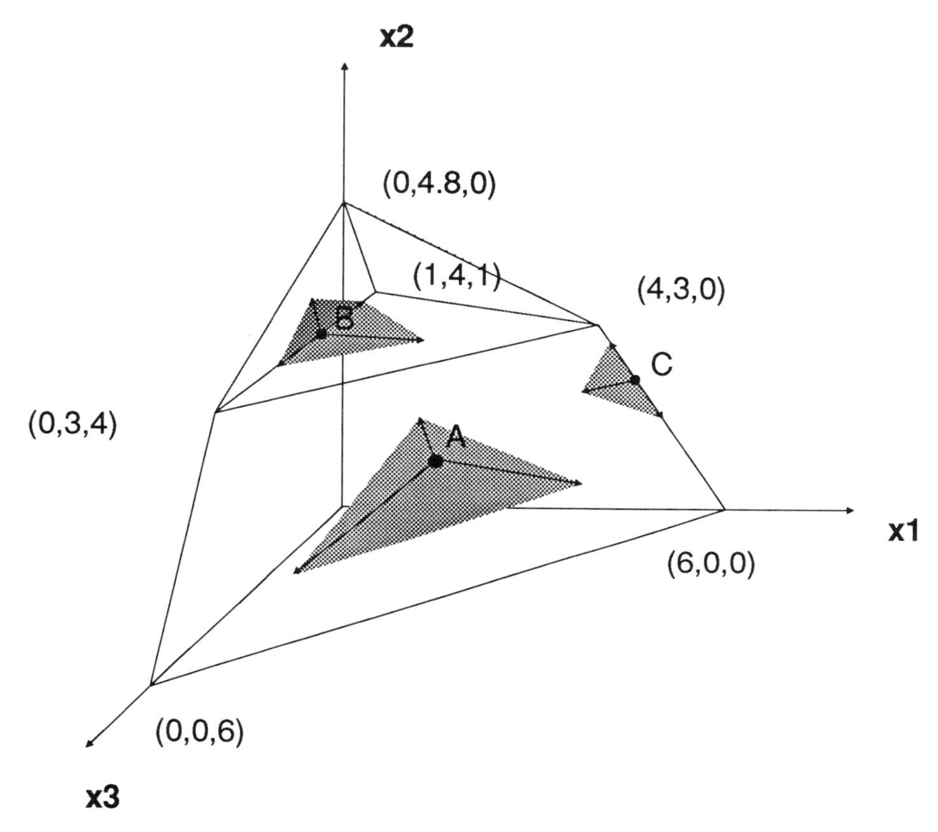

Figure 1: An Example Illustrating Representative Sets of Nondominated Tradeoffs

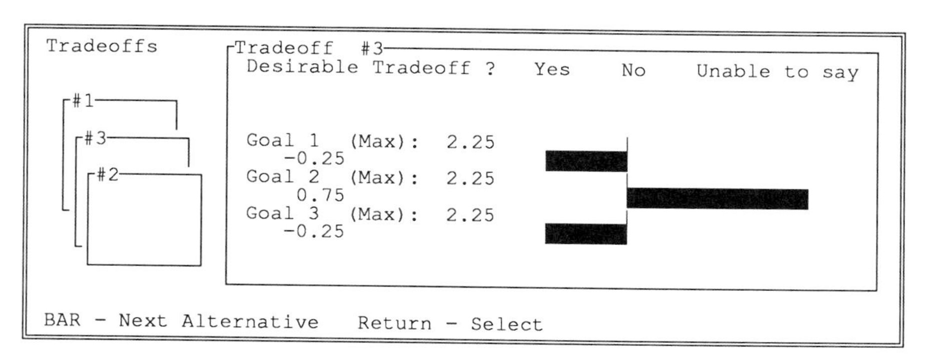

Figure 2: An Example of the Interface for Presenting the Representative Set of Tradeoff Vectors to the DM

Based on our previous considerations we are now ready to propose our approach.

An Approach

1 Find an initial nondominated solution

2 Generate a representative set of tradeoff vectors at the current solution, and ask the DM to evaluate them.

3 If the DM likes one of the tradeoff vectors go to Step 4; otherwise go to Step 6. If the DM likes several vectors offered choose the most preferred one and go to Step 4.

4 Use the tradeoff vector for generating a curve emanating from the current solution and traversing across the efficient frontier.

5 Present the solutions included in the curve visually and numerically in the spirit of Korhonen and Laakso (1986a). Ask the DM to indicate the solution he/she prefers most. Go to Step 2.

6 If the DM does not like any tradeoff vector, ask the DM if he/she is willing to carry out a neighborhood search using Pareto Race. If yes, then go to Step 7; otherwise stop.

7 If the DM using Pareto Race finds a new preferred solution, calculate the corresponding tradeoff vector and go to Step 4; otherwise stop.

Using the above approach, the DM has a possibility to find the most preferred solution, provided that a locally best solution is also globally best.

The approach can be adopted even for solving large-scale problems (see, Korhonen et al. (1989)). The neighborhood search at the current solution by Pareto Race either results in the best solution or provides an improved solution, which can be used in defining the curve traversing across the efficient frontier.

5. DISCUSSION

In this paper we have discussed the questions important to take into account in developing interactive procedures which also meet the requirements of behavioral realism. Related issues have recently been discussed by Vanderpooten (1990).

We also proposed a general framework for a procedure that supports the DM in finding the most preferred solution, provided a locally best solution is also globally best.

The advantages of the approach are:

- The DM is not overloaded with information, because the number of representative tradeoff vectors is usually not large. In the most general case it is equal to the number of objectives (Halme and Korhonen (1989)).

- A set of representative tradeoffs is, in the most general case, easily calculated in the simplex tableau corresponding to the current solution.

- New solutions are easily computable using parametric programming.

- General conditions for optimality can be given.

The disadvantages and remaining research problems are:

- If the locally best solution is not globally best, our approach does not lead to the global optimum. In this case, heuristics have to be introduced to define a "jump" to a better solution. Gardiner and Steuer (1990) propose a procedure for generating a representative set of nondominated solutions which can be used. Even this kind of a procedure does not work without assumptions about the DM's preference structure.

- The approach is path-dependent. During the search process the DM gradually costructs and revises his/her perception of the set of solutions. The solutions considered constitute his/her information base, which he/she will use as "reference information" in later evaluations. How this information affects the final choice, is one topic for further research.

Acknowledgements

The research was supported, in part, by grants from the Foundation of the Helsinki School of Economics, the Kordelin Foundation and the Y. Jahnsson Foundation, Finland.

The authors wish to thank the other members of our research group "*Multiple Criteria Decision Support*" at the Helsinki School of Economics for helpful comments.

References

Benayoun, R., de Montgolfier, J., Tergny J., and Larichev O. (1971), "Linear Programming with Multiple Objective Functions: Step Method (STEM)", Mathematical Programming, Vol. 1, No. 3, pp. 366-375.

Evans, J. P. and Steuer, R. E. (1973), "Generating Efficient Extreme Points in Linear Multiple Objective Programming: Two Algorithms and Computing Experience", in J. L. Cochrane, and

M. Zeleny (eds.): Multiple Criteria Decision Making, University of South Carolina Press, Columbia, South Carolina.

Fishburn, P. (1989), "Foundations of Decision Analysis: Along the Way", Management Science, Vol. 35, No. 4, pp. 387-405.

Gardiner, L. and Steuer, R. (1990), "Range Equalization Scaling and Solution Dispersion in the Tchebycheff Method: A Pleminary Study", The paper was presented at the IX-th International MCDM-Conference at Fairfax, Virginia, U.S.A., August 5-8, 1990.

Geoffrion, A., Dyer, J. and Feinberg, A. (1972), "An Interactive Approach for Multi-Criterion Optimization, with an Application to the Operation of an Academic Department", Management Science, Vol. 19, No. 4, pp. 357-368.

Halme, M. (1990), "Finding Efficient Solutions in Interactive Multiple Objective Linear Programming", Licentiate thesis, Helsinki School of Economics.

Halme, M. and Korhonen, P. (1989), "Nondominated Tradeoffs and Termination in Interactive Multiple Objective Linear Programming", in Lockett, A. G. and Islei, G. (Eds.): Improving Decision Making in Organizations, Springer-Verlag, pp. 410-423.

Kahneman, D. and Tversky, A. (1979),"Prospect Theory: An Analysis of Decisions Under Risk", Econometrica, Vol. 47, pp. 262-291.

Korhonen, P. and Laakso, J. (1986a), "A Visual Interactive Method for Solving the Multiple Criteria Problem", European Journal of Operational Research, Vol. 24, N:o 2 (February 1986), pp. 277-287.

Korhonen, P. and Laakso, J. (1986b), "Solving Generalized Goal Programming Problems Using a Visual Interactive Approach", European Journal of Operational Research, Vol. 26, N:o 3 (September 1986), pp. 355-363

Korhonen, P., Moskowitz, H. and Wallenius, J. (1990), "Choice Behavior in Interactive Multiple Criteria Decision-Making", Annals of Operations Research, Vol. 23, pp. 161-179.

Korhonen, P. and Wallenius, J. (1988), "A Pareto Race", Naval Research Logistics, Vol. 35, N:o 6, pp. 615-623.

Korhonen, P., Wallenius, J. and Zionts, S. (1989), "A Computer Graphics-Based Decision Support System for Multiple Objective Linear Programming", unpublished manuscript.

Steuer, R. E. (1977), "Multiple Objective Linear Programming with Interval Criterion Weights", Management Science, Vol. 23, No. 3, pp. 305-316.

Steuer, R. (1986), Multiple Criteria Optimization: Theory, Computation, and Application, John Wiley & Sons, New York.

Vanderpooten, D. (1990), "Three Basic Conceptions Underlying Multiple Criteria Interactive Procedures", the paper was presented at the IX-th International MCDM-Conference at Fairfax, Virginia, U.S.A., August 5-8, 1990.

Wierzbicki, A. (1980), "The Use of Reference Objectives in Multiobjective Optimization", in G. Fandel and T. Gal (Eds.), Multiple Criteria Decision Making, Theory and Application, Springer-Verlag, Berlin, pp. 468-486.

Zionts, S. and Wallenius, J. (1976), "An Interactive Programming Method for Solving the Multiple Criteria Problem", Management Science, Vol. 22, No. 6, pp. 652-663.

Zionts, S. and Wallenius, J. (1980), "Identifying Efficient Vectors: Some Theory and Computational Results", Operations Research, Vol. 28, pp. 785-793.

MULTICRITERIA PROBLEMS WITH DESIGNED DECISION ALTERNATIVES

Larichev O.I.
Professor, Institute for Systems Studies (VNIISI)
Pr. 60 let Octyabrja,9, MOSCOW 117312, USSR

Pavlova L.I.
Professor, Moscow Civil Engineering Institute
Jaroslavskoe shosse,26, MOSCOW 129337, USSR

Osipova E.A.
Junior researcher, Institute for Systems Studies (VNIISI)
Pr. 60 let Octyabrja,9, MOSCOW 117312, USSR

ABSTRACT

The paper considers the multicriteria mathematical programming problems whose alternatives are not defined by determining a set of feasible solutions but rather constructured with the components of this set chosen in conformity with the requirements set to criteria estimates. That is, criteria interrelated in a sense that the improvement of estimates for one and the same alternative is gained at the expense of degradation of others. Some criteria are qualitative. The practical example is an urban development problem where alternative versions consist of a number of sites convenient for residential construction. An interactive system is developed helping designers in creating alternative versions of urban development, in improving successively each of these versions and in comparing the improved versions with one another. A case is provided to illustrate the system proposed.

1. Introduction

There is a wide range of multicriteria linear programming problems that can be represented as (1):

$$\max \{ C_1(x), \ldots, C_K(x) \} \qquad \text{Eq.(1)}$$

$$x \in D \qquad \text{Eq.(2)}$$

where k = number of criteria;
C = linear function of k-dimensional vector X;
D = feasible decision domain (it is generally assumed that $D = \{ F(x) \geqslant 0, \ x \geqslant 0\}$, where F(x) is a linear function of x). In this case, decision is on the domain Pareto boundary.

The majority of papers on multicriteria linear programming (see overview Larichev & Polyakov (1980)) assume that feasible domain D has form of a linear polygon.

There are, however, problems with an ill-defined domain. Imagine, for example, a problem of territory development in a region. The region area considerably exceeds the one that can be developed with the available scarce resources. Hence, there is not one choice step but rather a combination thereof:

(1) Preliminary (conceptual) choice of the site to be developed in the region.It is clear that a number of alternatives are subject to comparison.

(2) Refined choice of the site against many criteria. Note that a multiattribute choice is applicable to a number of the earlier selected alternatives, for it is only this kind of choice that can expose the genuine strengths and weaknesses thereof.

(3) Choice from several better alternatives also performed against many criteria. Given this procedure, the feasible domain is as if designed on the basis of a number of requirements to it. It is noteworthy that the requirements are modified and refined in the process of choice.

Let us call the new problem a problem with designed decision alternatives under scarce resources.

Write the problem in the following form (2):

$$C^{**} = \max_{i}^{DM} \{ C^{*1}, \; \ldots \;, C^{*N} \} \qquad \text{Eq.(3)}$$

$$\text{where } C^{*i} = \max_{x,R} \{ C_1(x,R), \quad \ldots \quad, C_K(x,R) \} \quad \text{Eq.(4)}$$

$$x \in F_i \;; \quad x = (x_1 \;, \; \ldots \;, x_n) \qquad \text{Eq.(5)}$$

$$R \in G_i \;; \quad R = (R_1 \;, \; \ldots \;, R_t) \qquad \text{Eq.(6)}$$

where N = number of choice alternatives, i.e. feasible domains of n-dimensional vector x designed on the basis of general constraints D;
F_i = feasible decision domain assigned for the i-th conceptual choice alternative;
K = number of criteria;
C_j (x,R) = criterion, linear function of n-dimensional vector x and t-dimensional vector R of resources allocated with a view to improving criteria estimates;
R = additional resources allocated for improving criteria estimates (in the general case, t-dimensional vector);
G_i = constraint on alloted resources assigned to the i-th conceptual choice alternative (in the general case, t-dimensional vector);
$\max^{DM}$ = best decision from the standpoint of the a priori unknown decision maker preferences.

If functions C_j (x,R) are linear and F_i are assigned as a system of linear equations and inequalities, we deal with a multicriteria linear programming problem with designed decision alternatives under scarce resources.

There are two generally accepted versions of problem (1) statement concerning the feasible domain. One is known as systems optimization, Glushkov et al. (1982). It assumes that the domain modifications are possible when the decision maker sets a desired goal in the criteria space, Glushkov at al.(1982).

Another version is associated with a morphological approach and the metrization of morphological space, Dubov at al.(1986). The basic idea boils down to the metrization in the morphological space where the distances correspond to the differences between the alternatives.

The essential differences between (2) and systems optimization problems and morphological approach are quite obvious. Statement (2) does not assign the feasible domain, it only outlines the conceptual alternatives that must be met by many decision choice alternatives. The purpose of decision (as the desired criteria value vector) is refined as the decision maker gets increasingly familiar with the conceptual versions F_i as well as tradeoffs between the criteria accomplished with versions F_i . In contrast to the, morphological approach, the versions proper are designed in a quite different manner.

2. Description problem statement

The development of industrial production, primarily concentrated in cities, of new working forms and conditions (due to new technologies and computerization), of transport and communal services, population growth, as well as the creation of conservation areas result in the periodic expansion of the city and its coalescence with suburbs.

Today, under the impact of the ever increasing urbanization, the urban areas grow twice as quickly as the city population. The territorial expansion breeds a large range of problems placing demand on research into the laws of formation of habitation conglomerations, into the forms and structures of intracity and suburb settlement, into the location of centers of human habitation, as well as into rational nature management.

Characteristic of the current urban population growth is a relative stability of urban population and rural migration. This fact considerably modifies the familiar formula: growth of the city is the growth of its population. As a matter of fact, the territorial expansion is due to the growth of all estimated standards - from floor area to the units of service and cars per capita - which, naturally, leads to the expansion of the city and its coalescence with suburbs thereby forming a metropolitan cluster. Hence,the urban processes are connected with territorial formation of habitation clusters, and the growth problem boils down to that of redistribution of population and centers of its activity.

Due to the fact that the reserves of free sites are nearly exhausted, there remain the following alternatives for the territorial expansion of the majority of cities:

(1) utilization of developed areas by way of replacing low-storied buildings and renovation of urban environments;

(2) development of sites requiring special excavation work leading to higher costs of construction;

(3) alienation of agricultural lands for urban construction;

(4) development of special areas, for example those occupied in some cities by airfields. This measure results not only in the restitution of free areas to the city but also in elimination of a noise zone covering the central districts of many cities.

The variations of the four states, with regard to concrete socioeconomic and natural conditions of the city, provide a wide-variety of compositional arrangement approaches to integrating construction sites into a concept of spatial urban development. They serve as a basis for the layout alternatives and frequently, in spite of land shortage, make it possible to study a large number of urban development alternatives. Besides, the practical urban design takes account of the factors such as spatialcompositional and historico-cultural heritage which form the context of the idea, the foundation of the concept. These factors are expressively embodied in the descriptions of architectural monuments, natural relics, and established housing i.e. architectural environment. The feasibility studies and the general layout concept fix the miniature future construction and panorama, the connection of new areas with the central part and architectural environment, the ratio between reconstruction and new construction, developed and free space. Thus far these factors, as manifestations of artistic and psychological principles in the perception of environment, remained a prerogative of designer and made up the set of tools. The latter are behind the development of a conceptual version. These factors, no doubt, are not formalizable. They constitute an extensive factual material required for a comprehensive approach to urban research integrating the traditional methods of analysis and accurate computation techniques.

The analysis and comparison procedure starts as early as the phase of conceptual version formation involving juxtaposition of subjective factors of the city peculiarities and the objective quantitative characteristics of reasonability of its development. The development of conceptual alternatives is accompanied by building of an information model and a system of constraints, determining the form of their presentation.

The bright expressiveness of urban construction idea, integrating the newly developed areas (prepared at the feasibility study phase) distinguishes one conceptual version from the other, at the same time a great deal of alternatives can be developed within the frameworks of one and the same urban construction concept. The development of the basic conceptual versions is an objective of creative search. It would be logical to exercise the refined choice of an area with a man-machine procedure permitting the concurrent generation and choice of alternatives by analyzing various combinations of construction sites which can be used in the considered conceptual version.

Thus, the concepts of expansion trends and opportunities for urban development embodied in layout alternatives are translated into a concrete problem of spatial organization and approaches to its solution.

The quality of problem solution alternatives is evaluated, in the general case, against four criteria - transportation, ecological,landscape-compositional, and economic, Osipova (1988). The transportation criterion is indicative of the degree of people's satisfaction with their jobs and time they spend to get to work, Pavlova (1986; 1987). It is measured by computing the size of the population dissatisfied with their jobs and time taken to get to work. The ecological criterion expresses the smoke content over residential areas. It is measured by computing the size of the population residing in areas with smoke content in excess of tolerable concentrations. The landscape-compositional criterion reflects the richness of landscape and proximity to the historical center of the city. It is measured by computing the number of people living amid poor landscape and far from the center of the city. The economic criterion reflects costs of the version implementation.

The problem criteria are far from independent.

The designer may undertake some steps with a view to improving some criteria at the expense of others. For example, the development of sewage treatment facilities will deteriorate the economic criterion estimate, thereby reducing the smoke content over the developed areas. The introduction of additional bus routes, connecting residential areas with enterprises, while increasing the version implementation costs, reduces the time needed to get to and from the job thereby increasing the population's satisfiability with their jobs. The layout of parks and creation of artificial reservoirs, while increasing costs improve the estimate of landscape-compositional criterion.

Along with criteria, the following constraints are assigned: on population settlement on different sites; on demand for workforce; on total size of active population.

Emphasize once again the specifics of the given problem:

(1) each conceptual version of urban development is characterized by a combination of sites which can be used for reconstruction and new construction;

(2) each version consists of a concrete set of sites for reconstruction and new construction;

(3) each site and each version are characterized by the estimates against the above criteria;

(4) the criteria are dependent; one of them (landscapecompositional) is purely qualitative.

3. Formal problem statement

Specify the formal statement of the above problem. Let the urban development project management select N conceptual alternatives of urban development. The suggested approach boils down to optimization of each conceptual alternative against the vector criterion followed by selection of some one alternative considered best from the standpoint of decision maker's preferences. Throughout selection, the decision maker's role is performed by a group of persons responsible for urban development including, no doubt, the project manager. Each of the N conceptual alternatives involves finding the best decision against the four criteria listed above.

The transportation and ecological criteria have form:

$$C_1(x) = \sum_{j=1}^{m} C_j^1 x_j \rightarrow \max \qquad \text{Eq.(7)}$$

$$C_2(x) = \sum_{j=1}^{m} C_j^2 x_j \rightarrow \max \qquad \text{Eq.(8)}$$

where m = number of sites in conceptual alternative; x_j = number of people inhabiting the j-th site; C_j^1 = share of population satisfied with the j-th site location; C_j^2 = share of ecologically pure area of the j-th site. If one stems from an assumption on the uniformity of people distribution on the site area, then coefficients C_j^2 reflect the share of the population residing in favorable ecological environment. Special emphasis should be placed on the landscape-compositional criterion. It is a qualitative criterion reflecting expert judgements of the site environments:

$$C_3(x) = \sum_{j=1}^{m} C_j^3 x_j \rightarrow \max \qquad \text{Eq.(9)}$$

where coefficients C_j^3 are defined by experts. The following is correct for coefficients C_j^i:

$$0 \leq C_j^i \leq 1 \qquad \text{Eq.(10)}$$

The last is the cost criterion determining the cost of construction:

$$C_4(x) = \sum_{j=1}^{m} C_j^4 x_j \rightarrow \min \qquad \text{Eq.(11)}$$

where C_j^4 = specific costs per head for the j-th site.

Since the urban development project covered two -three five-year periods, the cost constraints cannot be determined precisely. Therefore an analysis of urban development plan improvement through additional investments is of immediate interest.

Let R_1 resources be allocated in addition to the resources required for the settlement of H thousand people under the i-th conceptual alternative. Denote in terms of y_j^i the modifications introduced in coefficients C_j^i (i=1,2,3) when improving the site quality against the first three criteria. Then the formal problem statement may have the following form (3):

$$\sum_{j=1}^{m} (C_j^i + y_j^i) x_j \rightarrow \max, \quad \text{where } i = 1,2,3. \qquad \text{Eq.(12)}$$

$$\sum_{j=1}^{m} C_j^4 x_j + R_1 \rightarrow \min \qquad \text{Eq.(13)}$$

$$x_j \geq 0 \qquad \text{Eq.(14)}$$

$$y_j^i = f (R_1) \qquad \text{Eq.(15)}$$

$$\sum_{j=1}^{m} x_j \leq H \qquad \text{Eq.(16)}$$

$$A x \leq b \qquad \text{Eq.(17)}$$

The latter condition reflects numerous constraints imposed on terms of site development and settlement. As applied to the considered problem, constraints $A x \leq b$ will have the following form:

(a) upper constraints on site settlement:

$$x_i \leq P_i^2, \quad i = \overline{1, J}, \qquad \text{Eq.(18)}$$

where J is total number of sites;

(b) constraints on provision of jobs to population at one of the city's enterprises:

$$\sum_{i=1}^{m} x_i \leq L^c, \qquad \text{Eq.(19)}$$

(c) constraints on the total size of active population:

$$\sum_{i=1}^{m} x_i = L. \qquad \text{Eq.(20)}$$

By assigning various additional resources: $R_2, R_3 \ldots R_t$, ($\max R_i \leqslant G_i$), one may study the conceptual alternative sensitivity to additional investments. It is easily seen that problem (3) consists of two problems: (a) problem of the best population settlement on the sites against four criteria; (b) problem of additional resource R distribution to improvement of the sites criterion estimates. Note that upon solving problem (b), problem (a) turns into a multicriteria linear programming problem.

4. Solution technique

Problem (3) constitutes a multicriteria problem of nonlinear parametric programming. It can be considerably simplified through its meaningful properties and by solving the two aforestated problems in turn.

The allocation of additional resources R improves the estimates by the first three criteria. An analysis shows that the improvement ordinarily takes form of a number of steps, such as: (1) laying high-speed transportation lines connecting the city center with the developed suburbs; (2) installation of additional filters on air pollution sources; (3) landscape improvement by digging artificial ponds, afforestation, etc.

The decision maker sequentially studies these steps and determines, with the experts' assistance, their cost. As a rule, such measures affect several sites at a time. The decision maker earmarks a set of resource allocation alternatives not exceeding the assigned value of G_i. Then, the modified coefficients C_j^i are introduced in the selected alternatives, and problem (3) is subjected to solution.

It seems reasonable, we believe, to solve it by STEM method, Benayoun at al. (1980) as one of the best, Larichev, Polyakov & Nikiforov (1987). Solution is exercised through interaction with the decision maker.

5. Comparison of conceptual alternatives

Following a detailed analysis and potential improvement of each conceptual alternative, the alternatives are compared with a view to selection of the best one. From methodological point of view, it is necessary to compare a small number of (ordinarily 5 to 7) objects having estimates against four criteria. It would be expedient to exercise a pairwise comparison of the objects and use a psychologically correct method of estimate compensation, Larichev (1979).

6. A case

The suggested approach was employed in selecting between the Middle Russia urban development alternatives. Following is a description of the development alternative choice as applied to one of the cities. A forecast shows that the city's population will reach 550,000 people by a ceratin future date, and the expected increment during the planned period is 120,000 people. Construction and reconstruction was carried out in stages. The described computations relate to the second-stage construction. The projected population growth for that period amounts to about 35,000 people, whereas that of industrial personnel - to about 10,000; the number of jobs at enterprises is also expected to increase by 10,000. The entire area suitable for development with regard to layout, geological, architectural, and conservation constraints consisted of 50 sites in four districts of the city. Each site was capable of housing from 1.3 to 2.5 thousand people. The persons in charge of urban development advanced two radically differing concepts of urban development. Both concerned the future of the airfield. Initially, the airfield was situated on the outskirts, but due to urban development it happened to be in the city's center.

The first concept envisages removal of the airdrome from the city boundaries. As far as the operation requires additional resources, the investments are taken into consideration in reviewing different construction alternatives in line with that concept. Given acceptance of this development concept, the four dispersed city districts, former separate townships, are to grow together into a unified organism connected by the common layout, transportation, communication, and service structures.

The second concept suggests to leave the airfield where it is. In this case, the city will grow largely at the expense of peripheral area development.

The study of the first conceptual alternative started with an analysis of feasible decision domain. Attention was given to improvement of ecological conditions in the central areas through installation of filters at a number of enterprises and afforestation of peripheral areas with poor landscapes.

The following reviewed alternatives are worth mentioning:

1. It is largely the central and adjacent sites that people move to. Since one of enterprises is located in the city center, about 3.5 thousand people live in the ecologically unfavorable environment. Some part of the population live in peripheral areas with poor landscapes, therefore about 2.5 thousand people find themselves in areas with inadequate landscapes. Costs run to around 125 million roubles.

2. The ever more stringent requirements with respect to ecological criterion lead to redistribution of population to the ecologically safe areas but more distant from the city center. The ecologically unfavorable central sites become free. About 6 thousand people reside in inadequate, in terms of landscape, conditions. The costs grow to 150 million roubles. Now only 1 thousand people live in unsatisfactory conditions, in ecological terms.

3. Certain measures were taken with a view to improving ecological conditions in central sites. They involved planting of greenery in peripheral areas. The costs stand at about 140 million roubles. Around 1.5 thousand people live in ecologically unfavorable conditions. The entire population is satisfied with conditions against the landscape-compositional criterion. It is mostly the central city areas and peripheral sites close to the center that are inhabited.

Most appealing, from decision maker's standpoint, turned to be an alternative associated with the installation of filters at a number of enterprises, as well as afforestation of peripheral sites. This alternative is superior to the first one against ecological and landscape-compositional criteria, and costs 10 million roubles less than the alternative associated with redistribution of population to the relatively pure (ecologically) areas which are more distant from the center.

The search for the best alternative as applied

to the second urban development concept was conducted in a similar manner. Consideration was given to the following alternatives.

1. Population moves to the vacant central sites near the airfield as well as to peripheral district sites close to the center. About 4,000 people live in unfavorable conditions, with respect to ecological criterion, 5.500 - in poor landscape areas. The cost by this alternative amounts to about 140 million roubles.

2. The improvement in terms of ecological criterion through redistribution of population to different sites raises expenses to 210 million roubles. This is connected with the development of inconvenient, from construction point of view, but ecologically pure areas, the need for gasification, heat- and water supply , and construction of roads. Around 3,000 people live in ecologically inappropriate conditions, and about 5.500 people - in insufficiently good conditions with respect to landscape.

3. Additional 30,000 roubles were allocated to afforestation of poor landscape peripheral sites.The requirements to ecological criterion get more and more strict; not more than 3,000 people may live in ecologically unfavorable conditions. The population resides mostly on ecologically favorable sites. The total costs incurred herewith amount to 290 million roubles.

Yet another alternative chosen by the decision maker from the second concept implies additional resources to be committed to afforestation. Under this alternative, the population lives in relatively favourable conditions, in ecological terms. All three peripheral sites develop equally intensively. The airfield noise reduces.

The selection of the best alternative with respect to each concept was followed by the final alternative choice. Most acceptable, from decision maker's point of view, was the first concept alternative associated with additional investments in construction of treatment facilities for central site sewage, and layout of parks in the more distant areas. Its advantages are apparent against several criteria:

1. In terms of cost - 140 million roubles as against 290 million with respect to the best second concept alternative.

2. In terms of ecology - only 1.500 people live in ecologically unfavorable conditions as against 3, 000 with respect to the second concept.

3. In social terms, for people lagely live in areas close to the center and not far from industrial enterprises.

Thus, the decision maker chose the first concept associated with the replacement of the airfield.

REFERENCES

1. Benayoun, R., Montgolfier, J., Tergny, J. & Larichev, O. "Linear Programming with Multiple Objective Functions: STEP Method (STEM)", Math. Progr., Vol. 1, N 3, pp. 366-375, 1971.
2. Dubov, Yu.A., Travkin, S.I. & Yakimets, V.N. Multicriteria Models of System Alternative Development and Choice. Moscow, Nauka Puiblishers, 1986.
3. Glushkov, V.M., Mikhalevich, V.S., Volkovich, V.L. & Dolenko, G.A. "On Systems Optimization in Multicriteria Linear Programming Problems", Kibernetika, No. 3, 1982.
4. Larichev, O.I. The Art and Science of Decision Making, Nauka,Moscow, 1979.
5. Larichev, O.I. & Polyakov, O.A. "Man-Machine Procedures of Multicriteria Mathematical Programming Problem Solution", Ekonomika i Mat. Metody, Vol. 16, Issue 1, pp. 129-145, 1980.
6. Larichev O., Polyakov O., Nikiforov A. Multictiterion linear programming problems (Analytic Survey), Journal of Economic Psycology, N 8, 1987.
7. Osipova, E.A. "A Problem of Urban Development Alternative Formation", Man-Machine Decision Procedures. Collected Papers. Issue 11. Moscow, VNIISI Press, 1988.
8. Pavlova, L.I. "Identification of Preference Factors and their Impact on Settlement", Applied Problems of Macrosystem Control Transactions of the 1st All-Union School-Workshop. Moscow, VNIISI Press, 1986.
9. Pavlova, L.I. "Settlement and Employment Model", Current and Future Trends in Large City Development in the USSR and Abroad. Express Information. Issue 7, Moscow, 1987.

MULTIOBJECTIVE CONTROL OF THE RISK OF EXTREME EVENTS IN DYNAMIC SYSTEMS

Duan Li
Research Assistant Professor
Department of Systems Engineering, University of Virginia
and
Yacov Y. Haimes
Lawrence R. Quarles Professor and Director
Center for Risk Management of Engineering Systems
University of Virginia, Charlottesville, Virginia, USA

Abstract

Several different risk control strategies for dynamic systems within a multiobjective framework are discussed in this article, with emphasis on the control of risk of extreme events in dynamic systems. A taxonomy is introduced whereby systems are classified according to the effects that their controls have on various risk functions. In addition, the trade-off relationships between the cost function and the identified risk functions are explored.

1. Introduction

The expected value of adverse effects, which has been the most commonly used measure of risk, is in many cases inadequate, since this scalar representation of risk commensurates events that correspond to all levels of losses and their associated probabilities. The common expected-value approach is particularly deficient for addressing extreme events, since these events are concealed during the amalgamation of events of low probability and high consequence and events of high probability and low consequence.

The partitioned multiobjective risk method (PMRM) developed by Asbeck and Haimes (1984) separates extreme events from other noncatastrophic events, and thus provides the decisionmaker with more valuable and useful information. In the literature, the PMRM has been extended to linear dynamic systems by Leach and Haimes (1987). The analytical relationships between the conditional expectation and the statistics of extremes have also been derived by Karlsson and Haimes (1988a; 1988b) and have been further extended in Mitsiopoulos and Haimes (1989) and in Mitsiopoulos, Haimes and Li (1991). Risk management in a hierarchical multiobjective framework is discussed in Li and Haimes (1987).

We consider here a multiobjective control strategy for risk of extreme events for a class of dynamic systems. By introducing several definitions on the control of extreme events, a new taxonomy is introduced whereby systems are classified according to the effects that their controls have on the risk functions. A system is defined to be "a system with neutral risk control" if the control has the same effects on the unconditional (common) expected value as on the conditional expected value for the extremes; otherwise, the system is termed "a system with risk-manipulatable control." According to the effects that their controls have on the uncertainty of the system, risk-manipulatable systems can be further divided into two classes: systems with mean-variance consistency risk control and systems with mean-variance inconsistency risk control.

2. Control of Extreme Events

Consider a general nonlinear class of discrete-time dynamic systems as follows:

$$x_{k+1} = h(x_k, u_k, w_k) \qquad k = 0,1,\cdots,T-1, \tag{1}$$

where k is the indicator of stage, x is the state vector, u is the control vector, and w is the process disturbance vector. We assume that the random sequence $\{w_k\}$ is independently distributed with a known probability density function $\{p_k(w_k)\}$.

We also assume that the perfect state information of x_k can be obtained by measurement at stage k. These assumptions cause the stochastic process $\{x_k\}$ generated by Eq. (1) to be Markovian. Thus, the conditional probability density function of x_{k+1}, based on the information set available at stage k, is of the form $p(x_{k+1}|x_k,u_k)$. In this study, we use a scalar function of the terminal state vector, $f(x_T)$, to represent the degree of risk of the system. Obviously, $f(x_T)$ is a random variable. Its probability density function $p[f(x_T)]$ can be obtained from the probability density function of x_T, $p(x_T)$. Similarly to the PMRM, we consider several risk functions on various partitioned probability ranges. As consistency with the PMRM, we partition the probability axis into three ranges. The partition points are denoted by α_1 and α_2 with $0 < \alpha_1 < \alpha_2 < 1$. For each given partitioned point α_i and given control u, we assume there exists a unique $s_i(u)$ such that

$$P[f(x_T) \le s_i(u)] = \alpha_i \qquad i = 1, 2. \tag{2}$$

Having found the risk partitioning points, the conditional expectations at various partitioned ranges are given as follows (we reserve f_1 for the cost function):

$$f_2(u) = \frac{\displaystyle\int_{-\infty}^{s_1(u)} f(x_T)\ p[f(x_T)|u]df(x_T)}{\displaystyle\int_{-\infty}^{s_1(u)} p[f(x_T)|u]df(x_T)} \tag{3}$$

$$f_3(u) = \frac{\displaystyle\int_{s_1(u)}^{s_2(u)} f(x_T)\ p[f(x_T)|u]df(x_T)}{\displaystyle\int_{s_1(u)}^{s_2(u)} p[f(x_T)|u]df(x_T)} \tag{4}$$

$$f_4(u) = \frac{\displaystyle\int_{s_2(u)}^{\infty} f(x_T)\ p[f(x_T)|u]df(x_T)}{\displaystyle\int_{s_2(u)}^{\infty} p[f(x_T)|u]df(x_T)} \tag{5}$$

Here $f_2(\cdot)$, $f_3(\cdot)$, and $f_4(\cdot)$ provide measures for risk with low consequence, medium consequence, and high consequence, respectively. One observation is that the denominator of Eq. (3) [or Eq. (4), or Eq. (5)] is independent of the control u. Adhering to the notation in the PMRM, we use $f_5(\cdot)$ to represent $E[f(x_T)]$. In this paper, the control of $f(x_T)$ is performed either by minimizing $E[f(x_T)]$, the expectation of $f(x_T)$, or by minimizing $E_{ext}[f(x_T)]$, the conditional expectation $f_4(\cdot)$ of $f(x_T)$ at the range of catastrophic risk. The cost function $f_1(\cdot)$ for implementing a risk control policy $\{u_k\}$ is denoted by J(u). The risk management policies are evaluated in a multiobjective framework. Specifically, we are concerned with the following two multiobjective risk control problems:

P1 [control of expectation]:

$$\min \begin{bmatrix} f_1(\cdot) = J(u) \\ f_5(\cdot) = E[f(x_T)] \end{bmatrix} \tag{6}$$

subject to $x_{k+1} = h(x_k, u_k, w_k)$

P2 [control of extreme events]:

$$\min \begin{bmatrix} f_1(\cdot) = J(u) \\ f_4(\cdot) = E_{ext}[f(x_T)] \end{bmatrix} \tag{7}$$

subject to $x_{k+1} = h(x_k, u_k, w_k)$

The risk-based models given in Eqs. (6) and (7) represent different risk management considerations. In many cases, a decisionmaker concerned with risk management problems will be more interested in the extreme values of random variables than in the expected value. Public perception of catastrophic risks is an important consideration. Although the concept of expectation has physical meaning and mathematical elegance, one deficient point remains: What would be a possible "worse" realization of a random variable if we only control its expectation? Intuitively, we would say that, in order to reduce the catastrophic risk, we should reduce the variance of the risk function at the same time we minimize its mean value. The concept of the PMRM in Asbeck and Haimes (1984) furnishes some answers to control expected extreme events. The conditional expectation $f_4(\cdot)$ separates extreme events from other noncatastrophic events, and thus provides the decisionmaker with significant information about extreme events.

3. Classification of Risk Control Systems

In this section, several examples will be studied which will give us insight into the effects different risk management criteria have on the selection of the best risk control strategy. Systems will be classified according to the effects that their controls have on risk functions.

Example 1: Consider the following linear-quadratic-Gaussian problem:

$$\text{P1: } \min \begin{bmatrix} f_1(\cdot) = \sum_{k=0}^{T-1} c_k u_k^2 \\ f_5(\cdot) = E(x_T) \end{bmatrix} \qquad \text{P2: } \min \begin{bmatrix} f_1(\cdot) = \sum_{k=0}^{T-1} c_k u_k^2 \\ f_4(\cdot) = E_{ext}(x_T) \end{bmatrix}$$

$$\text{subject to } x_{k+1} = ax_k + bu_k + w_k \qquad k = 0,1,\cdots,T-1,$$

where $x_k \in R$ is the state, $u_k \in R$ is the control, $c_k > 0$, and $\{w_k\}$ is an independent normal sequence with mean μ_k and variance σ_k^2. For this linear system, it is well known that x_T is of a normal distribution, with the mean and the variance given as follows:

$$E(x_T) = a^T x_0 + \sum_{i=0}^{T-1} a^{T-1-i}(bu_i + \mu_i)$$

$$Var(x_T) = \sum_{i=0}^{T-1} a^{2(T-1-i)} \sigma_i^2$$

We should note that the variance of x_T is independent of the control u. The conditional expectation on the catastrophic-damage region $[s_T, \infty)$ is by definition

$$f_4 = \frac{\displaystyle\int_{s_T}^{\infty} \frac{x_T}{\sqrt{2\pi \mathrm{Var}(x_T)}} \exp\{\frac{-[x_T - E(x_T)]^2}{2\mathrm{Var}(x_T)}\}\, dx_T}{\displaystyle\int_{s_T}^{\infty} \frac{1}{\sqrt{2\pi \mathrm{Var}(x_T)}} \exp\{\frac{-[x_T - E(x_T)]^2}{2\mathrm{Var}(x_T)}\}\, dx_T},$$

where $s_T = E(x_T) + s_T'[\mathrm{Var}(x_T)]^{1/2}$ and s_T' is a constant. After some manipulation [see Leach and Haimes (1987) for a detailed derivation], we obtain

$$f_4(\cdot) = E(x_T) + \beta_1[\mathrm{Var}(x_T)]^{1/2},$$

where

$$\beta_1 = \frac{\displaystyle\int_{s_T'}^{\infty} \frac{\tau}{\sqrt{2\pi}} e^{-\tau^2/2}\, d\tau}{\displaystyle\int_{s_T'}^{\infty} \frac{1}{\sqrt{2\pi}} e^{-\tau^2/2}\, d\tau}.$$

Since $\beta_1[\mathrm{Var}(x_T)]^{1/2}$ is a constant term that is independent of the control u, problems P1 and P2 generate the same set of noninferior solutions. Moreover, the trade-offs between $f_1(\cdot)$ and $f_5(\cdot)$ and between $f_1(\cdot)$ and $f_4(\cdot)$ are equal for all noninferior solutions. Thus, the control has the same impact on the expectation and on the conditional expectation. We call this class of systems "systems with neutral risk control." In other words, the control is "neutral" in minimizing the expected value and in minimizing the expected extreme.

Example 2: Consider the following scalar system:

$$\text{P1: } \min \begin{bmatrix} f_1(\cdot) = \sum_{k=0}^{T-1} c_k u_k^2 \\ f_5(\cdot) = E(x_T^2) \end{bmatrix} \qquad \text{P2: } \min \begin{bmatrix} f_1(\cdot) = \sum_{k=0}^{T-1} c_k u_k^2 \\ f_4(\cdot) = E_{ext}(x_T^2) \end{bmatrix}$$

subject to $x_{k+1} = ax_k + bu_k + w_k \qquad k = 0,1,\cdots,T-1$

$$x_0 > 0$$
$$u_k \in U_k = \{u_k | ax_k + bu_k > 0\},$$

where the process disturbance is of an exponential distribution,

$$p_W(w_k) = \exp(-w_k) \qquad \text{if } w_k \ge 0$$
$$p_W(w_k) = 0 \qquad \text{if } w_k < 0.$$

The control target in Example 2 is to drive the terminal variable x_T close to zero without excessive expenditure of control effort. Given x_{T-1}, the probability density function of x_T is also exponential,

$$p_X(x_T) = \exp[-(x_T - ax_{T-1} - bu_{T-1})] \qquad \text{if } x_T \ge ax_{T-1} + bu_{T-1}$$
$$p_X(x_T) = 0 \qquad \text{if } x_T < ax_{T-1} + bu_{T-1}.$$

It is easy to derive that

$$E(x_T) = ax_{T-1} + bu_{T-1} + 1$$
$$Var(x_T) = 1.$$

The probability distribution function of $y = f(x_T) = x_T^2$ is

$$P_Y(y) = Pr(Y \le y) = Pr\{0 \le X_T \le \sqrt{y}\} = \int_0^{\sqrt{y}} p_X(x_T)dx_T.$$

Thus, the probability density function of y [i.e., $f(x_T)$] is

$$p_Y(y) = \frac{1}{2\sqrt{y}} p_X(\sqrt{y}) = \frac{1}{2\sqrt{y}} \exp[-(\sqrt{y} - ax_{T-1} - bu_{T-1})]$$
$$\text{if } y \ge (ax_{T-1} + bu_{T-1})^2$$

$$p_Y(y) = 0 \qquad \text{if } y < (ax_{T-1} + bu_{T-1})^2.$$

The expectation function $f_5(\cdot)$ is given by

$$f_5(\cdot) = E(x_T^2) = Var(x_T) + E^2(x_T) = 1 + E^2(x_T).$$

The conditional expectation on the catastrophic-damage region $[s_T, \infty)$, where $s_T = [E(x_T) + s_T']^2$ with a constant $s_T' > 0$, is, by definition

$$f_4(\cdot) = \frac{\displaystyle\int_{[E(x_T)+s'_T]^2}^{\infty} y\, p_Y(y)dy}{\displaystyle\int_{[E(x_T)+s'_T]^2}^{\infty} p_Y(y)dy}$$

$$= \frac{\displaystyle\int_{E(x_T)+s'_T}^{\infty} x_T^2 \exp[-(x_T-E(x_T)+1)]dx_T}{\displaystyle\int_{E(x_T)+s'_T}^{\infty} \exp[-(x_T-E(x_T)+1)]dx_T} .$$

After some simplification, we find that

$$f_4(\cdot) = 1 + [E(x_T) + s'_T +1]^2.$$

The trade-offs between $f_1(\cdot)$ and $f_4(\cdot)$ and between $f_1(\cdot)$ and $f_5(\cdot)$ are different. Thus, the control has different effects on the expectation, $f_5(\cdot)$, and on the conditional expectation, $f_4(\cdot)$. We call this class of systems "systems with risk-manipulatable control." For this specific problem, it can be easily seen that every control effort will make a greater reduction in the measure of the expected extreme, $f_4(\cdot)$, than in the measure of the expected value, $f_5(\cdot)$. The variance of $f(x_T)$ in this example can be derived, which is given as

$$\mathrm{Var}[f(x_T)] = \int\{f(x_T) - E[f(x_T)]\}^2 p_X(x_T)dx_T$$

$$= 4E^2(x_T) + 8E(x_T) + 8.$$

It is easy to see that the minimization of $E[f(x_T)]$ is consistent with the minimization of $\mathrm{Var}[f(x_T)]$. We further call this type of systems "systems with mean-variance consistency risk control."

Example 3: Consider the following scalar one-step horizon nonlinear system:

$$\text{P1: } \min \begin{bmatrix} f_1(\cdot) = u_0^2 \\ f_5(\cdot) = E(x_1) \end{bmatrix} \qquad \text{P2: } \min \begin{bmatrix} f_1(\cdot) = u_0^2 \\ f_4(\cdot) = E_{ext}(x_1) \end{bmatrix}$$

subject to $x_1 = 0.5x_0 + bu_0 + w_0$,

where x, u, and w are the same as in Example 1; however b is an unknown parameter with mean $\hat{b}$ and variance B. It can be shown that

$$E(x_1) = 0.5x_0 + \hat{b}u_0 + \mu_0$$
$$Var(x_1) = u_0^2 B + \sigma_0^2.$$

In this nonlinear system, the variance of x_1 is affected by the control u. There is also a conflict between the minimization of the expectation and the minimization of the variance. Similar to the results obtained in Example 1, we derive

$$f_4(\cdot) = E(x_1) + \beta_1[Var(x_1)]^{1/2}$$

Note however that in this example both the expectation and the variance are functions of the control u. The control u has different impacts on the expected value, $f_5(\cdot)$, and the expected extreme, $f_4(\cdot)$. Thus, this system is "a system with risk-manipulatable control." Because of the inconsistency between the mean and the variance, the expected value, $f_5(\cdot)$, and the expected extreme, $f_4(\cdot)$, may conflict with each other. We further call this class of systems "systems with mean-variance inconsistency risk control."

The concept of "systems with mean-variance inconsistency risk control" is closely related to the concept of dual control introduced by Feldbaum (1965). Consider the nonlinear system given in Eq. (1) with an observation equation

$$y_k = g(x_k, u_k, v_k), \tag{8}$$

where y_k is the observation vector and v_k is the observation noise vector. Denote by M(r,i,k) the rth central moment of the ith component of x at time k conditioned upon Y^k and U^{k-1}, where $Y^k = \{y_0, \ldots y_k\}$ and $U^{k-1} = \{u_0, \ldots, u_{k-1}\}$. The control u is then said to have no dual effect of order r ($r \geq 2$) if

$$E[M(r,i,k)|Y^j, U^{k-1}] = E[M(r,i,k)|Z^j], \tag{9}$$

where $Z^j = \{Y^j, U^{j-1} = 0\}$ with $j \leq k$. Roughly speaking, a control has a dual effect if, in addition to its effect on the state, the present control might affect the future state uncertainty.

The difference between our study of the control of extreme events and the concept of dual control is that the dual effect is defined on the objective function in our study, while the dual effect is defined on the state variables in the study of dual control.

The control in Example 3 has a dual effect in both senses, since the variance of the state variable (which also represents the objective function) is a function of u. Because the conditional expectation depends on the variance, it is possible to reduce the value of $f_4(\cdot)$ by reducing the variance. To enhance our point, let us modify Example 3 with time-horizon T. Assume that unknown parameter b is initially of a normal distribution and is independent of process noise $\{w_k\}$. It is easy to see that the posterior distribution of b is still normal. Let $\hat{b}_k$ and B_k be the posterior mean and the posterior variance of b at stage k. Since both x and b are normal, it follows that

$$\begin{aligned} B_{k+1} &= \mathrm{Var}[b|x(k+1)] \\ &= \mathrm{Var}(b|x_k) - \mathrm{Cov}(b,x_{k+1})[\mathrm{Var}(x_{k+1})]^{-1}\mathrm{Cov}[x_{k+1},b] \\ &= B_k\sigma_k^2/[B_k u_k^2 + \sigma_k^2]. \end{aligned}$$

This clearly illustrates the control's "learning" ability. In addition to its effect on the objective function, the control can also enhance the future information accuracy by probing.

Table 1 summarizes the main characteristics of different classes of systems in our classification for the purpose of risk management.

Table 1. Classification of Risk Control Systems

Systems with neutral risk control	$\frac{\partial f_1(\cdot)}{\partial f_4(\cdot)} = \frac{\partial f_1(\cdot)}{\partial f_5(\cdot)}$, systems in which $\mathrm{Var}[f(x_T)]$ is independent of the control
Systems with risk-manipulatable control	$\frac{\partial f_1(\cdot)}{\partial f_4(\cdot)} \neq \frac{\partial f_1(\cdot)}{\partial f_5(\cdot)}$, systems with mean-variance consistency or irreducible systems uncertainty
	$\frac{\partial f_1(\cdot)}{\partial f_4(\cdot)} \neq \frac{\partial f_1(\cdot)}{\partial f_5(\cdot)}$, systems with mean-variance inconsistency

4. The Trade-off Relationship in the PMRM

The PMRM separates information about catastrophic events and low-damage events and provides a fuller description of risk than that of the traditional expected value. One new question arises: What rationality principles should guide risk-related decisionmaking based on such multidimensional risk information? A trade-off relationship may furnish an answer. There exists a relationship between the conditional and unconditional expected-value functions in the PMRM. Let θ_i denote the denominator of $f_i(\cdot)$, $i = 2,3,4$, in Eqs. (3), (4), and (5), respectively. Multiplying $f_i(\cdot)$ by θ_i, $i = 2,3,4$, and summing them together, we get, from the definition of the expected value,

$$f_5(\cdot) = \theta_2 f_2(\cdot) + \theta_3 f_3(\cdot) + \theta_4 f_4(\cdot). \tag{10}$$

Consider a family of multiobjective optimization problems, $\min[f_1(\cdot), f_i(\cdot)]'$, $i = 2,3,4,5$, under some constraints. Assume that u^* is a noninferior solution for all four multiobjective optimization problems. If, for the noninferior control u^*, the trade-off between the cost function $f_1(\cdot)$ and the ith risk function $f_i(\cdot)$, $\lambda_{1i}(u^*)$, is strictly positive, then the noninferior frontier in the objective space around the neighborhood of $[f_1(u^*), f_i(u^*)]'$ can be represented (Chankong and Haimes 1983) by

$$f_1 = f_1(f_i) \tag{11}$$

and

$$\lambda_{1i}(u^*) = -\partial f_1(\cdot)/\partial f_i(\cdot)|_{u^*}. \tag{12}$$

Taking the partial derivative with respect to f_1 on both sides of Eq. (10) and using the fact (Chankong and Haimes 1983) that $\lambda_{1j} = 1/\lambda_{j1}$, we obtain

$$\frac{1}{\lambda_{15}} = \frac{\theta_2}{\lambda_{12}} + \frac{\theta_3}{\lambda_{13}} + \frac{\theta_4}{\lambda_{14}}. \tag{13}$$

The trade-offs between the cost function $f_1(\cdot)$ and any risk function $f_i(\cdot)$, $i \in \{2,3,4,5\}$, allow decisionmakers to see the marginal cost of a small change in the risk objective, given a particular level of risk assurance for each of the partitioned risk regions and for the unconditional risk index. A knowledge of the relationship of all of these marginal costs will give the decisionmaker insights that are useful for determining the acceptable risk.

5. Epilogue

Various risk control systems have been classified in this article. The relationship between the control of the expected value and the control of the extremes for different classes of systems has been explored. The control algorithms have not been specified, since most of them can be achieved by dynamic programming. We should note, however, that dynamic programming may not be applicable for some control tasks associated with extreme events. This constitutes one important research direction for risk management that is currently being pursued.

Acknowledgments

Support for this research was provided, in part, by the National Science Foundation, Grant No. CES-8617984, under the title "Hierarchical-multiobjective management of large scale infrastructure," and Grant No. BCS-8912630, under the title "Integrating the statistics of extremes with conditional expectation."

References

Asbeck, E., and Y. Y. Haimes, "The Partitioned Multiobjective Risk Method," Large Scale Systems, Vol. 6, No. 1, pp. 13-38, 1984.

Chankong, V., and Y. Y. Haimes, Multiobjective Decision Making: Theory and Methodology, New York: Elsevier-North Holland, 1983.

Feldbaum, A. A., Optimal Control Systems, New York: Academic Press, 1965.

Leach, M. R., and Y. Y. Haimes, "Multiobjective Risk-Impact Analysis Method," Risk Analysis, Vol. 7, No. 2, pp. 225-41, 1987.

Li, D., and Y. Y. Haimes, "Risk Management in a Hierarchical Multiobjective Framework," in Toward Interactive and Intelligent Decision Support Systems, eds. Y. Sawaragi, K. Inoue, and H. Nakayama, Berlin: Springer-Verlag, Vol. 2, pp. 180-89, 1987.

Karlsson, P., and Y. Y. Haimes, "Risk-Based Analysis of Extreme Events," Water Resources Research, Vol. 24, No. 1, pp. 9-20, 1988a.

Karlsson, P., and Y. Y. Haimes, "Probability Distributions and Their Partitioning," Water Resources Research, Vol. 24, No. 1, pp. 21-29, 1988b.

Mitsiopoulos, J., and Y. Y. Haimes, "Generalized Quantification of Risk Associated with Extreme Events," Risk Analysis, Vol. 9, No. 2, pp. 243-54, 1989.

Mitsiopoulos, J., Y. Y. Haimes, and D. Li, "Approximating Catastrophic Risk Through Statistics of Extremes," to appear in Water Resources Research, 1991.

HIERARCHICAL MULTIOBJECTIVE PROGRAMMING AN OVERVIEW

Elliot R. Lieberman
Visiting Assistant Professor
Department of Management Science and Systems
State University of New York at Buffalo
Buffalo, New York 14260
August 1990

Abstract: This paper reviews the major currents in hierarchical programming research to date. Attention is drawn to some blind spots in the research corpus: both individual works that have undeservedly received little attention and divided camps that apparently have failed to recognize each other's significant contributions in the field. The paper concludes by identifying areas that are likely to prove fruitful for future research.

Abstract: This paper reviews the major currents in hierarchical programming research to date. Attention is drawn to some blind spots in the research corpus: both individual works that have undeservedly received little attention and divided camps that apparently have failed to recognize each other's significant contributions in the field. The paper concludes by identifying areas that are likely to prove fruitful for future research.

The subject of this paper is the area of research alternatively called hierarchical or multilevel analysis/modeling/optimization/programming. For simplicity, we will use the term hierarchical programming. Strictly speaking, this is not a survey paper, but a broad stroked overview, intended to focus attention on the major centers of research activity and to point out research directions overlooked in earlier, finer grained surveys. (Excellent surveys of various segments of the field can be found in Burton and Obel 1977; Grana and Torrealdea 1986; Haimes 1973a; Mahmoud 1977; Nachane 1984; and Singh, Drew, and Coales 1975.) At the same time, it is hoped that this paper will stimulate potentially fruitful interchanges between what appear to be two somewhat insular "camps" that have been concurrently pursuing hierarchical programming research.

Definitions

We begin with some key definitions.

Hierarchical approach is a concepual representation of a complex system whose own structure and rules of behavior reflect the ordered, multilevel structure and processes of the system being modeled. Such representations are appropriate and, in fact, necessary for systems whose

- input-output behavior cannot be understood without
  - partitioning the whole system into subsystems, and then
  - aggregating the subsystems to reconstitute the totality.
- control structure is beyond the capacity of a single controller.
- operation involves a number of interdependent units with
  - particularized functions
  - shared resources
  - interrelated goals
  - interrelated constraints

Two basic processes are inherent in the hierarchical approach:

Decomposition: The conceptual partitioning of a system into independent, interrelated modules or subsystems in an effort to

- reduce the dimensionality of the overall system,
- increase model-reality verisimilitude, and
- simplify modeling and computational requirements.

The results are the

- specification of a collection of uncoupled system, and the

- definition of their coupling relationships.

Coordination: The process whereby a supremal subsystem in a hierarchy causes the harmonious functioning of subordinate subsystems by

- manipulating interactions,
- resolving conflicts,
- adjusting goals, and
- adjusting the model structure.

Coordination is effected through the choice of **coordination parameters** and a **coordination principle**.

Depending the nature of the coordination parameters and principle chosen, two alternative types of coordination will result:

Goal Coordination: The supremal unit influences subordinate subsystems by adjusting their goals. An example of a coordination principle that produces goal coordination is the Interactive Balance Principle.

Model Coordination: The supremal unit influences subordinate subsystems by modifying the structure of the subsystem model. An example of a coordination principle that produces model coordination is the Interactive Prediction Principle.

Before proceeding to our overview of hierarchical programming, one last concept needs to be defined.

Coordinatability: The existence of feasible values for given coordination parameters, such that the chosen coordination principle will be satisfied.

We now turn to the overview of some of the major tendencies in hierarchical programming.

Overview

Broadly speaking, there have been two general orientations to hierarchical programming. On the one hand, there has been an optimal control/systems engineering orientation, where the emphasis has been placed on large scale process control systems (e.g., water resource, electric power, manufacturing, and traffic control systems). On the other hand, there has been an operations research/mathematical programming orientation, which has devoted attention primarily to production planning and scheduling problems and other industrial engineering problems. Research in the former category generally grows out of the seminal work of Mesarovic, Macko, and Takahara [1972], whereas research in the latter category most often has its roots in the work of Hax and Meal [1975].

While at times the line between the two orientations blurs, almost without exception, practitioners of one of the two orientations have evidenced little awareness of the work of those following the other orientation. The grounds for this disjuncture are not entirely clear, although they may rest in differing penchants toward theory, methodology, and applications. The optimal control/systems engineering research has established an elegant and extensive theoretical foundation for multilevel analysis of large scale systems. However, by self-admission (see, for example, Mahmoud 1977 and Singh, Drew, and Coales

1975), this research has produced less extensive results in terms of applications and software implementations. By contrast, the mathematical programming research appears to be much more applications driven. Its important theoretical breakthroughs have not been impelled by a mathematician's quest to extend, refine, and elaborate a body of theory but by a pragmatic interest in developing models that are imminently suitable in a given industrial setting.

We now take a closer look at some the major currents within each of the two orientations. Table 1 summarizes the main currents in hierarchical programming among those researchers pursuing an optimal control/systems engineering orientation. The table is evidence of the breadth of the endeavor, especially from the standpoint of theoretical developments. The work encompasses deterministic, stochastic, dynamic, and both open and closed loop systems.

In the optimal control/systems engineering approach to hierarchical multiobjective programming, individual subsystems are considered to have multiple objective functions, and the supremal subsystem's vector objective function is a function of the subordinate subsystem objectives. (See Figure 1.) This is a rather stringent interpretation of what constitutes a multiobjective problem in hierarchical programming, since even when each subsystem possesses but one objective, the overall system can be considered to have a multiplicity of possibly competing objectives. In the strictly defined case shown in Figure 1, one can either attempt to generate a representation of the non-inferior set or develop an interactive method which identifies a preferred alternative. In the former situation it is necessary to develop procedures that reduce the problem of finding the non-inferior set for both the supremal and infimal objective functions $[\bar{F}, \bar{f}]$ to the more manageable problem of just finding the non-inferior set for the subordinate system objectives $[\bar{f}]$. Haimes, *et al.* 1990 report interesting work in this regard, although further research is clearly needed.

Within the mathematical programming orientation to hierarchical multiobjective programming, we witness three discernible areas of inquiry. By far the most extensive has been the work in Hierarchical Production Planning originated by Bitran, Haas, Hax, and Meal at MIT. Their approach employs a three-level formulation that is used to model hierarchical production planning processes. The upper level problem determines a medium range (typically, one year) production plan for aggregate product types; it is formulated as a linear programming problem whose objective is to minimize total production, labor, and inventory costs. The middle and lower level problems take the form of continuous knapsack problems whose respective objectives are to minimize the sum of set-up costs and the sum of squared deviations from assigned production levels. Decision variables link the supremal units to their subordinate subsystems. Considerable innovativeness has been manifested in developing new solution procedures for this problem [Bitran, Haas, Hax 1981]. Another fruitful area of research has been the extension of the Hierarchical Production Planning approach to include production scheduling and finance/marketing subsystems. In fact, plugging non-optimization submodules (e.g., material requirements planning, forecasting, or simulation modules) into the overall optimization hierarchy has proven to be a very productive area for this school of research.

Another current within the mathematical programming orientation has been the research on bilevel and multilevel programming problems. Essentially multilevel extensions to Stackleberg games, this research is applicable to systems in which an upper level subsystem implicitly determines the feasible region of a subordinate subsystem. Figure 2 depicts the problem formulation.

The final major direction of research associated with the mathematical programming

Table 1: Optimal Control Orientation in Hierarchical Programming.

Locale	Principal Researchers	Areas of Investigation	Application Areas
Case Western Reserve University Cleveland, Ohio, USA	Mesarovic, Macko, Haimes, Lasdon, Lefkowitz *et al.*	Theory of Multilevel Systems Systems Systems Identification and Parameter Estimation	Water Resources
	Tarvainen	Hierarchical Multiobjective Optimization	
Cambridge University, England	Singh, *et al.*	Dynamic Hierarchical Control Theory and Solution Techniques	Traffic Control, River Pollution Control
Cairo University Egypt	Mahmoud, Hassan, Darwish		
Institute of Automatic Control, Technical University, Warsaw, Poland	Findeisen, Brdys, Malinowski, Wozniak	Theory of Hierarchical Optimal Control Feedback Structures and Closed Loop Optimization Dynamic Hierarchical Control	Industrial Processes
McMaster University, Canada	Abad	Business Applications of Hierarchical Optimal Control	Production, Marketing, and Financial Planning in Firms

$$DM_0: \quad \min \begin{bmatrix} F_1(\bar{f}^1, \ldots, \bar{f}^N) \\ \vdots \\ F_n(\bar{f}^1, \ldots, \bar{f}^N) \end{bmatrix}$$

$$DM_1: \quad \min \begin{bmatrix} f_1^1(\bar{x}_1, \bar{m}_1, \bar{y}_1) \\ \vdots \\ f_{n_1}^1(\bar{x}_1, \bar{m}_1, \bar{y}_1) \end{bmatrix}$$

$$\vdots$$

$$DM_N: \quad \min \begin{bmatrix} f_1^N(\bar{x}_N, \bar{m}_N, \bar{y}_N) \\ \vdots \\ f_{n_N}^N(\bar{x}_N, \bar{m}_N, \bar{y}_N) \end{bmatrix}$$

subject to

(Process Constraints:) $\bar{y}_i = H_i(\bar{x}_i, \bar{m}_i) \quad i = 1, \ldots, N$

(Coupling Constraints:) $\bar{x}_i = \sum_{j=1}^{N} C_{ij}\bar{y}_j \quad i = 1, \ldots, N$

(State Constraints:) $g_i(\bar{x}_i, \bar{m}_i, \bar{y}_i) \leq 0, \quad i = 1, \ldots, N$

$$\sum_{i=1}^{N} \bar{q}_i(\bar{x}_i, \bar{m}_i, \bar{y}_i) \leq 0$$

where $f_i^j(\bar{x}_i, \bar{m}_i, \bar{y}_i)$ the i-th objective function of subsystem j,
$F_l(\bar{f}^1, \bar{f}^2, \ldots, \bar{f}^N)$ is l-th objective function of the overall system,
$\bar{x}_i$ is the vector of inputs to the i-th subsystem from other subsystems,
$\bar{m}_i$ is the i-th subsystem's vector of decision variables, and
$\bar{y}_i$ is the i-th subsystem's vector of outputs.

Adapted from Haimes, Tarvainen, Shima, and Thadathil 1990

Figure 1: The Hierarchical Multiobjective Programming Problem

Table 2: Mathematical Programming Orientation in Hierarchical Programming.

Locale	Principal Researchers	Areas of Investigation	Application Areas
Massachusetts Institute of Technology, Cambridge, Mass., USA	Bitran, Haas, Hax, Meal *et al.*	Hierarchical Production Planning Extensions to Production Scheduling and Marketing	Industrial Production and Operations
	Candler, Townsley, Bard, Wen, Cruz, Falk, *et al.*	Multilevel Programming Problems (Multilevel Variant of Stackle-berg Game)	Organizational Decision Making
	Bialis, Karwan, Gallo, Ülkücü, Konno, *et al.*	Bilevel Programming	
Institute of Cybernetics, Kiev, USSR	Mikhalevich, Volkovịch, *et al.*	Multilevel Optimization with Prioritized Lateral Linkages and Expansion of Subsytem Feasible Domains Using Systems Optimi-zation	Organizational Decision Making

Multilevel Programming Problem

Definition: A nested sequence of optimization problems, solved in a predetermined order, where the feasible region of each subsequent problem is implicitly determined by the solution of the preceding problem.

$$\left.\begin{array}{c}\max f_1(\bar{x}^1, \bar{x}^2, \ldots, \bar{x}^n)\\ \bar{x}^1 \in X^1\\ \bar{g}^1(\bar{x}^1) \geq \bar{0}\end{array}\right\} \text{where } \bar{x}^2 \text{ solves}$$

$$\left.\begin{array}{c}\max f_2(\bar{x}^1, \bar{x}^2, \ldots, \bar{x}^n)\\ \bar{x}^2 \in X^2\\ \bar{g}^2(\bar{x}^1, \bar{x}^2) \geq \bar{0}\end{array}\right\} \text{where } \bar{x}^3 \text{ solves}$$

$$\vdots$$

$$\left.\begin{array}{c}\max f_{n-1}(\bar{x}^1, \bar{x}^2, \ldots, \bar{x}^n)\\ \bar{x}^{n-1} \in X^{n-1}\\ \bar{g}^{n-1}(\bar{x}^1, \bar{x}^2, \ldots, \bar{x}^{n-1}) \geq \bar{0}\end{array}\right\} \text{where } \bar{x}^n \text{ solves}$$

$$\begin{array}{c}\max f_n(\bar{x}^1, \bar{x}^2, \ldots, \bar{x}^n)\\ \bar{x}^n \in X^n\\ \bar{g}^n(\bar{x}^1, \bar{x}^2, \ldots, \bar{x}^n) \geq \bar{0}\end{array}$$

where f_i is objective function for subsystem i,
X^i are the decision variables over which subsystem i has control, and
Set $\{\bar{x}^i : \bar{g}^i(\bar{x}^1, \bar{x}^2, \ldots, \bar{x}^{i-1}, \bar{x}^i) \geq \bar{0}\} \cap X^i$ depends on the settings of $\bar{x}^1, \bar{x}^2, \ldots, \bar{x}^{i-1}$.

(Note: This is a multilevel variant of the Stackleberg game.)

Figure 2: Multilevel Programming Problem Formulation

Table 3: Major Issues facing Optimal Control Orientation.

- Choice of optimization procedure:
  - Feasible decomposition
  - Nonfeasible decomposition
  - Predictive-Corrective decomposition
  - Mixed decomposition
- Coordination principles
  - Goal coordination
  - Model coordination
  - Multiple coordination
- Coordinatability
- Convergence properties
- Sub-optimality
  - When is it acceptable?
  - When is it preferable?
  - How to measure extent of sub-optimality?
- Multiple objectives associated with a individual subsystems
  - Decomposition procedures for such cases
  - Displaying non-inferior solution sets
- Multiple decision makers (controllers)
  - Competitive as well as cooperative situations
  - Negotiating schemes
- Closed loop structures
- Perturbations and stability properties

Table 4: Major Issues facing Mathematical Programming Orientation.

- Infeasibility problems caused by aggregation
  - Feasible upper level solutions result in infeasible lower level solutions, because factors are ignored at the upper level in order to avoid excessive detail.
  - Example: Monthly production levels, which are feasible at the upper level, may prove infeasible at the lower level, because only the lower level knew that all the required product had to be delivered in the first week.
- Inconsistency problems caused by conflicting objectives at different levels
  - Example: Upper level wants to maximize service level, while lower level wants to minimize inventory costs.
- Suboptimality problems caused by aggregation
  - One of the main advantages offered by a hierarchical approach is being able to avoid excessive detail at the upper level, thereby reducing computational costs, forecasting inaccuracies, and the need to model difficult managerial interactions. However, ignoring certain factors at the upper level often leads to solutions that are suboptimal in terms of the overall system.
  - Example: Ignoring machine set-up costs in deriving the upper level's annual aggregate production plan can lead to suboptimal overall plans.
- Innovative methods that respond to applications specific considerations
  - Example: Modified knapsack method developed for case of high set-up costs (Bitran, Haas, Hax 1981).
- Integrating models and methods with the unique organizational culture and structure of a particular business concern
  - Example: Monthly review committees played a central role in fostering strong, bi-directional communications, which proved essential for implementing a new hierarchically integrated production and distribution system at American Olean Tile Company (Miller and Liberatore 1988).

Table 5: Opportunities for Cross-Fertilization between the Two Orientations in Hierarchical Programming.

Contribution from Optimal Control Orientation		Contribution from Mathematical Programming Orientation
Theory and methods for handling subsystems with multiple objectives.	$\Leftarrow$ $\Rightarrow$	Experience with diverse subsystem modules (e.g., simulation and MRP in concert with optimization models).
Research on feedback effects (both open and closed loop)	$\Leftarrow$ $\Rightarrow$	Applications specific theoretical and methodological breakthroughs (e.g., revised Knapsack Method to handle the production planning requirements of middle and lower subsystems).
Well developed conceptual framework for pursuing hierarchical programming research.	$\Leftarrow$ $\Rightarrow$	Intensive theoretical work in particular methodological niches (e.g., Bilevel and Multilevel Programming, Systems Optimization, etc.).
Elegant theoretical work (particularly in dynamic systems).	$\Leftarrow$ $\Rightarrow$	Considerable experience with implementation issues, computational considerations, and model-organization congruence.

orientation is found at the Ukrainian Academy of Sciences Institute of Cybernetics in Kiev, where researchers have incorporated V. M. Glushkov's earlier work on Systems Optimization and V. S. Mikhalevich's work on the Method of Sequential Analysis into a hierarchical setting. Two dinctive features of this approach are that

1. It employs the concept of interaction priorities to formalize interconnections between subsystems. That is, iRj means that subsystem i has priority in decision making over subsystem j.

2. It includes a procedure, known as the System Optimization Method, that modifies subsystem feasible regions in order to overcome situations in which no solution exists that is simultaneously feasible for all subsystems.

In Tables 3 and 4 we summarize the major issues addressed by researchers within each orientation toward hierarchical programming. Table 5 points out areas of potentially fruitful cross-fertilization between the two research orientations.

References

No attempt is made here to give a comprehensive listing of publications on hierarchical programming. Instead, three types of sources are cited: survey articles where the interested reader can find extensive lists of sources, full book-length sources, and selected listings of rarely cited articles .

Abad, P. L. 1982. "An Optimal Control Approach to Marketing-Production Planning." *Optimal Control Applications and Methods* 3: 1-13.

———. 1982. "Approach to Decentralized Marketing Production Planning." *International Journal of Systems Science* 13: 227-235.

———. 1985. "A Two-Level Algorithm for Decentralized Control of a Serially Connected Dynamical System." *International Journal of Systems Science* 16: 619-624.

———. 1987. "A Hierarchical Optimal Control Model for Coordination of Functional Decisions in a Firm." *European Journal of Operations Research* 32, no. 1: 62-75.

Bard, Jonathan F. 1982. "An Explicit Solution to the Multi-level Programming Problem." *Computers and Operations Research* 9, no. 1: 7-10.

———. 1984. "Optimality Conditions for the Bilevel Programming Problem." *Naval Research Logistics Quarterly* 13: 13-26.

———. 1985. "Geometric and Algorithmic Developments for a Hierarchical Planning Problem." *European Journal of Operations Research* 19: 372-383.

Bialis, W. F. and M. H. Karwan. 1982. "On Two-Level Optimization." *IEEE Transactions on Automatic Control* 27, no. 1: 211-214.

Bitran, G. R., E. A. Haas, and A. C. Hax. 1981. "Hierarchical Production Planning: A Single Stage System." *Operations Research* 29, no. 4: 717-743.

———. 1982. "Hierarchical Production Planning: A Two-Stage System." *Operations Research* 30, no. 2: 232-251.

Bitran, G. R. and A. C. Hax. 1981. "Disaggregation and Resource Allocation Using Convex Knapsack Problems." *Management Science* 27: 431-441.

Burton, R. M. and B. Opel. 1977. "The Multilevel Approach to Organizational Issues of the Firm: A Critical Review." *OMEGA* 5, no. 4: 395-444.

Candler, W. 1988. "A Linear Bilevel Programming Algorithm: A Comment." *Computers and Operations Research* 15: 297-298.

Candler, W. and R. Townsley. 1982. "A Linear Two-Level Programming Problem." *Computers and Operations Research* 9, no. 1: 59-76.

Findeisen, W., F. N. Bailey, M. Brdys, K. Malinowski, P. Tatjewski, and A. Wozniak. 1980. *Control and Coordination in Hierarchical Systems.* Chichester: John Wiley and Sons.

Gabbay, H. 1976. "A Hierarchical Approach to Production Planning." Ph.D. dissertation, Cambridge, Massachusetts: Massachusetts Institute of Technology.

Grana, M. and F. J. Torrealdea. 1986. "Hierarchically Structured Systems." *European Journal of Operations Research* 25, no. 1: 20-26.

Haimes, Y. Y. 1973a. "Decomposition and Multilevel Approach in the Modeling and Management of Water Resource Systems." *Decomposition of Large-Scale Problems* ed. D. M. Himmelblau, 8, no. 3: 347-368.

———. 1973b. "Integrated System Identification and Optimization." In *Advances in Control Systems,* ed. C. T. Leondes, 435-518. New York: Academic Press.

———. 1973c. "Multilevel Dynamic Programming Structure for Regional Water Resource Management." In *Decomposition of Large-Scale Problems,* ed. D. M. Himmelblau, 369-378. Amsterdam: North Holland.

———. 1977. *Hierarchical Analysis of Water Resources Systems.* New York: McGraw-Hill.

Haimes, Y. Y., K. Tarvainen, T. Shima, and J. Thadathil. 1990. *Hierarchical Multiobjective Analysis of Large-Scale Systems.* New York: Hemisphere Publishing Corp.

Hax, A. C. 1977. "Integration of Strategic and Tactical Planning in the Aluminium Industry." In *Applied Mathematical Programming,* eds. S. P. Bradley, A. C. Hax, and T. L. Magnanti, Chapter 6. Reading, Massachusetts: Addison-Wesley.

Hax, A. C. and G. R. Bitran. 1979. "Hierarchical Planning Systems — A Production Application." In *Disaggregation Problems in Manufacturing and Service Organizations,* eds. L. P. Ritzman, *et al.* Boston: Martinus Nijhoff.

Hax, A. C. and J. J. Golovin. 1978. "Computer Based Operations Management System (COMS)." In *Studies in Operations Management,* ed. A. C. Hax, 429-461. New York: North Holland-American Elsevier.

———. 1978. "Hierarchical Production Planning Systems." In *Studies in Operations Management,* ed. A. C. Hax, 400-428. New York: North Holland-American Elsevier.

Hax, A. C. and H. C. Meal. 1975. "Hierarchical Integration of Production Planning and Scheduling." In *Studies in Management Science. Vol. 1: Logistics,* ed. M. A. Geisler, 53-69. New York: North Holland-American Elsevier.

Lasdon, L. 1970. *Optimization Theory for Large Scale Systems.* London: Macmillan.

Liberatore, M. J. and T. Miller. 1985. "A Hierarchical Production Planning System." *Interfaces* 15, no. 4: 1-11.

Mahmoud, M. S. 1977. "Multilevel Systems Control and Applications: A Survey." *IEEE Transactions on Systems, Man, and Cybernetics* 7: 125-143.

Mahmoud, M. S., M. F. Hassan, and M. G. Darwish. 1985. *Large-Scale Control Systems: Theories and Techniques.* New York: Marcel Dekker, Inc.

Mesarovic, M. D., D. Macko, and Y. Takahara. 1970. *Theory of Hierarchical Multilevel Systems.* New York: Academic Press.

Mikhalevich, V. S. and V. L. Volkovich. 1982. *Computational Methods for the Research and Design of Complex Systems.* Moscow: Nauka.

Mikhalevich, V. S., V. L. Volkovich, and G. V. Kolenov. 1988. "An Algorithm for Coordination of Solutions in a Distributed System of Independent Problems with Linear Models." *Cybernetics* 24, no. 3: 271-280.

Miller, T. and M. J. Lieberatore. 1988. "Implementing Integrated Production and Distribution Planning Systems." *International Journal of Operations and Production Management* 8, no. 7: 31-41.

Nachane, D. M. 1984. "Optimization Methods in Multilevel Systems: A Methodological Survey." *European Journal of Operations Research* 21, no. 1: 25-38.

Sandell, N. R., P. Varaiya, and M. Athans. 1975. "Survey of Decentralized Control Methods for Large Scale Systems." In *Systems Engineering for Power: Status and Prospects. Proceedings of the Engineering Foundation Conference on Systems Engineering, Henniker, New Hampshire, 17-22 August 1975,* 334-355. Washington, D.C.: U.S. Energy Research and Development Administration.

Singh, M. G. 1982. *Dynamical Hierarchical Control.* Amsterdam: North Holland.

Singh, M. G., S. A. W. Drew, and J. F. Coales. 1975. "Comparison of Practical Hierarchical Control Methods for Interconnected Dynamical Systems." *Automatica* 11: 331-350.

Singh, M. G. and A. Titli. 1978. *Systems: Decomposition, Optimization, and Control.* Oxford: Pergamon Press.

Tarvainen, K. 1981. "Hierarchical Multiobjective Optimization." Ph.D. dissertation, Cleveland, Ohio: Case Western Reserve University.

———. 1986. "On the Generation of Pareto Optimal Alternatives in Large Scale Systems." *Proceedings of the 4th IFACS Symposium on Large Scale Systems, Zurich, Switzerland*, n.p.

Volkovich, V. L., A. N. Berezhnoi, G. V. Kolenov, and Yu. P. Chaplinskii. n.d. "A View on Decision Making in Multilevel Organizational Systems." (Collection title unknown.), n.p.

Volkovich, V. L. and Yu. P. Chaplinskii. 1987. "Algorithms for System Optimization in Linear Models when the Directive Region is Given by Balance Relationships." *Soviet Journal of Automation and Information Sciences* no. 6: 46-54.

Volkovich, V. L., G. V. Kolenov, and S. O. Mashchenko. 1988. "An Algorithm of Search for an Admissible Solution in Linear Distributed Systems." *Soviet Journal of Automation and Information Sciences* no. 4: 70-77.

———. 1989. "An Algorithm for Solving a Linear Optimization Problem in a Distributed System." *Soviet Journal of Automation and Information Sciences* no. 1: 40-49.

Volkovich, V. L. and V. M. Voinalovich. n.d. "O koordiniruemosti dvukhurovenykh ierarkhicheskikh sistem." (On Coordinatability of Two-Level Hierarchical Systems.) In *Modelirovanie v ekonomicheskikh issledovaniyakh,* n.p.

Wen, U. P. 1981. "Mathematical Methods for Multilevel Linear Programming." Ph.D. dissertation, Buffalo, New York: State University of New York at Buffalo.

Wismer, D. A. 1971. *Optimization Methods for Large Scale Systems with Applications.* New York: McGraw-Hill.

MULTICRITERIA DECISION MAKING FOR SELECTING FREEWAY INTERCHANGE LOCATIONS IN TAIWAN

Chien-Yuan Lin
Associate Professor, Inst. of Building and Planning
National Taiwan University
Junn-Yuan Teng
Inst. of Traffic and Transportation
National Chia-Tung University
Taipei, Taiwan

ABSTRACT

Due to the rapid socio-economic development, freeway transportation demand has increased dramatically in the past few years in Taiwan. To improve the accessibility of freeway system, transportation planners are facing to a critical decision problem in selecting locations for both additional interchanges along existing freeway and new interchanges for a new freeway. To help planners in tackling such a complicated decision making problem, a computerized two-phase multicriteria evaluation model is developed. In this paper, we will describe the structure of the proposed two-stage evaluation model and its application in a case study.

I. INTRODUCTION

Since the completion of the first freeway in 1978, freeway transportation has not only improved the economy development in Taiwan, but also saved tremendous travel time for various freeway users. Since the setting of freeway interchange will bring both positive and negative impacts on the selected area, selection of freeway interchange locations has to be carefully dealt with, no matter if a new interchange is to be added to an existing freeway or a set of interchange locations are to be designed for a new freeway system. Along with the rapid economic development and population growth in the past few years, especially the increase of car ownership, demand for freeway transportation has sharply increased. Adding new interchange facility to the existing freeway has been requested by both local governments and interest groups, however, land owners in the suggested interchange locations are against due to the under-priced compensation for land acquisition. In dealing with this complicated decision problem, transportation planners are crucially in need for an effective decision making tool.

Based on factors usually considered in selecting freeway interchange locations, a two-phase multicriteria evaluation model is developed to serve as an operational tool for transportation planners. The result system is an integration of a location model and a multicriteria evaluation model. In this paper, contents and framework of the two-phase model are described. Besides, a case study of selecting an additional location for the existing freeway is demonstrated.

II. CONSIDERATION FACTORS OF FREEWAY INTERCHANGE LOCATION

Freeway is highway with limited access for vehicles so that interchanges are the only entrances and exits for vehicles to enter or leave a freeway. Impacts and consideration factors related to the selection of freeway interchanges are described as follows:

1. Positive Impacts of Freeway Interchange

A. Direct Benefits

Areas around freeway interchanges will have better accessibility to freeway service. Potential benefits include savings in travel expense and travel time, alleviation of local traffic bottleneck, and better transportation service to communicate with other areas.

B. Indirect Benefits

Indirect benefits include better opportunities for rationalization of production and distribution system, promotion of local industry and resource development, and balancing regional development.

2. Negative Impacts of Freeway Interchange

A. Negative Impacts on Local Area
Aside of its potential positive influences on regional economy and development, the provision of interchange will have negative influences such as traffic increase upon local roadways, environmental pollution, and the increase of noise.

B. Negative Impacts on Freeway Operation
Potential negative impacts on freeway operation include downgrade of freeway service level due to traffic increase, toll loss because of inappropriate location, disturbance on freeway mainline traffic.

3. Consideration Factors in Selecting Interchange Locations

Some factors that are often considered in selecting locations for freeway interchanges in practice are described as follows:

A. Local Transportation Demand:
When long-distance transportation demand increases to a certain extent, improvement of better accessibility to freeway system are always considered. If freeway service is not provided, long and mid-long distance trips will be forced to travel along local roadways.

B. Local Development Promotion
The provision of freeway interchange will improve the accessibility of an area. For the promotion of local development, local socio-economic and environmental impacts have to be cautiously considered.

C. Comprehensive Transportation Plan
One of the purposes in comprehensive transportation planning is to coordinate investments of transportation infrastructures. Locations that have been planned for setting freeway interchanges should be favorably considered.

D. Industrial and Tourism Development
Provision of transportation infrastructure is always required to stimulate local industrial and tourism development. To serve existing transportation demands generated from major industrial area or to promote tourism development for an area, connection to freeway interchanges will be a plus.

E. Military Purpose
National defense has played a key role in the development of freeway system in Taiwan. It has been a long time that relationship between Taiwan and Mainland China are in a tense situation. In the foreseeable future, military consideration can not be totally discarded.

III. Two-Phase Evaluation Model

In a plural society, decision has to be traded-off among many conflicting objectives and different value judgements have to be dealt with at the same time. Therefore, MCDM are widely applied in various resource allocation problems, such as transportation planning, energy planning, urban planning and water resource planning [Keeney and Raiffa, 1976]. The complicated decision task of a freeway interchange location system exactly fits with the property of MCDM applications.

To apply MCDM in selecting freeway interchange locations, four basic elements need to be prepared. They are alternative set, criterion set, expected outcomes, and preference structure. In a typical MCDM application, the task will be processed in four steps, namely, structuring of the decision problem, assessing possible impact of alternatives, constructing alternative preferences of decision makers, evaluating and ranking alternatives [Keeney 1984]. Such a process assumed that all candidate designs are feasible and objectives can be traded-off. However, this is not true to freeway interchange location selection problem in Taiwan, since most candidate interchange location proposed by local governments are technically infeasible and few are definitely worth for setting with no need of further negotiation. To cope with this special decision making problem, an integrated two-phase evaluation model is developed to facilitate planners in freeway interchange planning process. Based on the two-phase evaluation model, when candidate locations are proposed, feasibility of these candidates are evaluated first against technical constraints, such as legal and technical considerations. Infeasible alternatives are purged before MCDM evaluation methods are applied. Only those that have passed the first phase are qualified for further evaluation in order to find out the highest ranked solution. Therefore, the whole decision process split into two stages.

Many single-phase choice models have been developed previously [MacCrimmon 1974] and integration of single-phase choice model into MCDM framework was once investigated either [Fishboun 1974]. Wright and Barbour uses MCDM to screen the alternative in the first phase, and repeat another MCDM method in the second phase evaluation. [Wright and Barbour 1977]. However, we found no application similar to the model in this paper, not mention any identical computer system has been presented. To facilitate the operation of two-phase evaluation model, a location-allocation model (LOCATOR III) [Lin 1987] is integrated in the second phase to provide locational performances for MCDM evaluation.

1. Evaluation in First-Phase

Candidate locations proposed by local agencies and planning agencies are taken as alternative for evaluation. In the first phase, they are evaluated against their technical feasibility. Four criteria are included in this phase, namely, connection to international airport, connection to international seaport, topologically constraint and minimal gap between interchanges. By use of Lexicographic Ordering Method, alternatives are screened. All those that are topologically constrained(i.e., steep slope segment or river bank and those that are too close to other interchanges are not accepted for further consideration). All those that are connected to international airport or seaport are absolutely recommended without further evaluation. Those that are not screened out in the first phase are proceeded for further evaluation.

2. Evaluation in Second-Phase

The second evaluation phase is to identify the most satisfied alternatives resulted from the first evaluation process. In this phase, the group decision making method [Saaty 1988]. Various experts and scholars are included to process the preference difference analysis in order to gain consensus weights. In addition, impact analysis of various alternatives will be prepared for two types of criteria. In one type, performance are provided by location analysis model, including average travel time and standard deviation of travel time; performances in the other type are provided by planner by different evaluation methods in order to formulate impact matrix, or decision matrix. Based on decision matrix and criterion weight matrix, MCDM can be used to ranking the different alternatives. In this paper, we will use Simple Additive Weighting Method, (SAW) to rank alternatives.

IV. EVALUATION CRITERIA AND WEIGHTS

First of all, an objective hierarchy structured in the second evaluation phase, and than weights for each different criterion are prepared by a group decision making method.

1. Objective Hierarchy Structure

Two levels of objective hierarchy structure, shown as Figure 1., are included in the proposed model. At the objective level, four objectives are included, namely, operational economy, socio-economic development, transportation and environment. Criteria included for each objective and corresponding measurement methods are described as follows:

A. Operational Economy

C1. Total land acquisition cost

Land acquisition cost includes costs for land and compensation for all betterments on it. Total land acquisition cost is measured by monetary unit Less land cost indicates that a candidate should be more favorably considered.

C2. Total construction cost

The construction cost has to do with interchange's geometry. Geometries of freeway interchange are subjected to factors related to traffic engineering, traffic performance and system maintenance. Construction cost are measured by monetary unit. Less construction cost indicates that a candidate should be more favorably considered.

B. Socio-Economic Development

C3. Ease of land acquisition

Land acquisition problem is a serious problem for selecting freeway interchange locations in Taiwan, since land price has been rocketed continuously and officially assessed land compensation price is always far behind. No land owner will be happy if his land is bought out by government at the officially assessed price. To measure the difficulty of land acquisition, number of total households living near the area means more objections may be happen during the acquisition process.

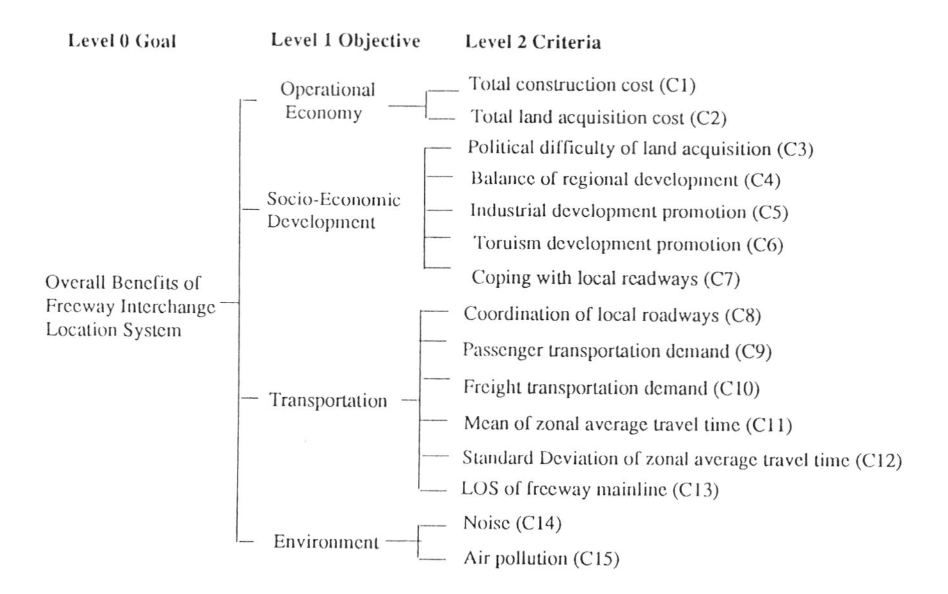

Fig 1. Objective Hiearchy Structure of Selecting Freeway Interchange Location

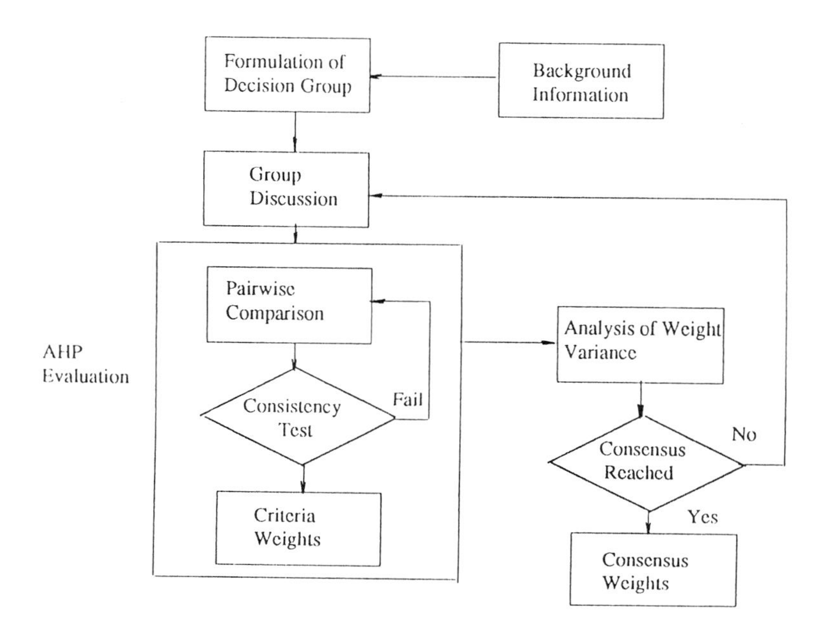

Fig 2. Determination Process of Consensus Weights

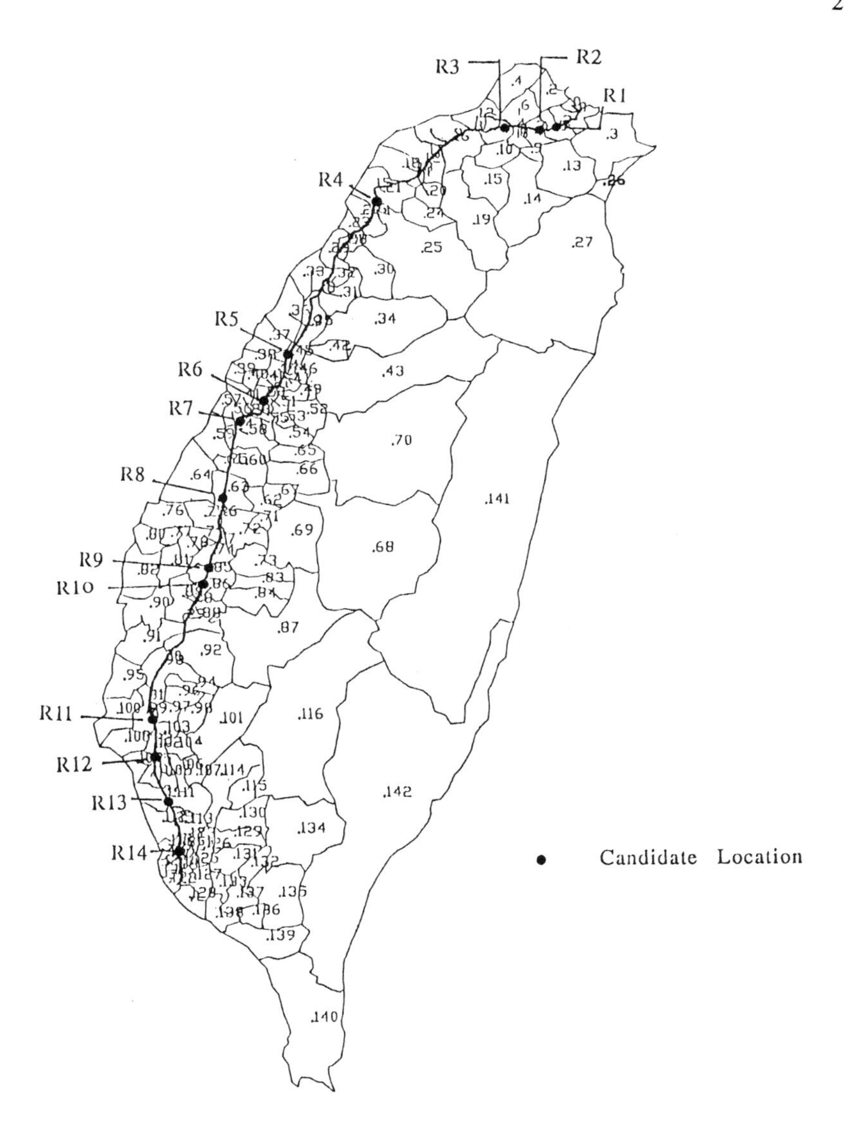

Fig 3 Locations of Candidate Freeway Interchanges

C4. Balancing regional development
One of the goals of Comprehensive Plan for Taiwan Area is to balance the regional developments. Since freeway is a required means for underdeveloped area to improve its attractiveness, its potential contribution should be considered in the selection of freeway interchange locations.

C5. Promotion of industry development
Promotion of industry development has long been one of the objectives of national development . Location near existing industrial area should be favorably considered for freeway interchange location. Candidate location with more industrial area nearby should be favorably considered.

C6. Promotion of tourism development
To promote tourism development, transportation facilities are used to better the accessibility of a proposed recreation area. The more tourists visiting an area near a proposed location, the more favorable the proposed location should be considered.

C7. Local government facilities
Areas with local public facilities, such as prefecture's government center, often need freeway interchange to improve its accessibility. Existence of prefecture administration center near the candidate area is used as indicator.

C. Transportation

C8. Local roadway coordination
Coordination of local roadway system will influence the interchange's operation performance. Locations having better local roadway coordination deserves favorable consideration.

C9. Coping with passenger transportation demand
One of freeway's objectives is to serve long distance passenger transportation. The more passengers depart or arrive in an area, the more urgent an area needs interchange.

C10. Coping with freight transportation demand
The more freights need to be transported near a candidate location, the more urgent it needs freeway interchange.

C11. Mean of average travel time
The accessibility of freeway interchanges are determined by their spatial distribution. Theoretically, more evenly distributed interchanges will reduce average travel time for all traffic zones. Average travel time are measured by minutes. Less mean of average travel time indicates better arrangement of interchange location.

C12. Standard deviation of average travel time
Mean value are the expected value and standard deviation indicates variation in statistical analysis. Small standard deviation indicates average travel time in each interchange's service area are more evenly distributed, this is favorable to equity consideration in selecting freeway interchange location. Standard deviation has no measurement unit.

C13. LOS of freeway mainline
If freeway interchanges are densely distributed, more short-distance transportation trips will be attracted to freeway. As a result, traffic on freeway may be increased and the level of service (LOS) of the freeway segment near candidate location are considered.

D. Environment

C14. Noise
Noise is one of the adverse effects resulted from freeway traffic on its adjacent area. In this study, number of residents within a radius of two kilometers from the candidate location is used as the indicator of noise impact .

C15. Air pollution
Air pollution along freeway interchanges are resulted from traffic flow. CO, NOx, HC are the typical pollutants. Again, activity population within two kilometers of the candidate location are used as indicator of air pollution impact.

2. Determination of Criteria Weights

Weights in MCDM represent the preference structure of decision makers. Higher weight value for a criterion indicates more importance are placed on that criterion. Although weights can be determined by decision makers directly, results may be subjective and consistency can not be validated. To remedy this shortcoming, in this study, Analytic Hierarchy Process (AHP) is applied. Through AHP, weights for each criterion are determined in a more consistent way and constancy can be checked. Members in the decision group are invited from different specialization areas, such as traffic engineers, local representatives. transportation planners and environmental scientists. The process of generating consensus weight scheme is shown as Figure 2. Based on the objective hierarchy structure in Figure 1, AHP is then used to evaluate the preference of each member in the decision group.

V. CASE STUDY

Due to the rapidly increasing demand for freeway transportation, how to select one from many candidate locations for adding a new interchange to the existing freeway system is a real world problem encountered in Taiwan. Fifteen candidate freeway interchange locations are included for evaluation in this case study. First of all, they are evaluated in the first stage to screen out locations that are technically infeasible or absolutely required for socio-economic purpose. In this case, R1, R5 and R10 are purged out since they are located too close to existing interchanges and technically infeasible in terms of traffic engineering. R7 and R15 are beside river and infeasible in terms of geometry. As a result of the first stage evaluation, only ten candidate locations are feasible and competitive enough for evaluation in the second phase.

To prepare decision matrix and weights for the second evaluation stage, performances such as mean of zonal average travel time(C11) and its standard deviation (C12) are generated by the location model. Performances for other criteria are gained from socio-economic data and related studies, contents in the decision matrix are shown as Table 1. AHP method is applied to obtain consensus weights in this case study.

In the weights deriving process, a group meeting is conducted for participants to discuss with each other. Consistency test and variance analysis are then followed. In this study, a consensus weighting scheme is reached after three rounds of evaluation. Average weights for the ten members are shown as Table 2. Based on the decision matrix and consensus weights, priorities of alternative locations are ranked. Normalization method used in this paper is described as follows [Nijkamp 1977]:

(1).Cost Criteria

$$\hat{e}_{ji} = 1 - \frac{e_{ji} - e_j^-}{e_j^+ - e_j^-} \qquad (1)$$

and

$$e_j^+ = \max_i \{ e_{ji} \} \qquad (2)$$

$$e_j^- = \min_i \{ e_{ji} \} \qquad (3)$$

where

$\hat{e}_{ji}$: normalized evaluation value of ith alternative for jth criterion

e_{ji} : evaluation value of ith alternative for jth criterion

e_j^+ : maximal evaluation value for jth criterion

e_j^- : minimal evaluation value for jth criterion

(2) Benefit Criteria

$$\hat{e}_{ji} = 1 - \frac{e_j^{+} - e_{ji}}{e_j^{+} - e_j^{-}} \qquad (4)$$

Normalized evaluation scores for each alternative are multiplied by consensus weights of each criterion. Higher score indicates higher priority.. Scores for each alternative are computed by the equation shown below:

$$S_i = \sum_j W_j \hat{e}_{ji} \qquad (5)$$

where

S_i : overall score for ith alternative

W_j : consensus weight for jth criterion

Therefore, if Si>Sk(i=k), then Ai>Ak (Ai, Ak indicates alternative i and alternative k). Based on above method, normalized scores, overall scores and ranking for each alternative are computed and shown as Table 3. Alternative location R13 has an overall score of 0.819 and it is recommended for the first choice.

VI. CONCLUSION

In this paper, we have demonstrated how an integrated decision support system based on multicriteria evaluation model and location model is developed. This model can be applied as an effective tool to support transportation planners in selecting freeway interchange locations. There are three features in the integrated system in terms of MCDM application. Firstly, to deal with " technical " criteria, a two-phase evaluation process is designed. With this filtering device, only competitive and technically feasible alternatives will be included for further evaluation and ranking through typical MCDM process. Secondly, a location-allocation model for location analysis is integrated into the decision model so that once a candidate location is to be included for evaluation, locational performances will be automatically generated and transferred to form the impact matrix. By so doing, planners can save a lot of time in data preparation. Thirdly, the whole process, from the input of location design to the presentation of evaluation result, is programmed as an interactive computer graphic system. This has provided a friendly man-machine interaction environment for transportation planners in solving MCDM problem. These three features together with the embeded robustness of AHP method have made MCDM as an operational tool for practical application. Besides, it should be noted that the integrated two-phase MCDM method can also be used to select more than one locations at one time by including more candidate locations in one design.

To enhance the capability of the freeway interchange MCDM model, s couple of directions are proposed for further developments First of all, criteria included for evaluation should be defined more precisely so that evaluation results can be more easily used for communication in public hearing. Secondly, preparation of weight set is always a critical step in practical applications. This problem involves who are to be included in the decision group and what method is to be used to derive compromised weighting scheme for succeeding evaluation. Further development of weight extraction method will certainly make MCDM more useful in practical applications.

Table 2 Consensus Weights

Criteria / Decisio Maker		Operational Economy		Socio-economic Development					Transportation						Environment	
		C1	C2	C3	C4	C5	C6	C7	C8	C9	C10	C11	C12	C13	C14	C15
Δ	1	0.128	0.130	0.017	0.120	0.074	0.082	0.010	0.062	0.075	0.045	0.064	0.022	0.023	0.053	0.095
Δ	2	0.142	0.099	0.051	0.047	0.036	0.057	0.099	0.062	0.077	0.082	0.046	0.029	0.089	0.031	0.053
●	3	0.062	0.119	0.102	0.169	0.065	0.021	0.059	0.036	0.079	0.083	0.011	0.010	0.021	0.083	0.080
●	4	0.075	0.034	0.207	0.131	0.059	0.027	0.017	0.063	0.043	0.096	0.042	0.067	0.038	0.045	0.051
●	5	0.044	0.034	0.119	0.137	0.040	0.089	0.029	0.042	0.072	0.062	0.050	0.050	0.094	0.054	0.084
*	6	0.052	0.071	0.101	0.060	0.027	0.085	0.028	0.048	0.189	0.121	0.025	0.022	0.034	0.050	0.087
*	7	0.082	0.057	0.116	0.150	0.050	0.013	0.039	0.072	0.160	0.122	0.030	0.030	0.011	0.023	0.045
*	8	0.081	0.058	0.101	0.049	0.078	0.082	0.041	0.030	0.152	0.117	0.022	0.020	0.040	0.043	0.086
Ø	9	0.040	0.037	0.081	0.086	0.046	0.083	0.028	0.072	0.085	0.088	0.024	0.021	0.030	0.100	0.179
Ø	10	0.068	0.044	0.019	0.093	0.051	0.119	0.037	0.064	0.075	0.072	0.022	0.027	0.023	0.092	0.194
Sum of Weights		0.774	0.683	0.914	1.042	0.526	0.658	0.387	0.551	1.007	0.888	0.336	0.298	0.403	0.574	0.959
Average		0.077	0.068	0.091	0.104	0.053	0.066	0.039	0.055	0.101	0.089	0.034	0.030	0.040	0.057	0.096

Note : 1. Weights in Objective Level are Operational Economic (0.146), Socio-economic Development (0.353), Transportation (0.348), Environment (0.153).
2. Background : Δ Construction Engineer ● Local Representative * Transportation Planner Ø Environment Expert

Table 3 Normalized Evaluation Value and Overall Score

Criteria / Candidate	Operational Economy		Socio-economic Development					Transportation						Environment		Overall Score	Rank
	C1	C2	C3	C4	C5	C6	C7	C8	C9	C10	C11	C12	C13	C14	C15		
R2	0.208	0.400	0.429	0.250	0.000	0.000	0.000	1.000	0.000	0.000	0.000	0.231	0.167	0.548	0.548	0.261	10
R3	0.333	0.200	0.333	0.125	0.000	0.000	0.000	0.607	0.333	1.000	0.251	0.229	0.111	0.355	0.355	0.313	9
R4	0.750	0.000	0.683	0.500	0.674	0.667	1.000	0.214	1.000	0.548	0.630	0.888	0.694	0.774	0.774	0.647	3
R6	0.000	0.800	0.000	0.000	1.000	1.000	1.000	0.929	0.593	0.806	0.128	0.188	0.000	0.000	0.000	0.405	8
R8	0.958	0.600	0.952	0.750	0.398	0.000	0.000	0.357	0.333	0.581	0.524	0.668	1.000	0.839	0.839	0.614	4
R9	1.000	1.000	7.746	1.000	0.771	0.333	1.000	0.643	0.667	0.355	0.912	0.933	0.944	0.903	0.903	0.788	2
R11	0.583	0.300	0.714	0.375	0.560	0.000	0.000	0.000	0.815	0.226	0.009	0.000	1.000	1.000	1.000	0.494	6
R12	0.292	0.500	0.619	0.250	0.216	0.333	0.000	0.714	0.704	0.484	0.125	0.240	0.500	0.581	0.581	0.444	7
R13	0.667	0.600	1.000	1.000	0.448	0.667	1.000	0.429	0.741	0.968	1.000	1.000	0.833	0.935	0.935	0.819	1
R14	0.500	0.200	0.841	0.625	0.286	0.333	1.000	0.643	0.407	0.806	0.379	0.608	0.389	0.710	0.710	0.554	5

JUDGEMENTAL MODELLING AS A RESEARCH TOOL IN INDUSTRIAL MARKETING DECISIONS

Geoff Lockett and Pete Naude

Manchester Business School
Manchester, UK

Abstract

This paper presents a case study using Judgemental Modelling. It examines the attributes that were important to a group of decision makers in buying a large computer system, and how the various suppliers' systems scored on those attributes. This exercise was then repeated with the different suppliers, thus enabling an examination of the level of understanding that the potential suppliers had of the buyers' needs. The paper focuses on how the information needs of the buyers changed over the decision making process, and illustrates the consequences of how the misunderstanding of this change on the part of the suppliers influenced the outcome of the final decision.

Although only a single case study, we believe that the quality of the data collected illustrates the validity of this technique as a research tool in studying the decision making process in industrial marketing.

Introduction

The marketing concept calls for suppliers in the market place to understand the needs and wants of their target markets (Kotler, 1988). We would expect firms adopting this concept to have a higher level of understanding of their customer's needs than those that do not. In industrial markets, where sales are often the culmination of months or even years of negotiations between the buyer and potential suppliers, we would expect the understanding to be higher still. This research measures the extent of that understanding, and examines the consequences of a supplier misunderstanding how the buyer's needs change over the decision making process.

Industrial Marketing Models

Four models in industrial marketing deserve attention for their contribution to helping marketers understand the selling process better, and hence to act more effectively in their sales efforts.

The first, put forward by Robinson, Faris, and Wind (1967), was a model of the purchase process itself, rather than a general model of industrial buying behaviour. It classified purchases by buy-class (new purchase, modified rebuy, or repeat purchase), and by (one of eight) buy-phases, varying from recognition of the problem through to feedback and evaluation. The model has two particular benefits: it helps to understand how the information needs of the buyer vary over both the **class** and the **phase** of the decision.

The next two models were more general in their perspective, building up an understanding of the factors affecting the individuals in the decision making unit (DMU), and how the DMU can interact to reach a decision. Webster and Wind (1972) identified four variables as relevant to the buying decision: the environment, the organisation, the buying centre, and the role of the individuals. Sheth (1973) identified as important the psychological makeup of the individuals, the conditions which lead to joint decision making, and the methods of resolution of conflict among members of the DMU. However these models imply a rather confrontational attitude between buyer and seller. They do not capture the essentially **relationship based** nature of much of industrial marketing, where a buyer will interact and explore various purchasing possibilities with potential suppliers, often for months before a contract is awarded. Such a relationship "is often close. It may be long term and involve a complex pattern of interaction between the two companies" (Hakansson, 1982:1).

The shortcomings of these traditional models have lead to the interactional approach, based on the general philosophy of systems theory, and calling for the simultaneous study of as many individuals as may contribute to the decision. Three theoretical papers (Hakansson and Ostberg (1975), Nicosia and Wind (1977), and Bonoma and Johnston (1978)) have been instrumental in advocating this approach. An examination of networks within industrial markets, as typified by the work done by the IMP Group, also falls into this interactional approach (See, for example, Cunningham and Culligan (1989), Easton (1989), and Hakansson and Johanson (1989)).

This paper accepts the interactional approach as the most applicable paradigm. It examines the level of understanding that suppliers have of a buyer's needs, and as such is concerned with a part of the total system involved in a particular transaction.

The specific questions that will be addressed are:

* What attributes does the DMU use to evaluate suppliers, and how do they vary in importance? How do the DMU members perceive the various potential suppliers that they are evaluating?
* How do the suppliers asses the importance of these same attributes? How well do they understand their competitors' positions?
* As the decision making process progresses, how do the information needs of the DMU change, and what influence does this change have on the role of the suppliers?

In the next section we discuss these issues using research on a real life study.

Background to the Purchase Decision

The National Health Service in England is divided into fourteen regions, each run by its own Regional Health Authority. Each unit is in turn subdivided into a number of Districts. Over the past few years, a number of reports have addressed the collection and use of information in the Health Service (Korner (1984) and National Health Service (1989)), and this has resulted in increasing attention being paid to computerization. One particular Authority, one of the first to purchase new computer systems, decided to follow a two stage approach, with a few initial pilot sites (i.e. Districts) exploring the options in the market, and making this information available to subsequent Districts. This study concerns the decision making process that took place in the first pilot site within that Authority.

With over a dozen Districts, and with each system costing in excess of £250,000, the size of the market, even within this one Authority, was significant. Given the influence that the sale would have on subsequent decisions both within the particular Authority and possibly in others, there was a high level of interest within the Authority and amongst the suppliers as to which computer system would be selected.

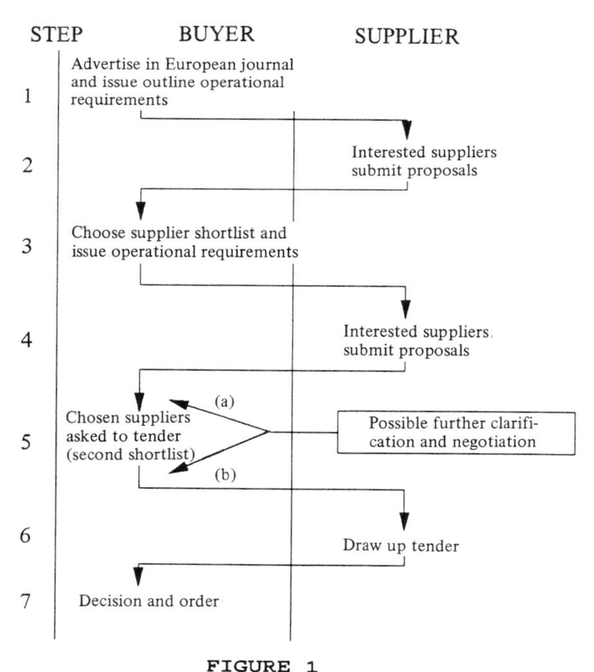

FIGURE 1
THE PURCHASING PROCESS

Although numerous people were tangentially involved in the purchase, there was a key DMU of four people within the hospital, all of whom would be heavy users of the new system. The decision making process is outlined in Figure 1. There were essentially three distinct phases: choosing the first shortlist from all firms responding to the initial advertisement; whittling this down to identify those suppliers that would be asked to tender (the second shortlist); and then deciding between the competing bids.

There were over 50 responses to the advertisement. After receiving the responses, a period of considerable information gathering was undertaken by the DMU. This resulted in a reduction of the initial list to just four firms (i.e. the result of Step 3 in Figure 1). The next step was to narrow this down further in order to identify which of the firms should be asked to submit formal tenders (i.e. Step 5).

For this decision, the DMU adopted a structured process whereby they first decided on the attributes that should be used to choose between the four suppliers and their systems, and then decided how these attributes varied in importance (A judgemental modelling package was used for collecting and analyzing the data (See Islei and Lockett (1989)). Finally, they scored the four suppliers on the various attributes.

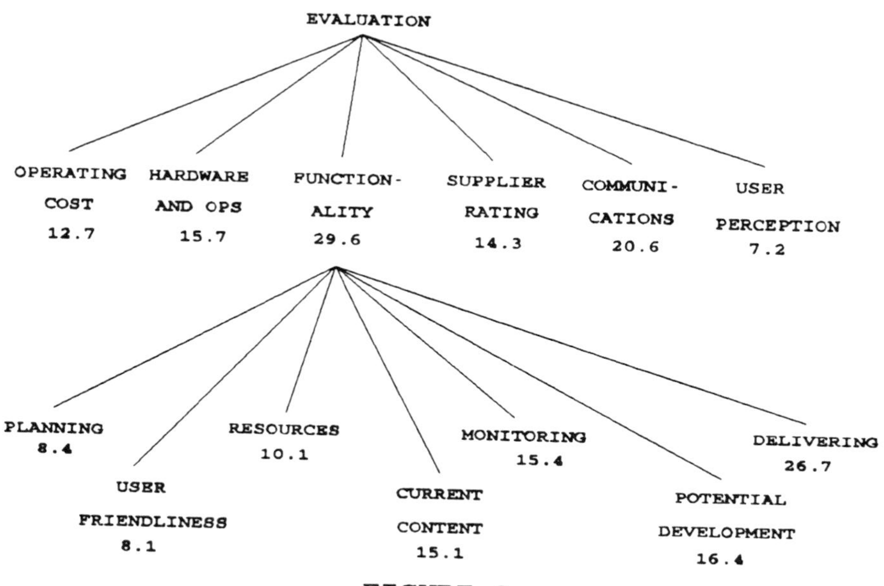

FIGURE 2
THE BUYERS' ATTRIBUTE HIERARCHY

Once the structure of the attribute hierarchy had been agreed upon, each of the DMU members rated the relative importance of each attribute. The hierarchy used and overall (averaged) scores are indicated in Figure 2. Appendix 1 gives the definitions used for each attribute.

By far the most importance was attached to Functionality, the only attribute with second level elements. It accounted for almost 30% of the DMU's weighting, with Communications was second with just over 20%. Three of the four individuals weighted these two attributes alone over 50%. User Perception was regarded as the least important attribute by all participants, averaging out at 7.2%. Within Functionality, Delivering was the most important element. Planning, User friendliness, and Resourcing were the least important.

Once the DMU had weighted each supplier on each attribute, the data were appropriately weighted and aggregated, and an overall score for each supplier obtained. This is shown in Table 1 below.

TABLE 1
THE BUYERS' ATTRIBUTE AND PRODUCT SCORES

Supp.	OpCost	HwOps	Func	SupRat	Comms	UserP	TOTAL
A	29.4	49.9	76.1	46.1	69.1	21.8	292
B	28.1	33.2	60.3	29.2	38.4	15.0	204
C	34.3	43.3	83.4	38.7	58.0	18.4	276
D	35.4	30.8	75.7	28.7	40.2	16.5	227

Supplier A achieved the highest total score. The suppliers are viewed distinctly, with none occupying the same market niche. Comparing the Suppliers, A scores well on four attributes: Hardware and Operations, Supplier Rating, Communications, and User Perception. The low score on Operating Cost is not seen as a serious problem, given the relative unimportance of the attribute (See Figure 2). However, Functionality is the most important attribute, and the relatively poor score on this attribute may well be cause for concern. Table 1 also indicates that Supplier C scored highest on Functionality and is associated with Communications. Although Supplier D scores third overall, it is seen as doing well on Operating Cost and on Functionality. Supplier B is not highly associated with either of the two most important attributes, Communications and

Functionality. Indeed, it does not score within the first two on any attribute.

At this stage of the Decision Making Process, it appeared that Supplier A was the overall favourite, and that Supplier B was considered to offer the least attractive of the four systems.

The Perceptions of the Suppliers

During the above phase of the decision making process it became evident that Supplier D would not be submitting a tender. The other three suppliers were approached and the senior sales person in charge of the particular account questioned as to their perception of the hospital's needs. Each supplier was asked to identify who they thought their shortlisted competitors for the particular tender were, and then asked to score the suppliers on the attribute hierarchy used by the DMU.

The competitors identified by the three suppliers are shown in Table 2. We see, for example, that Supplier A considered themselves as well as C and D as the competition; while Supplier B thought that themselves, A and C were in contention. Only Supplier C correctly identified all four potential suppliers.

TABLE 2
The Perceived Competition

		The Actual Supplier		
		A	B	C
	A	X	X	X
The	B	-	X	X
Perceived	C	X	X	X
Competition	D	X	-	X

Although numerous firms had responded to the initial advertisement, the short listed suppliers were obviously well aware of who their core competition was, and there was virtually no uncertainty as to where the boundaries lay (Easton, 1988). This is to be expected: the earlier reports (Korner (1984) and National Health Service (1989)) had resulted in excellent growth prospects for a particular and highly specialised market segment. While other suppliers were interested in exploiting this growth, the established competitors did not really regard them as a threat. Two of the suppliers (A and C), identified the fourth supplier as a potential rival, and were not to know that he had in fact withdrawn from the race.

A clear picture emerged of the extent to which the various suppliers understood the needs of the buyer when they undertook the scoring process using the DMU's attribute hierarchy. These scores appear in Table 3, which gives the score for the buyers (from Figure 2 above) and the three suppliers, with the ranking of the attribute shown in parentheses.

TABLE 3
THE SCORES OF THE SUPPLIERS

	OpCost	HwOps	Func	SupRat	Comms	UserP
DMU	12.7(5)	15.7(3)	29.6(1)	14.3(4)	20.6(2)	7.2(6)
A	9.2(6)	16.3(3)	28.5(1)	17.5(2)	15.8(4)	12.8(5)
B	5.7(5)	5.1(6)	28.3(1)	21.2(3)	11.3(4)	28.3(1)
C	10.3(6)	13.7(4)	17.8(3)	22.3(1)	13.7(4)	22.3(1)

For Supplier A we see a high degree of overlap, except for the attributes Supplier Rating and Communications, where the rankings are reversed and differ by more than one position. Supplier B clearly overestimated the importance of the last attribute, User Perception, to the buyers. Functionality, though, was correctly seen as being of major importance. Supplier C sees as important the 'softer' issues in the sale: Supplier Rating and User Perception tie for first place, whereas they were fourth and sixth respectively for the buyers. The perceived importance of these two pushes Functionality into third place, and Communications into joint fourth place.

Overall, it appears that Supplier A has the best perception of the buyer's needs. We can surmise that both B and C place too much importance on managing the perception of the users, rather than seeing the importance of the functionality. While all three suppliers underestimated the Communications attribute, A gave it the highest score. All correctly perceived Operating Cost to be of relatively minor importance.

The Predicted Decision

The picture that emerged at this stage was that the DMU considered Supplier A the best. They rated them highest (Table 1), and indeed Supplier A seemed to have the best idea of what the buyers needed (Table 3). All three suppliers gave Supplier A a good chance of victory.

As a result of this process, it was decided that only two of the suppliers would be asked to tender. Supplier D had already withdrawn, and it was decided not to ask Supplier B. This left Suppliers A and C at the end of Step 5 (Figure 1), with the DMU believing that both suppliers could offer an acceptable system. Supplier A, however, was clearly the favoured candidate at this stage.

The Final Decision

Two months later, the order was given to Supplier C. Although the price of Supplier C was lower than that of A, our observations of the process indicate that this was not the reason for the buyers' change of heart. Rather, Supplier A did not understand how the needs of the DMU changed during the decision making process. There is no doubt that the three distinct phases of the decision making process require the proactive seller to offer the buyer different information over the life of the decision. These can be summarised as follows:

Phase 1

Objective:

To get onto Buyer Shortlist

Required Response:

Give sufficient information to allow the buyer to be clear that your product will meet the initial specifications as outlined in the advertised Operational Requirements.

Actual Response of Supplier A:

As an international leader in the particular systems, with many working systems worldwide, the response at this stage was perfectly adequate.

Phase 2

Objective:

To get onto Second Shortlist

Required Response:

Start negotiations with the client regarding the desired system attributes. Far more specific information is required than before.

Actual Response of Supplier A:

Excellent. At the end of this stage the DMU preferred A's system, and A had the best understanding among all potential suppliers of the buyer's needs.

<u>Phase 3</u>

Objective:

To win order

Required Response:

This is the most crucial stage. The level of competition is far stronger than before, with all suppliers now realistic winners. Additionally, the sale is by tender, and no post-tender negotiation is permitted. The response should therefore be to work extremely closely with the buyer, giving as much information as possible, both regarding the breakdown of costs and the reasons for recommending a particular configuration. Be prepared to negotiate about both price and system specifications to get as close to the buyer's requirements prior to tendering.

Actual Response of Supplier A:

Poorly executed. In comparison to Supplier C, A adopted a more arms-length approach, promising to have the final tender document delivered on the date required. There was little to no negotiation as to either additional changes to the system specifications or the structure of the final price. In terms of Figure 1, Supplier A continued with further clarification and negotiation up to the stage of being asked to tender - line a. They let their efforts drop once they had been asked to tender - line b.

Supplier C, in contrast, continued with detailed negotiations as long as possible. Their continued emphasis on managing the softer issues (Supplier Rating and User Perception - see Table 3) was the correct strategy at this stage of the decision making process.

As a result, Supplier A lost the tender.

Conclusion

This case study observed a purchase decision unfold over a five month period. It attempted to understand how well potential suppliers understood the needs of their customer, and how the information needs of the buyer vary over the decision making process. What emerges is that in this restricted market of few buyers concentrated in one industry and few able suppliers, there is a relatively good understanding of the buyer's needs by the viable suppliers. It also provides an example of a supplier with a better than average chance of winning the order then losing it. This is due to not fully appreciating how both the nature of the relationship between buyer and seller and the information exchange change over the decision making process. It is not sufficient to understand the stated needs of the buyer and to be able to supply them: the winning supplier also has to understand how the needs of the buyer changes over the decision making process.

Although only a single case study, the results do point to the importance of understanding the marketing process and the associated organisational dynamics. A judgemental modelling package enabled us to study in some detail needs and perceptions of the buyers at a particular point in the decision making process, and also how these were viewed by the competing suppliers. The results suggest that judgemental modelling is useful both as an aid to managerial decision making and as a tool for industrial market researchers.

Appendix 1

Description of the DMU's Attributes

Top Level Attributes:

OpCost	The Operating Costs, including maintenance. This is scored inversely, so that a high score indicates a low Operating Cost
HwOps	The Hardware and Operating System factors
Func	The Functionality of the application software, as described below
SupRat	The users' Rating of the Supplier's ability to provide support and maintenance, and their plans for continuing development
Comms	The ability of the system to Communicate with other systems, particularly the Patient Administration System
UserP	The estimate of how each supplier is Perceived by the general Users within laboratories

The Attributes within Functionality

Plan	The ability of the system to contribute to Planning within the Particular Service
UserF	The User Friendliness of the system
Resour	The ability of the system to contribute to activities associated with Resourcing and maintaining the Particular Service
CurCon	The Current Contents of the system, compared with user requirements
Monit	The ability of the system to assist with Monitoring and Reviewing the Particular Service, including Quality Control
PotDev	The Potential for Development of the system to meet user requirements. This includes assessment of system design
Deliv	The ability of the system to assist with Delivering the Particular Service

BIBLIOGRAPHY

Bonoma, T.V., and Johnston, W.J.
The Social Psychology of Industrial Buying and Selling, Industrial Marketing Management, Vol 17 pp. 213 - 224, 1978

Cunningham, M.T., and Culligan, K.L.
Relationships, Networks and Competitiveness: A Study of the Impact of Electronic Data Interchange Systems, **in** Research in Marketing: An International Perspective, Proceedings of the 5th IMP Conference, D.T. Wilson, S-L Han, and G.W. Holler, eds., September 1989.

Easton, G.
Competition and Marketing Strategy, European Journal of Marketing, Vol 22:2, pp 31-49, 1988

--------,
Industrial Networks - A Review, **in** Research in Marketing: An International Perspective, Proceedings of the 5th IMP Conference, D.T. Wilson, S-L Han, and G.W. Holler, eds., September 1989.

Hakansson, H., and Ostberg, C.
Industrial Marketing: An Organisational Problem?, Industrial Marketing Management, Vol 4, pp 113 - 123, 1975

--------, (Ed)
International Marketing and Purchasing of Industrial Goods, (A Study by the IMP Group), John Wiley And Sons, 1982

-------- and Johanson, J.
Relationships in Industrial Networks, **in** Research in Marketing: An International Perspective, Proceedings of the 5th IMP Conference, D.T. Wilson, S-L Han, and G.W. Holler, eds., September 1989.

Islei, G., and Lockett, A.G.
Judgemental Modelling Based on Geometric Least Square, European Journal of Operations Research, Vol 36 No 1, July, 1988

Korner, E.
Fourth Report to the Secretary of State: A Further Report on the Collection and Use of Information about Activities in Hospitals and the Community in the National Health Services, by The Steering Group on Health Services Information ('The Korner Reports'), HMSO, 1984

Kotler P.
Marketing Management: Analysis, Planning, Implementation, and Control (6th Edition) Englewood Cliffs: Prentice-Hall, 1988

National Health Service
Working for Patients, Working Papers 1-10, HMSO, 1989

Nicosia, F.M., and Wind, Y.
Emerging Models of Organisational Buying Processes, Industrial Marketing Management Vol 6 pp 353 - 369, 1977

Robinson, P.J., Faris, C.W., and Wind, Y.
Industrial Buying and Creative Marketing. Boston: Allyn and Bacon, 1967

Sheth, J.N.
A Model of Industrial Buyer Behaviour, Journal of Marketing, Vol 37 (Oct), pp 50 - 56, 1973

Webster, F.E., and Wind, Y.
A General Model for Understanding Organisational Buying Behaviour, Journal of Marketing, Vol 36 (April), pp 12 - 19, 1972

Wilson, D.T.
Dyadic Interactions, in Consumer and Industrial Buying Behaviour, Edited by Woodside, A.G., Sheth, J.N., and Bennett, P.D., North-Holland, 1977

THE FRENCH AND THE AMERICAN SCHOOL IN MULTI-CRITERIA DECISION ANALYSIS

F.A. Lootsma
Faculty of Mathematics and Informatics
Delft University of Technology
P.O. Box 356, 2600 AJ Delft, The Netherlands

Abstract

A crucial problem in multi-criteria analysis, even when the choice of an alternative has a priori known consequences, is the transition from the objective evaluation of the decision alternatives to the subjective weighing. The French school represented by the ELECTRE systems of Roy, and the American school represented by Saaty's Analytic Hierarchy Process (AHP), incorporate subjective human judgement in different ways. In the present paper we revise the AHP to demonstrate that its potential to rank the alternatives in a decreasing or increasing order of preference is somewhat overestimated. Using the persistent pattern of human comparative judgement in unrelated areas such as history, planning, and psychophysics, we obtain a natural scale to quantify verbal preferential statements. Sensitivity analysis based on a variety of geometric scales shows that the American school, as well as the French school of thinking, may be unable to rank the alternatives unambiguously in a subjective order of preference.

1. Introduction

It is customary in Europe to distinguish a French and an American school in the field of multi-criteria decision analysis. The founding father of the French school is B. Roy who developed a series of ELECTRE methods (see Roy (1985)) and prompted many scientists, mainly in French-speaking regions, to design related methods. The American school is inspired by the work of Keeney and Raiffa (1976) on multi-attribute value functions and multi-attribute utility theory. A popular method, typically fitting into this framework, is the Analytic Hierarchy Process (AHP) of Saaty (1980). Both schools are concerned with the same problem: the evaluation of a finite number of alternatives $A_1, \ldots, A_n$ under a finite number of conflicting criteria $C_1, \ldots, C_m$, by a single decision maker or by a decision-making body. The uncertainty of the future state of the world is supposed to be negligible, so that the choice of an alternative has a priori known consequences.

ELECTRE starts with a pairwise evaluation of the alternatives under each of the criteria separately. Using the physical or monetary values $g_i(A_j)$ and

$g_i(A_k)$ of the respective alternatives A_j and A_k under a measurable criterion C_i, and introducing certain threshold levels for the difference $g_i(A_j) - g_i(A_k)$, the decision maker may declare that he is indifferent between the alternatives under consideration, that he has a weak or a strict preference for one of the two, or that he is unable to express any of these preference relations. If the alternatives are not measurable under C_i, their performance is expressed on a qualitative scale with increasing values such as 1, 2, ..., 10 assigned to the respective echelons; thereafter, threshold levels are introduced and employed in the same way to elicit the required preference information. Both, indifference between A_j and A_k, as well as a weak or a strict preference for A_j, are summarized in the statement that A_j is at least as good as A_k or, equivalently, that A_j outranks A_k. Thus, under each criterion there is a complete or incomplete system of binary relations between the alternatives, the so-called outranking relations. Next, the decision maker is requested to assign weights to the criteria in order to express their relative importance. On the (rare?) occasion that each pair of criteria has a practically constant substitution rate, the information to calculate the weights is obtained by asking the decision maker how much compensation under one criterion he would just need to accept a degradation under the other criterion. Otherwise, each relevant viewpoint in the decision process is heuristically expressed in a trial set of criterion weights (importance factors) to be used for an elaborate sensitivity analysis. Finally, there is an aggregation step. For each pair of alternatives A_j and A_k, ELECTRE calculates the so-called concordance index, roughly defined as the total amount of evidence to support the conclusion that A_j globally outranks A_k, as well as the discordance index, the total amount of counter-evidence. Balancing the two indexes, ELECTRE finally decides whether A_j outranks A_k, whether A_k outranks A_j, or whether there is no global outranking relationship between the two alternatives. Eventually, ELECTRE yields a global system of binary outranking relations between the alternatives. Because the system is not necessarily complete, ELECTRE is sometimes unable to identify the preferred alternative. It only produces then a core of leading alternatives. Moreover, ELECTRE cannot always rank the alternatives completely in a subjective order of preference.

The AHP also starts with a pairwise evaluation of the alternatives under each of the criteria separately. In the basic experiment, where the alternatives A_j and A_k are presented under the criterion C_i, the decision maker is requested to express his indifference between the two, or his weak, definite, strong, or very strong preference for one of them. His verbal judgement (the selected gradation) is subsequently converted into a numerical value $r_{jk}^{(i)}$ on the so-called fundamental scale. Using the matrix $R^{(i)} = \{r_{jk}^{(i)}\}$, the AHP calculates the partial, single-criterion scores $\tilde{v}_i(A_j), j = 1, \ldots, n$, also referred to as the impact scores, approximating the subjective values of the alternatives under criterion C_i. It is worth noting that the partial scores are not unique. Because the ratio $\tilde{v}_i(A_j)/\tilde{v}_i(A_k)$ is defined for each pair (A_j, A_k) of alternatives, the partial scores have a multiplicative degree of freedom. They can accordingly be normalized in such a way that

$$\sum_{j=1}^{n} \tilde{v}_i(A_j) = 1; \; i = 1, \ldots, m. \tag{1}$$

Similar pairwise comparisons and similar calculations yield normalized weights $\tilde{w}(C_i), i = 1, \ldots, m$, for the respective criteria. Finally, there is an aggregation step generating the global, multi-criteria scores $\tilde{f}(A_j)$ via the arithmetic-mean rule

$$\tilde{f}(A_j) = \sum_{i=1}^{m} \tilde{w}(C_i)\tilde{v}_i(A_j); \; j = 1, \ldots, n. \tag{2}$$

By these quantities, usually referred to as the final scores, we have a global order (a global preference structure) defined on the set of alternatives. In the terminology of the American school, the partial and the final scores constitute partial value functions and a global value function respectively.

At first sight, the AHP yields stronger results than ELECTRE. The final scores can be used to identify the preferred alternative, to sort the alternatives into a limited number of categories, to rank the alternatives in a subjective order of preference, and to allocate resources to the respective alternatives on the basis of the relative preferences for them. Sensitivity analysis, however, shows that the rank order of the final scores varies under reasonable deviations from the fundamental scale, so that sorting procedures and resource allocation must be carried out with great care (Lootsma et al. (1990a)). Moreover, decision makers find it difficult to choose one of the verbal qualifications (indifference, weak, definite, strong, or very strong preference) in order to express their relative preference for one of the two alternatives in a pairwise comparison. This is particularly true under a measurable criterion C_i, when the performance of the alternatives A_j and A_k can be expressed in physical or monetary values $g_i(A_j)$ and $g_i(A_k)$. In ELECTRE, the treatment of measurable criteria is definitely more direct and more transparent than in the AHP. It seems to be easier for a decision maker to accept the AHP when the performance of the alternatives cannot be measured (when the colours of the alternatives are compared, for instance, or the design, the elegance, and the style), and when the criteria are compared on the basis of their relative importance in the actual decision problem.

This question leads us straightaway to the heart of the matter in the present paper. The physical or monetary values $g_i(A_j)$ and $g_i(A_k)$ are usually obtained by a more or less objective evaluation of the alternatives, that is, by scientific measurement or by cost calculations. The impact scores and the final scores are due to a subjective weighing of the alternatives, via human judgement expressed in verbal terms. The transition from the objective evaluation to the subjective weighing, as well as the quantification of verbal judgement, are still poorly explained. Saaty (1980) introduced on doubtful grounds a "fundamental scale" and an "eigenvector method" to calculate the impact scores and the criterion weights; his arithmetic-mean aggregation rule (2) does not really apply, because only the ratios $\tilde{v}_i(A_j)/\tilde{v}_i(A_k)$ are properly defined. In recent years, we introduced a class of geometric scales and a geometric-mean aggregation rule (Lootsma (1987)), but in real-life applications we are still not satisfied with the underlying theory: we cannot properly explain it to the decision makers. This prompted us to carry out some additional research on the nature of comparative human judgement.

2. Categorization of a price range

We start with the example which is frequently used to illustrate multi-criteria analysis: the evaluation and the selection of a car. This provides a heuristic introduction to illustrate the transition from car prices to the subjective judgements whereby cars are referred to as "cheap", "somewhat more expensive", "more expensive", or "much more expensive".

Usually, low costs are important for the decision maker so that he carefully considers the consumer price, and possibly the annual expenditures for maintenance and insurance. The consumer price as such, however, cannot tell us whether the car in question would be more or less acceptable to him. That depends on the context, that is, on his spending power and on the alternative cars which he seriously has in mind. In general, there is a minimum price $C_{\min}$ which he is prepared to pay, and a maximum price $C_{\max}$ which he can afford and which he does not really want to exceed. Intuitively, he will subdivide the price range $(C_{\min}, C_{\max})$ into a number of price categories by the introduction of price levels partitioning the range into sub-intervals which are felt to be of the same order of magnitude. We take $e_0, e_1, e_2, \ldots$ to stand for the so-called echelons of the category scale under construction, and $C_{\min} + e_0, C_{\min} + e_1, \ldots$ as the associated price levels. In order to model the requirement that the sub-intervals must subjectively be equal, we recall Weber's law (1834) in psycho-physics, stating that the just noticeable difference Δs of stimulus intensities must be proportional to the actual stimulus level s. Thus, taking here the price increment above $C_{\min}$ as the stimulus intensity, that is, assuming that the decision maker is not really sensitive to the price as such but to the excess above the minimum price $C_{\min}$ which he has to pay anyway, we set

$$e_\delta - e_{\delta-1} = \varepsilon e_{\delta-1} \ , \ \delta = 1, 2, \ldots,$$

which yields

$$e_\delta = (1+\varepsilon)e_{\delta-1} = \ldots = (1+\varepsilon)^\delta e_0.$$

Obviously, the echelons constitute a sequence with geometric progression. The basic step is e_0, and $(1+\varepsilon)$ is the progression factor. The number of categories is rather small, because the linguistic capacity of human beings to describe the categories in verbal terms is limited. We introduce the following qualifications to identify the subsequent price categories:

cheap,
cheap/somewhat more expensive,
somewhat more expensive,
somewhat more/more expensive,
more expensive,
more/much more expensive,
much more expensive.

Thus, we have four major, linguistically distinct categories: cheap, somewhat more, more and much more expensive cars. Moreover, there are three so-called threshold categories between them, which can be used if the decision maker hesitates between the neighbouring qualifications.

The next section will show that human beings follow the same pattern in many unrelated areas when they categorize an interval. They introduce three to five major categories, and the progression factor $(1+\varepsilon)^2$ is roughly 4. By the interpolation of threshold categories, they have a more refined subdivision of the given interval. Then there are six to nine categories, and the progression factor $(1+\varepsilon)$ is roughly 2. In the present section, we will use these results in advance, in order to complete the categorization of a price range. Let us, for instance, take the range between Dfl 20,000 (ECU 9,000) for a modest Renault 5 and Dfl 40,000 (ECU 18,000) for a well-equipped Renault 21 in the Netherlands. The length of the range is Dfl 20,000. Hence, setting the last price level $C_{\min} + e_6$ roughly at $C_{\max}$ we have

$$e_6 = C_{\max} - C_{\min},$$
$$(1+\varepsilon)^6 e_0 = 20,000; \quad 1+\varepsilon = 2,$$
$$e_0 = 20,000/64 \approx 300.$$

It is sometimes more convenient to associate the above-named qualifications, not with the sub-intervals, but with the price levels. Thus, cheap cars are roughly found at the price $C_{\min}+e_0$, somewhat more expensive cars at $C_{\min}+e_2$, etc. This will eventually lead to the following subdivision:

$C_{\min}+e_0$	Dfl 20,300	cheap cars,
$C_{\min}+e_1$	Dfl 20,600	cheap/somewhat more expensive cars,
$C_{\min}+e_2$	Dfl 21,200	somewhat more expensive cars,
$C_{\min}+e_3$	Dfl 22,500	somewhat more/more expensive cars,
$C_{\min}+e_4$	Dfl 25,000	more expensive cars,
$C_{\min}+e_5$	Dfl 30,000	more/much more expensive cars,
$C_{\min}+e_6$	Dfl 40,000	much more expensive cars.

We can now give a more precise interpretation for the qualifications. A somewhat more expensive car has a price increment e_2, which is 4 times the price increment e_0 of a cheap car, etc. We will use this observation to identify the so-called modifiers "somewhat more", "more", and "much more" with ratios 4:1, 16:1, and 64:1 respectively. Note that, by this convention, a car of Dfl 25,000 is somewhat more expensive than a car of Dfl 21,200 because the price increments also have the ratio 4:1. By the same token, a car of Dfl 21,200 is somewhat cheaper than a car of Dfl 25,000.

When the alternative cars are judged under the consumer-price criterion, the target is at the lower end $C_{\min}$ of the interval of possible prices. From this point the decision maker looks at less favourable alternatives. That is the reason why the above categorization, in principle an asymmetric subdivision of the interval under consideration, has an upward orientation. When the cars are judged under the reliability criterion, the orientation is downwards. Numerical data to estimate the reliability are usually available. Consumer organizations collect information about many types and models of cars which follow the prescribed maintenance procedures, and they publish the frequencies of technical failures in the first three or five years. Let us suppose that the decision maker only considers cars with a reliability of at least 95%, so that we are restricted to the interval $(R_{\min}, R_{\max})$ with $R_{\min} = 95$ and $R_{\max} = 100$. Following the mode of operation just described, we obtain the major echelons

$R_{\max}-e_0$	99.9%	reliable cars,
$R_{\max}-e_2$	99.7%	somewhat less reliable cars,
$R_{\max}-e_4$	98.7%	less reliable cars,
$R_{\max}-e_6$	95.0%	much less reliable cars,

because $e_0 = (100-95)/64 \approx 0.08$, with the progression factor such that $e_6/e_0 = 64$. In summary, the alternatives are compared with respect to a certain target. The relative performance is inversely proportional to the distance from the target. The reader can easily verify this in the two examples just given. If we take R_j and R_k to denote the reliability of the alternative cars A_j and A_k, for instance, then the inverse ratio

$$(R_{\max} - R_k)/(R_{\max} - R_j)$$

represents the relative performance of A_j and A_k under the reliability criterion. The qualifications "somewhat cheaper" and "somewhat more reliable" imply that the **inverse ratio** of the distances to the target (the echelons) is 4:1.

3. Category scales in other areas

It is surprising to see how consistently human beings categorize certain intervals of interest in totally unrelated areas.

a. **Historical periods.** The written history of Europe, from 3000 BC until today, is subdivided into a small number of major periods. Looking backwards from 1989, the year when the Berlin Wall was reopened, we distinguish the following turning points marking off the start of a characteristic development:

1947	42 years before 1989	beginning of cold war,
1815	170 years before 1989	beginning of industrial dominance,
1500	500 years before 1989	beginning of world-wide trade,
450	1550 years before 1989	beginning of middle ages,
-3000	5000 years before 1989	beginning of ancient history.

These major echelons, measured by the number of years before 1989, constitute a geometric sequence with the progression factor 3.3. We obtain a more refined subdivision when we introduce the years

1914	75 years before 1989	beginning of world-wars period,
1700	300 years before 1989	modern science established,
1100	900 years before 1989	beginning of high middle ages,
-800	2800 years before 1989	beginning of Greek/Roman history.

With these turning points interpolated between the major ones, we find a geometric sequence of echelons, with progression factor 1.8.

b. **Planning horizons.** In industrial planning, we usually observe a hierarchy of planning cycles where decisions under higher degrees of uncertainty and with more important consequences for the company are prepared at increasingly higher management levels. The planning horizons constitute a geometric sequence, as the following list readily shows:

1 week		weekly production scheduling,
1 month	4 weeks	monthly production scheduling,
4 months	16 weeks	ABC planning of tools and labour,
1 year	52 weeks	capacity adjustment,
4 years	200 weeks	production allocation,
10 years	500 weeks	strategic planning of company structure.

The progression factor of these major horizons is 3.5. In practice, there are no planning horizons between the major ones.

c. **Size of nations.** The above categorization is not only found on the time axis, but also in spatial dimensions, when we categorize the nations on the basis of the size of their population. The major echelons in the list to follow reveal a somewhat European bias:

small nations	4 million	DK, N, GR,
medium-size nations	15 million	NL, DDR,
large nations	60 million	D, F, GB, I,
very large nations	200 million	USA, USSR,
giant nations	1000 million	China, India.

We find again a geometric sequence, with progression factor 4.0. It seems to be reasonable to interpolate the following threshold echelons:

small/medium size	8 million	A, B, H, S,
medium size/large	30 million	E, PL,
large/very large	110 million	Japan.

The refined sequence of echelons has the progression factor 2.0.

d. **Loudness of sounds.** Vigorous research in psychophysics has revealed that there is a functional relationship between the intensity of physical stimuli (sound, light,...) on the one hand and the sensory responses (the subjective estimates of the intensity) on the other. Psychophysics starts from Weber's law (1834), stating that the just noticeable difference Δs of stimulus intensities must be proportional to the actual stimulus level s itself. In Fechner's law (1860), the sensory response $\Delta\Psi$ to a just noticeable difference Δs is supposed to be constant, which implies that $\Delta\Psi$ is proportional to $\Delta s/s$. Integration yields a logarithmic relationship between Ψ and s. Additional experience has finally shown that Fechner's law does in general not hold. Brentano (1874) suggested that the sensory response $\Delta\Psi$ might be proportional to the response level Ψ, so that $\Delta\Psi/\Psi$ would be proportional to $\Delta s/s$. By integration, one obtains that Ψ would be a power function of s. Empirical evidence in many areas of sensory perception prompted Stevens (1957) eventually to postulate the power law as a general psychophysical law. Thus, with s_1 and s_2 representing intensity levels of a particular stimulus such as sound or light, the sensory and the physical intensity ratios are connected by

$$\frac{\Psi(s_1)}{\Psi(s_2)} = \left(\frac{s_1}{s_2}\right)^{\beta}. \tag{3}$$

The exponent β has been established for many sensory systems under precisely defined circumstances. For a 1000 Hz tone it is roughly 0.3. It is customary in acoustics to use a dB-scale for sound intensities. Thus, the intensity s with respect to a reference intensity s_0 is represented by

$$dB(s) = 10 \log(s/s_0).$$

A difference of 10 dB between sound intensities s_1 and s_2 can henceforth be written as

$$dB(s_1) - dB(s_2) = 10,$$

which implies

$$s_1/s_2 = 10,$$

$$\Psi(s_1)/\Psi(s_2) = (s_1/s_2)^{\beta} \approx 2.$$

In other words, by a step of 10 dB the sound intensity is felt to be doubled. The interesting result for our purposes is that the range of audible sounds has roughly been categorized as follows:

40 dB	very quiet; whispering,
60 dB	quiet; conversation,
80 dB	moderately loud; electric mowers and food blenders,
100 dB	very loud; farm tractors and motorcycles,
120 dB	uncomfortably loud; jets during take-off.

Although the precision should be taken with a grain of salt because we have a mixture of sound frequencies at each of these major echelons, we obviously find here a geometric sequence of subjective sound intensities with the progression factor 4.

e. **Brightness of light.** Physically, the perception of light and sound proceed in different ways, but these sensory systems follow the power law with practically the same value of the exponent β. Hence, a step of 10 dB in light intensity is felt to double the subjective brightness. The range of visible light intensities has roughly been categorized as follows:

30 dB	star light,
50 dB	full moon,
70 dB	street lightning,
90 dB	office space lightning,
110 dB	sunlight in summer.

Under the precaution that the precision should not be taken too seriously because we have at each of these major echelons a mixture of wave lengths, we observe that the subjective light intensities also constitute a geometric sequence with the progression factor 4.

4. A natural scale for relative preferences

In a basic experiment of pairwise-comparison methods for multi-criteria analysis, two stimuli S_j and S_k (two alternatives A_j and A_k under a particular criterion, for instance) are presented to the decision maker whereafter he is requested to express his indifference between the two, or his weak, definite, strong, or very strong preference for one of them. We assume that the stimuli have unknown subjective values V_j and V_k for him, inversely proportional to the distances from a certain upper limit of attractiveness. The purpose of the basic experiments and the subsequent analysis is to approximate these values under the assumption that they have been normalized. The verbal comparative judgement, given by the decision maker and converted into a numerical value r_{jk}, is taken to be an estimate of the ratio V_j/V_k. The conversion is based on the results of the preceeding sections, that is, we use a geometric scale to quantify the verbal statements. Such a scale is conveniently characterized by a scale parameter γ, the logarithm of the progression factor $(1+\varepsilon)$. Thus, we set

$$r_{jk} = \exp(\gamma\delta_{jk})$$

where δ_{jk} is an integer designating the gradation of the decision maker's judgement as follows:

0	indifference between S_j and S_k,
+ 2	weak (mild, moderate) preference for S_j versus S_k,
- 2	weak (mild, moderate) preference for S_k versus S_j,
+ 4	strict (definite) preference for S_j versus S_k, etc.

Obviously, weak ("somewhat more") preference for S_j with respect to S_k is converted into $\exp(2\gamma) = (1+\varepsilon)^2$, strict ("more") preference into $\exp(4\gamma) = (1+\varepsilon)^4$, etc. When the progression factor $(1+\varepsilon)$ is set to 2, we have precisely the ratios for comparative judgement announced at the end of section 2. We set δ_{jk} to 1 if the decision maker hesitates between indifference and weak preference for S_j, etc. In summary, we use the even values of δ_{jk} to designate the major echelons (the major gradations) of comparative judgement, and the odd values for the threshold echelons (the threshold gradations).

The results of section 3 prompt us to propose a geometric scale with $(1+\varepsilon) = 2$ and $\gamma = 0.7$ as a natural scale for the quantification of the gradations just mentioned. In earlier experiments, we used a short or normal scale ($\gamma = 0.5$) and a long scale ($\gamma = 1$), for reasons to be explained at the end of this section. Those scales are still recommended for a sensitivity analysis because the refined sequences of major and threshold echelons in section 3 have a progression factor which is **roughly** equal to 2!

We approximate the vector $V = (\ldots, V_j, \ldots, V_k, \ldots)$ of subjective stimulus values via logarithmic regression, that is, we approximate V by the normalized vector $\overline{v}$ which minimizes the expression

$$\Sigma_{j<k}(ln\ r_{jk} - ln\ v_j + ln\ v_k)^2, \tag{4}$$

where the summation is further restricted to the pairs (j, k) judged by the decision maker. He does not really have to consider each pair of stimuli, an advantage which the eigenvector method of Saaty (1980) signally fails to offer. Minimization of (4) is carried out by solving the associated, linear system of normal equations, with variables $w_j = ln\ v_j$. Obviously, the w_j have an additive degree of freedom. The v_j will accordingly have a multiplicative degree of freedom, which is used to single out the normalized vector $\overline{v}$, with components summing up to unity.

By this procedure we calculate stimulus weights for an individual decision maker. We obtain a vector $\overline{v}$ of group weights, possibly a compromise, by minimizing

$$\Sigma_{j<k}\Sigma_{d\in D_{jk}}(ln\ r_{jkd} - ln\ v_j + ln\ v_k)^2, \tag{5}$$

where D_{jk} stands for the set of decision makers who judged the pair (j, k), and r_{jkd} for the estimate of V_j/V_k expressed by decision maker d. We are clearly assuming that the values V_j and V_k of the respective alternatives are the same for all decision makers. We solve the variables $w_j = ln\ v_j$ from the associated, linear system of normal equations, and we use the multiplicative degree of freedom in the v_j to obtain the normalized minimum solution of (5). In earlier papers, we have shown that the rank order of the calculated stimulus weights does not depend on the scale parameter γ. The leading stimulus remains number one.

The above procedure is applied m times to calculate the normalized impact scores $\overline{v}_i(A_j), i = 1, \ldots, m$, of the alternatives $A_j, j = 1, \ldots, n$, and only once to calculate the normalized weights $\overline{w}(C_i)$, $i = 1, \ldots, m$, of the criteria (it will be obvious that the criteria can also be taken to stand for stimuli which are considered in pairs). Thus, each decision maker carries out at most $m[\frac{1}{2}n(n-1)]$ pairwise comparisons to judge the alternatives under the respective criteria, and at most $\frac{1}{2}m(m-1)$ comparisons to assess the criteria themselves. As we have seen, not every possible pair has to be presented to each decision maker, but in order to reduce the notorious inconsistency of human judgement, they should consider as many pairs as possible.

We conclude this section with a note on our earlier choice of a value for the scale parameter γ. In real-life experiments with groups of decision makers (Lootsma et al. (1986)) we used the verbal statements (indifference, weak, strict, strong, very strong preference) in two different ways: (a) we converted them into numerical values on various geometric scales, with trial values assigned to γ, whereafter we applied logarithmic regression (formula (5)) to calculate stimulus weights, and (b) we converted weak, strict, strong, and very strong preference into preference without further gradations, whereafter we calculated the stimulus weights via the method of Bradley and Terry (1952) which does not have a particular scale. For practical purposes, the results of (a) and (b) were sufficiently close when γ varied between 0.5 and 1. The idea is obvious. If we assume that the members of the group are in principle subject to identically distributed perturbations and that they have the same stimulus values in the back of their mind, we may compare the results of (a) and (b) in order to match the scale parameter γ. The analysis of the present paper enables us to choose γ more precisely. We generate the natural scale by setting γ to the value of 0.7. Sensitivity analysis is carried out via the short scale ($\gamma = 0.5$) and the long scale ($\gamma = 1$).

5. Aggregation, final scores, and rank reversal

Aggregation may present unexpected results to the decision makers, when the underlying assumptions are ignored. First, we have to find a common nominator for the operation. We shall be assuming that the decision makers express their relative **preference** for the alternatives, under each of the respective criteria. Thus, they are not supposed to choose the qualification "somewhat cheaper" when they compare cars under the consumer-price criterion, but "weak preference" for one of them, etc. Next, we assume that the expressions such as "weak preference", "strict preference", and "strong preference", correspond to inverse ratios of echelons in a **common** range (D_{min}, D_{max}) on the one-dimensional axis of desirability. The orientation of the categorization is downwards from the maximum desirability D_{max} in the actual decision problem. Thus, if D_j and D_k denote the desirabilities of alternatives A_j and A_k under a given criterion, the inverse ratio

$$V_j/V_k = (D_{\max} - D_k)/(D_{\max} - D_j) \tag{6}$$

models the preference for A_j with respect to A_k. These assumptions enable us to operate with preference ratios. Let us now consider two alternatives A_j and A_k with their calculated profiles, the vectors $\overline{v}_i(A_j), i = 1, \ldots, m$, and $\overline{v}_i(A_k), i = 1, \ldots, m$, respectively. For each i the ratio

$$\overline{v}_i(A_j)/\overline{v}_i(A_k) \tag{7}$$

expressing the relative preference for A_j with respect to A_k under criterion C_i, is unique. Since we are dealing with ratios, it is natural to model the global preference for A_j with respect to A_k by the expression

$$\Pi_{i=1}^{m}(\overline{v}_i(A_j)/\overline{v}_i(A_k))^{\overline{c}_i}, \tag{8}$$

where $\overline{c}_i$ simply denotes the calculated weight $\overline{w}(C_i)$ of the i-th criterion. In an attempt to express the global preferences for the respective alternatives by final scores $\overline{f}(A_j)$ and $\overline{f}(A_k)$, we set the ratio

$$\overline{f}(A_j)/\overline{f}(A_k) \tag{9}$$

to (8), whence

$$\overline{f}(A_j) = \Pi_{i=1}^{m}(\overline{v}_i(A_j))^{\overline{c}_i}. \tag{10}$$

The final scores have a multiplicative degree of freedom. They can accordingly be normalized to sum up to unity.

The geometric-mean aggregation rule (10) has the interesting property of "infinite compensation for zero preference". Suppose that the decision maker is indifferent between A_j and A_k, which implies that the ratio (8) is roughly equal to 1. Imagine now that A_j remains fixed but that A_k can be varied continuously. If we take one of the impact scores in the profile of A_k to converge to 0, then indifference between A_j and A_k can only be maintained if at least one of the remaining scores of A_k goes to infinity. It is easy to verify that the arithmetic-mean aggregation rule (2) has "finite compensation for zero preference" only. Hence, the geometric-mean rule does not make it urgent to introduce a veto mechanism as in ELECTRE, which rules out certain alternatives with an extremely poor performance under one of the criteria. Note that formula (8) is based on ratios which do not depend on the units of measurement! It would be difficult to measure preference as such, but in the present method this is not a point of major concern. The ratios are typically dimensionless quantities.

6. Final remarks

Table 1 shows how the weights and the rank order of the criteria as well as the final scores and the rank order of the alternatives, in a project reported by Lootsma et al. (1990a), vary with the scale parameter γ. The scale variation is considerable, but the sensitivity of the weights and scores remains within reasonable limits. The rank order of the final scores happens to be stable. When rank reversal occurs in a real-life project, by a sensitivity analysis which shows the weights and scores on the natural scale ($\gamma = 0.7$) and on two neighbouring scales ($\gamma = 0.5$ and $\gamma = 1.0$), one has to warn the decision maker that the rank order of the alternatives has not been established beyond reasonable doubt (Lootsma et al. (1990b)). In doing so, we are close to the results of the French school of thinking in multi-criteria analysis, which is not always able to rank the alternatives completely in a subjective order of preference. It is important to realize this, because rank reversal is a frequently occurring phenomenon when we vary the numerical scale for comparative judgement.

References

[1] Bradley, R.A., and Terry, M.E., "The Rank Analysis of Incomplete Block Designs. I. The Method of Paired Comparisons". Biometrika 39, 324-345, 1952.

[2] Keeney, R., and Raiffa, H., "Decisions with Multiple Objectives: Preferences and Value Tradeoffs". Wiley, New York, 1976.

[3] Lootsma, F.A., "Modélisation du Jugement Humain dans l'Analyse Multicritère au moyen de Comparaisons par Paires". R.A.I.R.O./Recherche Opérationelle 21, 241-257, 1987.

[4] Lootsma, F.A., Meisner, J., and Schellemans, F., "Multi-Criteria Decision Analysis as an Aid to the Strategic Planning of Energy R&D". European Journal of Operational Research 25, 216-234, 1986.

[5] Lootsma, F.A., Mensch, T.C.A., and Vos, F.A., "Multi-Criteria Analysis and Budget Reallocation in Long-Term Research Planning". European Journal of Operational Research 47, 293-305, 1990a.

[6] Lootsma, F.A., Boonekamp, P.G.M., Cooke, R.M., and van Oostvoorn, F., "Choice of a Long-Term Strategy for the National Electricity Supply via Scenario Analysis and Multi-Criteria Analysis". European Journal of Operational Research 48, 189-203, 1990b.

[7] Roy, B., "Méthodologie Multicritère d'Aide à la Décision". Collection Gestion, Economica, Paris, 1985.

[8] Saaty, Th.L., "The Analytic Hierarchy Process, Planning, Priority Setting, Resource Allocation". McGraw-Hill, New York, 1980.

[9] Stevens, S.S., "On the Psychophysical Law". Psychological Review 64, 153-181, 1957.

Table 1. Weights and rank order of the criteria, final scores and rank order of the alternative energy research programs in a budget-reallocation study. Comparative human judgement has been encoded on a short scale ($\gamma = 0.5$), a natural scale ($\gamma = 0.7$), and a long scale ($\gamma = 1.0$) to show the scale sensitivity of the final scores and their rank order.

		γ=0.5		γ=0.7		γ=1.0	
Scale values	Indifference	1.0		1.0		1.0	
	Weak Preference	2.7		4.0		7.4	
	Strict Preference	7.4		16.0		54.6	
	Strong Preference	20.1		64.0		403.4	
Weights and rank order of criteria	Security of Energy Supply	26.1	2	27.7	2	29.4	2
	Energy Efficiency	13.3	4	10.8	4	7.7	4
	Long-Term Contribution	27.7	1	30.2	1	33.2	1
	Environmental Protection	25.1	3	26.2	3	27.2	3
	Suitability for Comm.Action	7.8	5	5.1	5	2.6	5
Final scores and rank order of alternatives	Photovoltaic Solar Energy	11.8	3	12.6	3	13.7	3
	Passive Solar Energy	6.9	9	6.0	9	4.7	9
	Geothermal Energy	8.7	7	8.0	7	7.0	7
	Advanced Energy Saving	13.2	2	14.1	2	15.4	2
	Saving in Industry	13.6	1	15.3	1	18.0	1
	Hydrocarbons	9.1	6	8.5	6	7.6	6
	New Energy Vectors	7.7	8	6.8	8	5.5	8
	Biomass Energy	5.6	10	4.4	10	3.0	10
	Solid Fuels	11.7	5	12.0	5	12.0	5
	Wind Energy	11.8	4	12.4	4	13.1	4

MULTICRITERION AID TO MAKE A SCHEDULE OF A SURGICAL SUITE

Jean-M. MARTEL
Université Laval
Business School
Sainte-Foy (Québec)
G1R 7P4 Canada

Danièle THOMASSIN
Case Western Reserve University
Weatherhead School of Management
Mids Department
Cleveland, Ohio 44106

ABSTRACT

The managing of health network activities is becoming more and more important and is essential within most communities and/or societies. Hospitals play a vital role within this network, as much by the quantity of their services as by their quality. Every institution has to efficiently manage its own resources, especially those related to operating room activities.

The particular character of a surgical suite renders classical scheduling methods inapplicable. This problem is essentially of a multidimensional nature and should be treated with a multicriterion approach. An ELECTRE type method is used to rank the surgical operations according to their priority, taking into account the available resources, the surgeons' preferences, the medical requirements of every patient and several characteristics of each operation.

1. INTRODUCTION

Health system management is essential within any community, insofar as the quality of life depends upon individuals who make it up. Such a system fulfills its functions through a variety of intermediary contributors, e.g. hospitals, medical clinics, psychiatric institutions,... each in charge of certain aspects of a health program. Among these intermediaries, hospitals are of vital importance, as much because of the quantity than because of the diversity of the needs which they address. They possess a sizeable infrastructure in terms of rooms, equipment, personnel, and so on, in order to satisfy the medical needs of a large population at a time and in a society more and more conscious of the meaningful saying: a healthy body makes a healthy mind.

Each hospital must efficiently manage its resources in order to offer high quality care at the lowest possible price to a maximum number of patients. However, these noble objectives conflict and the search for an acceptable compromise has been the object of many debates over the last few decades. Recognizing the importance of these debates, the objective of our research is to propose a way of improving the functioning and performance of a hospital's nerve center, that of its surgical suite.

The surgical suite consists of a series of rooms and equipment where specialized personnel have the great responsibility of carrying out surgical operations. Its performance not only has a direct impact on the quality of the care that is given but also upon the performance of other functions within the hospital. Poor planning of operations can lead to delays and eventual postponements concerning the initial plan, causing, among other things: patient discomfort due to either a prolonged or a repeated period of fasting; an undue prolongation of certain patients' stay at the hospital (undue because it is not caused by medical complications), a stay which is sought to be minimized as much for the well-being of the patient as for the costs it generates; an increase in admission delays of patients waiting for hospitalization which is even harder on those patients who are immobilized in corridors of emergency rooms; serious complications for patients admitted for one-day operations who often have only been given a short period of time off by their employers thus rendering the postponements of their operations litigious; or an increase in the functioning cost of the surgical suite, if the decision is made to extend the opening hours in order to catch up, entirely or in part, on the delays.

These few problems alone are enough to justify the importance of adequately running the activities of a surgical suite as much for the unjustified costs that are incurred by poor planning as for the trouble and medical complications that may result therefrom.

Unfortunately it is almost impossible to conceive of a caseload schedule which would use the rooms 100% of the time (taking the cleaning periods into account); which would perfectly foresee the length of the operations and the slightest incident as well as emergencies; which would respect the surgeons' preferences and the patients' expectations (taking the material limitation of the room into account) in order to schedule the patients' operations where and when it is most convenient for them; and finally, which would entirely take into account the particular medical requirements of each patient.

Although this may prove to be realistic if there are only a few patients to schedule, difficulties become increasingly present as the number of patients increase. Various conflicts and unforeseen circumstances begin to appear such that it becomes simply impossible to satisfy all the demands. Often, the best one can hope for is to be able to anticipate a time segment for each operation. Even this more modest objective is generally unattainable as the demand for operations regularly exceeds the maximal capacity of the multiple operating room system. Choices must therefore be made between operations presenting various characteristics, at times quite similar, but at times very different, if not incomparable.

The environment in which the program must be applied becomes complicated mainly because of the uncertainty as to the length of the operations and the emergencies. Even under normal conditions the duration of an operation cannot be anticipated with certainty. It is therefore easy to imagine that when complications do occur, our estimations (already imprecise) become completely erroneous, upsetting the schedule of the operating program. Moreover, emergency operations occur quite frequently, sometimes completely upsetting the established schedule. The basic problematic is therefore reinforced by the unpredictable nature of the environment in which it is considered.

The making of a surgical suite is more than a typical application of a general model of a schedule. In fact, the numerous socio-economic impacts (as conflicting as they generally are) as well as the highly unpredictable character of certain essential parameters, amply justify its particular character and, consequently, the interest and importance attributed to it.

The paper is structured in the following manner: the problem is formulated in section 2; section 3 presents the choice of an appropriate multicriterion method. This method is applied to solve a case using the date of a large hospital from the province of Québec's region.

2. PROBLEM FORMULATION

We have so far introduced the main characteristics and difficulties of the problem of making of a suite of surgical theatres. However, a more formal definition turns out to be necessary if one wants to arrive at a concrete method of helping create a schedule suitable to the reality of the running of a surgical suites operations.

Eventhough, as we have just mentioned, our decision situation distinguishes itself from classic models of schedule making, we can nonetheless inspire ourselves from these models. According to Tersine [1980] "scheduling" can be defined as the managing of the activities of a system in order to regulate the flow of "incoming" and "outgoing" and this by determining when the so called activities are to take place. Following an equivalent formulation, our problem can be expressed as follows:

determine in which room and at what time to schedule surgical operations, taking into account the availability of personnel and rooms, the physical constraints of the rooms, the surgeons' preferences and the medical requirements of each patient in order to maximize the percentage of use of the rooms and to minimize delays and postponements.

Several hospitals run their surgical suite with the help of a block-booking system (Morgan, 1973). According to such a system, each specialization is assigned, on a weekly basis (Monday to Friday from 8:00 a.m.- 4:00 p.m.), a time slot and an exclusive corresponding room. An operating room priority list is thus defined on which one can see, for each day of the week, the specialization which is allocated to each room and for which time period. The respective specialized surgeons who will be practicing during this reserved time period are also listed. Surgeons of a same specialization determine amongst themselves the time period(s) to which each has a right, taking into account what is globally allocated to their specialization. It is in this way that they will determine for each of the period a list of surgeons going from the one with the highest priority to the one with the lowest priority. This list is brought up to date at the frequence of about once a year.

Thus, every day of the week, the surgeons who have a priority to fulfill the next day send the operating requisitions which they wish to carry out to the operating system (*i.e.*: before 3:00 p.m.). Once all the requisitions are handed in, the schedule is then made up. The scheduling of the requisitions depends on both medical and organizational criteria as well as pressure by certain surgeons who wish to favour a few of their cases.

In this research, we have not questioned the block-booking process as much because of its easy application than because of its strongly institutionalized character. Rather, we have looked for a way of improving the scheduling process itself, by looking at the environment in which it takes place, the physical support used for its putting into action, and the rules upon which it is based.

In such a system, the priority operations are first scheduled. The scheduling of the remaining operations is then done according to the importance level of these operations. The order of the operations to be carried out depends upon various constraints and criteria which may well be different from one hospital to the next. For the studied hospital the various constraints of the system are in order of diminishing importance:

- the absolute priority given to infants, the elderly and diabetic patients;
- the surgeons' operating priorities;

- the priority given to patients with one-day operations

Operations which are not subject to these constraints are then ranked according to several importance criteria. It is precisely this form of classification that multicriterion aid is applied.

The A.A.E. model (Actions, Attributes, Evaluations) seems to be a natural way to formalize a multicriterion decision. The formulation of the problem is defined in three parts: all of the potential altervatives (a_i, $i = 1,2,\ldots,m$), all of the attributes or criteria (c_j, $j = 1,2,\ldots,n$) and all of the performance evaluations of each of the actions according to each of the criteria (e_{ij}).

In the case with which we are concerned, the potential alternatives are none other than all of the operations to be scheduled for a given day and which are not subject to the preceding constraints. For the studied hospital, the criteria retained are:

- the number of postponements of an operation (c_1);
- the age of the patient (c_2);
- the type of anesthesia (c_3);
- the cleanliness of the case (c_4);
- the expected operation time (c_5).

Once each operation is evaluated according to its importance on each criterion, a performance matrix is obtained (table 1).

TABLE 1. PERFORMANCE MATRIX

Criteria / alternatives	c_1 ----	c_j ----	c_n
a_1		¦	
¦			
a_i	----	e_{ij}	----
¦		¦	
a_m			

Since many criteria are of a qualitative nature, the measuring scale associated to them is ordinal. For example, the criterion "type of anesthesia", can take two modalities: general or local. All else being equal, a higher priority will be given to an operation needing a general anesthesia rather than one involving a local anesthesia.

The classification method used must not only be able to qualitatively treat the expressed importances and take into account the uncertainty surrounding certain evaluations, but it must also take into account the relative importance allocated to each criterion.

3. THE CHOICE OF A MULTICRITERION METHOD

We are confronted with multiple objectives/criteria formulation. Yet, the vast majority of the scheduling methods are essentially of a monocriterion nature. To bring us to this form two approaches are possible: reformulate certain objectives as a supplementary constraint to consider; or, look for a common unit of measurement and express each objective according to this unit in order to obtain only one expression to be optimized. However, both of these approaches are quite disputable. 1) The first approach more or less changes the philosophy of the problem, given that a constraint represents (in any decision problem) a perspective quite different than an objective to be attained. Moreover, this formulation does not enable us to represent an order of importance, a preference to see certain constraints satisfied at the expense of others. On the contrary, a solution will only be considered acceptable if all the constraints, which it is subject to, are entirely and totally respected. 2) The second approach, besides from the fact that it is generally quite difficult to put into practice, yields a risk of leading to the aggregation of incommensurable units.

Almost all of the attempts in multiobjective programming initially starting from the maximization of several objectives (Arthur and Ravindran [1982], Deckro *et al.* [1982], Koelling and Bailey [1984], inevitably come down to the creation of an equivalent monocriterion program. This is obtained by explicitly introduce a value function on all the objectives. That is how Evans [1984], in an article on various existing multiobjective techniques, specifies that "any multiobjective program can be restructured into a problem containing only one objective-function".

Thus, despite the significance of these methods and their ability to generate so called optimal schedules, it seems that their use (at least in our particular context) is not suitable. In fact, the choice of the best surgical suite lies rather in the search of an acceptable compromise as far as the degree of satisfaction of conflicting multiple objectives is concerned.

To adopt a monocriterion approach is to stick to (explicitely or not) the vision according to which there is a general rule, an essential guiding principle which determines the right direction in which it is suitable to develop the system in which we are interested. On the other hand, the multicriterion paradigm acknowledges the importance of several conflicting guiding criteria which compels the search of a compromise and of a given state of equilibrium. There is therefore no more optimum to obtain, but rather compromises to find an arbitraments to invent (Roy [1987]). This latest vision of things seems to us to be closest to the surgical suite scheduling reality.

There is a great variety of multicriterion methods (Scharlig [1985], Vincke [1989]) which can be classified in categories according to their preconized approach of aggregation. The first category regroups methods based on one single synthesis criterion (*e.g.* multiple attribute utility theory, goal programming), which are essentially compensatory methods. The biggest difficulty in using these methods is to determine the value functions (or utility functions) required in aggregation, and this is particularly true in context of our study.

It seems, according to our case study, that the second category of approach, the synthesis outranking (Roy [1985]), is more appropriate. In this approach, each pair $(a_i, a_i') \in A \times A$ is compared in order to determine whether or not "a_i is at least as good as a_i'". Although no method is perfect, the ELECTRE methods, which belonging to this second category, seem highly appropriate to our situation. With the ELECTRE methods it is possible to deal by the ordinal data and to conclude at the incomparability between two actions. Particularly, ELECTRE II (Roy and Bertier [1971]) has been developed to deal with a ranking problematic and it is rather simple and easy to understand as well as to apply. Thus, it will accommodate the type of data available in our study.

The basic principal behind the ELECTRE methods is that one alternative outranks another if it is at least as good as than the other according to the majority of criteria without being too bad on the other criteria. Thus the outranking must verify two conditions: one concordance and one non-discordance conditions. A concordance index is built in the following way:

$$C(a_i, a_i') = \sum_{j \in Ca_i, a_i'} \Pi_j$$

where $Ca_i, a_i' = \{\text{criteria } j \,/\, e_{ij} \geq e_{ij'}\}$ and Π_j represents the relative importance of criterion j, with $\sum_{j=1}^{n} \Pi_j = 1$.

Likewise two sets of discordance indexes are defined according to the scales used for each criterion.

4. ILLUSTRATIVE CASE

In this research, we try to improve the scheduling process itself, that is, to better understand the environment in which it proceeds, to clearly define the rules it obeys by and to support it with an appropriate methodological tool. It seems important to note that what is reported in this paper is only a part of a much global effort, in which attention was also given to matters such as the precision of the data - especially regarding the estimated operating times and number of emergencies -, the automatic updating of the schedule upon the arrival of unpredicted events and the creation of a flexible three days schedule.

Eliciting the scheduling rules required multiple meetings with the chief nurse, surgeons and the administrative director of the hospital. The following rules were finally agreed upon:

1. Upmost priority (*i.e* even before surgeons, operating priorities) is given to babies, the elderly and patients with diabetes;
2. Respect of the surgeons' operating priorities;
3. Postponement of one-day surgeries (*i.e.* which does not necessitate an hospitalization) must be avoided;
4. The priority of an operation increases each time it is postponed(c_1);
5. Children of 12 years or less are given priority (c_2);
6. Surgeries necessitating general anesthesia are scheduled before the ones involving local anesthesia (c_3);
7. The cleanest operations are performed before the more dirty ones (c_4); and,
8. The priority on an operation is a direct function of this estimated operating time (c_5).

From these rules, we then developed the following procedure: the babies (*i.e.* less than 12 pounds), the elderly (*i.e.* more than 70 years old) and the patients with diabetes are scheduled first, in any available and adequate room; then, with respect to the surgeons' operating priorities, one-day surgeries are scheduled; finally, the other operations are scheduled, according to the following criteria: 1) number of postponements; 2) patient's age; 3) type of anesthesia; 4) cleanliness of the operation; and, 5) the estimated operating time. The weights - or relative importance - associated with each criterion were obtained, again through intensive discussion with the involved actors. They are, in the order of presentation of the criteria, 0.25, 0.20, 0.10, 0.25 and 0.20 respectively. We therefore have a hierarchy of criteria, where number of postponements and cleanliness are the most important one, followed by the patient's age and the operating time and, last, the type of anesthesia.

Then, the scale of each criterion had to be assessed. While the number of postponements can take any positive integer value, its most probable values are 0, 1 and 2. The patient's age and the type of anesthesia are both dichotomic criteria. That is, a patient's age is either more than 12 years or less, and the anesthesia is either general or local. The level of cleanliness criterion can be any one of the following: clean, clean in contaminated territory, contaminated and infected. Finally, the operating time criterion can take any real positive value. However, any surgery rarely requires less than 0.5 hour or more than 5 or 6 hours, with an average of about 2 hours.

Given the small number of modalities for each criterion, and their respective discriminatory power, we have fixed only one discordance threshold for each criterion expected for the estimated operating time which can, a priori, take an infinite number of values. These discordance thresholds were set with respect to the scale of each criterion. They represent the difference starting from which there is too much opposition to the outranking of an operation a_i over an operation a_i' which would otherwise (*i.e.* over the other criteria) be preferable. For the number of postponements, a difference of 2 was judged large enough to stop the outranking of an operation otherwise more important. In the cleanliness case, only the modality infected was judged severe enough to block an outranking. Values of 90 and 120 minutes were selected as the discordance thresholds for the operating time criterion. Finally, no discordance threshold was set for the dichotomic criteria age and anesthesia. That is, these criteria participate in determining the priority of an operation over another only via the concordance condition.

Finally, we set values for the concordance thresholds, which establish the values according to which an operation has priority over another. The thresholds have been fixed at 0.80, 0.70 and 0.60 respectively and bear the following implications: at the most demanding level (*i.e.* C1 = 0.80), at least four criteria, including the two most important one (*i.e.* number of postponements and cleanliness), must support the outranking operation for it to pass the concordance test: at the C2 = 0.70 level, at least one of the most important criteria must favor the outranking operation; finally, at the least demanding level (*i.e.* C3 = 0.60), one of the most important criteria only can support the outranking operation, as long as the operation is favored by the next two most important criteria (*i.e.* patient's age and operating time). There therefore is a logical progression from the most to the least demanding threshold, in order to make sure that only the most significant outranking will be identified.

We now have all the values and parameters needed to apply ELECTRE II. To do so, we created a data set that would lead us to test the whole spectrum of thresholds, that is, using the following values: 0 and 2 postponements, patients of either 10 or 30 years old, both types of anesthesia, clean and infected operations and estimated operating times of 0h30, 2h00 or 2h30, for a total combinations set of 48 (2x2x2x2x3) operations. A table of these 48 combinations is presented in Appendix and the results of the scheduling is presented in table 2.

The results of ELECTRE II's classification show that the operations have been prioritized, with some of them on the same level. In order to see how this classification could be utilized, let us consider a very simple example in which we suppose that we have three operating rooms' with the following list of priorities:

- Room 1: Speciality 1 (s1) from 8h00 to 12h00
  Speciality 2 (s2) from 12h00 to 16h00
- Room 2: Speciality 3 (s3) from 8h00 to 12h00
  Speciality 4 (s4) from 12h00 to 16h00
- Room 3: Speciality 5 (s5) from 8h00 to 12h00
  Speciality 6 (s6) from 12h00 to 16h00

Similarly, the relationship between each operation and its corresponding speciality is given in table 3.

TABLE 2: ELECTRE II CLASSIFICATION

COMBIN.#	FINAL	DIRECT	INVERSE
1	1	1	1
7	2	2	2
2	3	3	3
13	3	3	3
19	3	3	3
14	3	3	3
3	3	3	3
9	3	3	3
20	3	3	3
8	3	3	3
25	4	5	4
15	5	4	6
4	6	7	5
21	7	6	8
31	8	8	7
10	9	9	8
5	10	10	10
16	10	10	10
22	10	10	10
17	10	10	10
6	10	10	10
12	10	10	10
23	10	10	10
11	10	10	10
26	10	10	10
37	10	10	10
43	10	10	10
38	10	10	10
27	10	10	10
33	10	10	10
44	10	10	10
32	10	10	10
39	12	12	12
18	13	14	13
28	13	12	15
45	15	16	14
34	16	15	17
24	17	17	16
29	18	18	18
40	18	18	18
46	18	18	18
41	18	18	18
30	18	18	18
36	18	18	18
47	18	18	18
35	18	18	18
42	19	19	19
48	20	20	20

TABLE 3: EXAMPLE

COMBIN.#	SPECIALITY	COMBIN.#	SPECIALITY
1	S1	24	S6
2	S4	25	S1
3	S1	26	S2
5	S5	28	S6
6	S1	31	S2
7	S3	33	S2
10	S6	34	S5
12	S5	35	S5
13	S3	37	S2
15	S2	39	S6
16	S4	41	S5
18	S6	43	S4
20	S3	44	S4
21	S3	45	S3
23	S1	48	S4

The operations under speciality 1 are 1, 3, 25, 6-23, with respective times of 2h30, 0h30, 2h30, 0h30 and 2h00, for a total estimated operating time of 8 hours. Given that only 4 hours are available, operations 1 and 3 will first be scheduled. The next most important operation is number 25, but there is absolutely no way that it can be scheduled within the still available 1h00 time period. Operation 6 would then probably be preferred.

For speciality 2, the list of operations, by order of importance, is 15, 31, 26-33-37. Their respective estimated operating time are 0h30, 2h30, 2h00, 0h30 and 2h30, for a total of 8 hours. Operations 15 and 31 will therefore be scheduled. Given the time still available and the estimated time of the remaining operations - all ex-eaquo -, operation 33 should then be chosen, yielding a total operating time of 3h30. The same situation holds for speciality 6, for which operations 10, 39 and 18 will be scheduled, for a total estimated operating time of 3h30.

Specialities 3, 4 and 5 yield to a problem similar to the one encountered for speciality 1, that is: if we strictly follow the order given by ELECTRE II, we will be left with an open time slot in which the next most important operation, in terms of priority, can not be scheduled. We therefore have a choice to make, as to whether the operating suite hours should be extended, and for which room, so as to accommodate some or all of these operations, or t instead schedule less important operations that would, however, fill the day's schedule.

It should be clear from this example that the objective is not to force a classification, but rather to make suggestions, to support the decision process. As such, the proposed classification can and should be analyzed judiciously, in each case, and possibly supplemented by ad hoc rules. However, since the suggested classification rest on the application of rules that were explicitly stated accepted, it provides a formal baseline against which each scheduling decision can now be judged.

REFERENCES

Arthur, J.L. and A. Ravindran, "A Multiple Objective Nurse Scheduling Model" IEE Trans., Vol. 13, 1, 1981 (55-60).

Deckro, R.F., J.E. Hebert and E.P. Winkofshy, "Multiple Criteria Job-Shop Scheduling", Computers & Operations Research, Vol. 9, 1982 (279-85).

Evans, G.W., "An Overview of Techniques for Solving Multiobjective Mathematical Program", Management Science, Vol. 30, no. 11, November 1984, pp. 1268-1282.

Koelling, C. and J.E. Bailey, "A Multiple Criteria Decision Aid for Personnel Scheduling", IEE Trans., Vol. 16, 4, 1984 (299-307).

Morgan, D.W., "Improved Scheduling Through Block Booking", Canadian Hospital, Vol. 50, 1973 (41-55).

Roy, B. & P. Bertier, "La méthode ELECTRE II: une méthode de classement en présence de critères multiples", note de travail no. 142, Direction scientifique, Groupe METRA, 1971.

Roy, B., Méthodologie multicritère d'aide à la décision, Economica, Paris, 1985.

Roy, B., "Des critères multiples en recherche opérationnelle: Pourquoi?", dans Cahier du Lamsade, no. 80, Sept. 1987, Université Paris-Dauphine.

Sharling, A., Décider sur plusieurs critères - Panorama de l'aide à la décision multicritère, Presses Polytechniques Romandes, Lauzanne, Suisse, 1985.

Tersine, R.J., Production/Operations Management: Concepts, Structure & Analysis, North Holland, 1980.

Vincke, P., L'aide multicritère à la décision, Editions de l'Université de Bruxelles, Bruxelles, 1989.

APPENDIX: THE SET OF 48 OPERATIONS

POSSIBLE COMBINATIONS					
COMBIN.#	# POSTP. (0.25)	AGE (0.20)	ANESTH. (0.10)	CLEANL. (0.25)	TIME (0.20)
1	2	≤12	general	clean	150
2	2	≤12	general	clean	120
3	2	≤12	general	clean	30
4	2	≤12	general	infected	150
5	2	≤12	general	infected	120
6	2	≤12	general	infected	30
7	2	≤12	local	clean	150
8	2	≤12	local	clean	120
9	2	≤12	local	clean	30
10	2	≤12	local	infected	150
11	2	≤12	local	infected	120
12	2	≤12	local	infected	30
13	2	>12	general	clean	150
14	2	>12	general	clean	120
15	2	>12	general	clean	30
16	2	>12	general	infected	150
17	2	>12	general	infected	120
18	2	>12	general	infected	30
19	2	>12	local	clean	150
20	2	>12	local	clean	120
21	2	>12	local	clean	30
22	2	>12	local	infected	150
23	2	>12	local	infected	120
24	2	>12	local	infected	30
25	0	≤12	general	clean	150
26	0	≤12	general	clean	120
27	0	≤12	general	clean	30
28	0	≤12	general	infected	150
29	0	≤12	general	infected	120
30	0	≤12	general	infected	30
31	0	≤12	local	clean	150
32	0	≤12	local	clean	120
33	0	≤12	local	clean	30
34	0	≤12	local	infected	150
35	0	≤12	local	infected	120
36	0	≤12	local	infected	30
37	0	>12	general	clean	150
38	0	>12	general	clean	120
39	0	>12	general	clean	30
40	0	>12	general	infected	150
41	0	>12	general	infected	120
42	0	>12	general	infected	30
43	0	>12	local	clean	150
44	0	>12	local	clean	120
45	0	>12	local	clean	30
46	0	>12	local	infected	150
47	0	>12	local	infected	120
48	0	>12	local	infected	30

A FUZZY/STOCHASTIC MULTIOBJECTIVE LINEAR PROGRAMMING METHOD

Raymond NADEAU , professor
Sciences de l'administration, Université Laval, Québec, CANADA, G1K 7P4

Bruno URLI , professor
Université du Québec à Rimouski, Rimouski, Québec, CANADA, G5L 3A1

Abstract. In the context of multiobjective linear programming (MOLP) problems where there is indetermination around some parameters, we suppose that the decision maker knows only the limits of variation of these parameters and eventually one of their central values. For such situations, we propose a new MOLP method which is inspired by fuzzy and stochastic linear programming methodologies. By means of adequate transformations, the fuzzy/stochastic MOLP problem is transformed into a deterministic one. The latter program is then solved by an interactive approach derived from the STEM method. Our methodology is illustrated by a didactital example.

1. Introduction

In many real situations which can be modelled by multiobjective linear programming (MOLP) techniques, there is indetermination around some parameters. As Roy (1987) explains, there are many kinds of imprecision, uncertainty and inaccurate determination. If this indetermination is caused mainly by external factors which are out of the control of the decision maker (D.M.), it is considered as uncertainty and usually modelled by probability distributions; in the context of MOLP problems, this approach was used, among others, by Goicoechea and al (1982) in the PROTRADE method and by Teghem and al (1986) in the STRANGE method. But the indetermination around the parameters of problems can also express inaccurate determination, imprecision, i.e. be the result of inexact measurement of data. Then this imprecision is generally modelled by fuzzy numbers; in the context of MOLP problems, this approach has been proposed, for example, by Slovinski (1986) in the FLIP method.

In reality, these two kinds of indetermination are often present simultaneously in the same problem. So, in strategic planning, for example, if one has to evaluate the cost of a high-voltage line for a hydroelectric power plant in the year 2000, there is obviously uncertainty but also imprecision about this cost. In this paper, we deal with problems where those two sources of undetermination are present together. More precisely, we suppose that the D.M. is in a situation of incomplete information about some parameters: for those parameters, he knows only their limits of variation and eventually one of their central values. For this fuzzy/stochastic problem, we propose a MOLP methodology which has recourse neither to probability distributions nor to fuzzy numbers. At first, by an approach inspired by goal programming and fuzzy numbers respectively, the fuzzy/stochastic objective functions and the fuzzy/stochastic constraints are transformed into deterministic ones. The obtained deterministic equivalent MOLP problem is then solved by an interactive procedure derived from the STEM method of Benayoun and al.(1971). Finally, our methodology will be illustrated by a didactic example.

2. The problem and a general idea of the proposed algorithm

We deal with the following general fuzzy/stochastic MOLP problem:

$$\begin{aligned} \text{Max} \quad & Z_k(x) = \Sigma_{j=1}^{n} c_{kj}.x_j \\ \text{s.t.} \quad & \Sigma_{j=1}^{n} a_{ij}.x_j \leq b_i \ , \ x_j \geq 0 \ , \ i=1,...,m \ ; \ j=1,...,n \ ; \ k=1,...,K, \end{aligned} \tag{1}$$

where $x \in R^n$ and the parameters c_{kj} , a_{ij} and b_i are non-deterministic variables for which the D.M. possesses only the following incomplete information:

- for each coefficient c_{kj}, the limits of variation c_{kj}^- and c_{kj}^+ and eventually a central value c_{kj}^o are known;
- for each coefficient a_{ij} and b_i, the limits of variation a_{ij}^-, a_{ij}^+ and b_i^-, b_i^+ are known.

In order to solve (1), we have developed a new interactive fuzzy/stochastic MOLP method. The general idea of our method can be summed up as illustrated in Figure 1. First, from the fuzzy/stochastic multiobjective program (1), we obtain a multiobjective deterministic program by appropriated transformations of the non-deterministic objective functions and constraints. Then we propose an interactive algorithm to solve this deterministic program. The first phase of this algorithm consists of obtaining a first compromise. Afterwards, through interactives phases, new compromises are presented to the D.M. until he finally considers one to be satisfactory. The principal steps of our algorithm are summed up in the schema of Figure 1: each one of those steps is explained with more details in what follows.

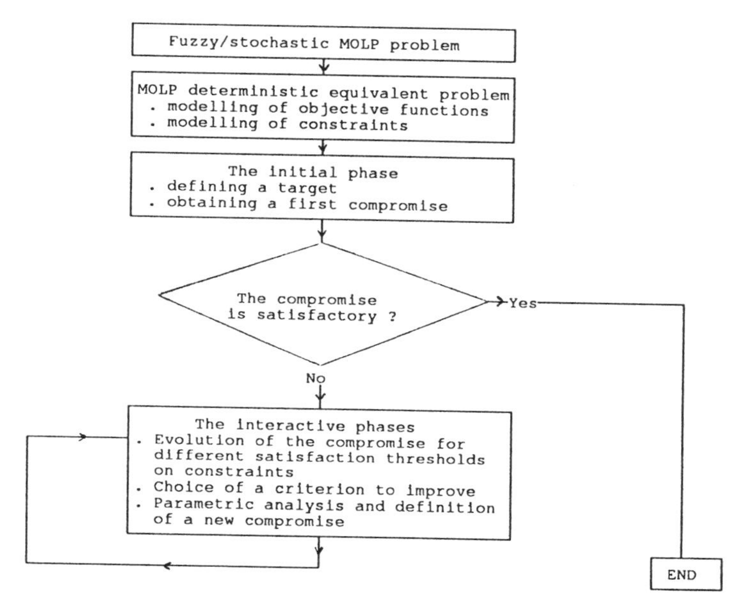

Figure 1: the general algorithm

3. The multiobjective deterministic equivalent program

3.1. *Modelling fuzzy/stochastic objective functions.*

To deal with the fuzzy/stochastic objective functions $Z_k(x)$, $k=1,\dots,K$, we will turn to a "Goal Programming" approach by transforming the initial problem (1) into a new problem modelled by means of non-deterministic inequalities in the form:

$$Z_k(x) \geq g_k, \tag{2}$$

where g_k represents the goal to be attained relatively to the objective $Z_k(x)$. It is supposed that $c_{kj} \in [c_{kj}^-, c_{kj}^+]$ and $g_k \in [g_k^-, g_k^+]$; g_k^+ represents the goal that the D.M. wishes to attain for $Z_k(x)$, whereas g_k^- represents the inferior bound which the D.M. is ready to accept for the goal g_k. In other words, the problem then consists of looking for an admissible solution x* such that the attained values for the different objective functions $Z_k(x)$ are as close as possible to the goals g_k^+ while at the same time satisfying the minimal goals g_k^-. For each objective function

$Z_k(x)$, the solution is compared to the goal g_k in such a way that the D.M. is asked to express his preferences with regard to the deviations relative to the goals, that is relative to $(g_k-Z_k(x))$. In this context, the greater this deviation is, the less satisfying it is to the D.M. and the weaker his degree of satisfaction is for such a solution.

The degree of satisfaction of the D.M. relative to the attainment of the goal g_k is expressed by a function of the difference $(g_k-Z_k(x))$. This degree of satisfaction translates the following assertion: "the D.M. is even less satisfied that $Z_k^-(x)=c_k^-.x$ is closed to g_k^- and even more satisfied that $Z_k^+(x)=c_k^+.x$ is closed to g_k^+". To do this, two slack variables $d_k^+=g_k^+-Z_k^+(x)$ and $d_k^-=g_k^--Z_k^-(x)$ are used, so the degree of satisfaction $P_k(d_k^+,d_k^-)$ is expressed by the linear function (3) which is represented graphically by Figure 2.

$$P_k(d_k^+,d_k^-)=\begin{cases} 0\,, & \text{if } d_k^-\geq 0 \\ 1-d_k^+/(d_k^+-d_k^-)\,, & \text{if } d_k^+\geq 0\geq d_k^- \\ 1\,, & \text{if } d_k^+\leq 0 \end{cases} \tag{3}$$

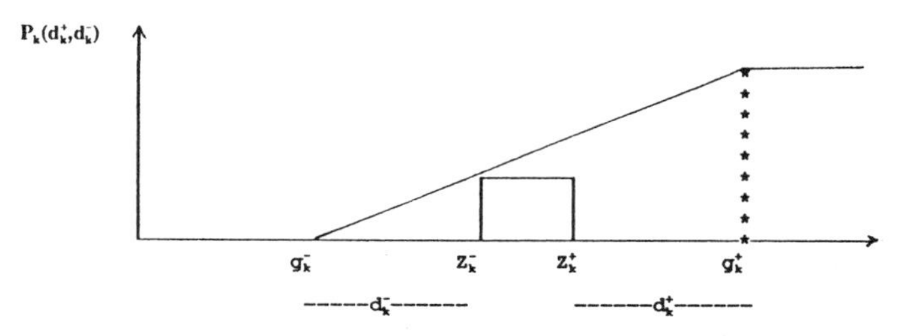

Figure 2: the degree of satisfaction $P_k\ (d_k^+,\ d_k^-)$

Consequently, the initial problem which consisted of "maximizing" the objective $Z_k(x)$ is replaced by the problem of maximizing the D.M.'s degree of satisfaction P_k relative to the attainment of goals g_k. Thus, the fuzzy/stochastic MOLP problem (1) is transformed into the following equivalent multiobjective program which, except for the constraints $\Sigma_{j=1}^{n} a_{ij}.x_j \leq b_i$, is deterministic:

$$\begin{aligned} \text{Max} \quad & P_k(d_k^+,d_k^-)=1-d_k^+/(d_k^+-d_k^-) \\ \text{s.t.} \quad & d_k^+=g_k^+-Z_k^+(x)\,,\ d_k^-=g_k^--Z_k^-(x)\,,\ d_k^+\geq 0\,,\ d_k^-\leq 0 \\ & \Sigma_j\, a_{ij}.x_j \leq b_i\,,\ x_j\geq 0 \qquad i=1,...,m\ ;\ j=1,...,n\ ;\ k=1,...,K. \end{aligned} \tag{4}$$

However, it is necessary in this approach to determine the goals g_k. The bounds g_k^- and g_k^+ for g_k can either be obtained from the D.M. or by means of an automatic procedure. Through this last procedure, these goals are determined from what we identify as a goal matrix and which, in a way, consists of a generalization of the classic payoff matrix in order to take into account the non-deterministic aspect in the constraints of problem (1). For each objective function $Z_k(x)$, we resolve the single-criteria deterministic programs obtained by replacing the fuzzy/stochastic parameters of problem (1) by their best values and by their worst values respectively, i.e.

$$\begin{array}{lll} \max\ Z_k^+(x) & \text{and} & \max\ Z_k^-(x) \\ \text{s.t. } a_i^-.x \leq b_i^+ & & \text{s.t. } a_i^+.x \leq b_i^- \end{array} \qquad (5)$$

The least constraining situation will be identified by β=0, the most constraining by β=1 and the optimal solution of the corresponding program will be designed by x_k^β, β=0,1, k=1,..K. Then the goals will be chosen as follows: $g_k^+ = Z_k^+(x_k^0)$ and $g_k^- = \min_{\beta,j} Z_k^-(x_j^\beta)$, β=0,1; j=1,..K. The goal g_k^+ is the ideal point because it is, in the space of objective functions, the point corresponding, for each objective function, to the best possible value of $Z_k(x)$ in the best of cases. The value g_k^- corresponds, for each objective, to the worst value of the goal matrix.

From this general model, a particular model corresponding with the case where, in addition to the extreme values, the D.M. also knows the central values c_{kj}^o of the fuzzy/stochastic parameters c_{kj}, can be developed. Then the fuzzy/stochastic objective function $Z_k(x)$ is transformed by using the inequality (2) where $Z_k^+(x)$ and $Z_k^-(x)$ are replaced by $Z_k^o(x)$, i.e $Z_k^o(x) \geq g_k$, where $g_k \in [g_k^-, g_k^+]$ and $Z_k^o(x)=c_k^o.x$. This inequality translates the following assertion: "the D.M. is less satisfied that $Z_k^o(x)$ is close to g_k^- and more satisfied that $Z_k^o(x)$ is close to g_k^+". So, as $d_k^+ - d_k^- = g_k^+ - g_k^-$ in this case, the D.M.'s degree of satisfaction $P_k(d_k^+, d_k^-)$ defined by (3) becomes:

$$P_k(d_k^+) = \begin{cases} 0\,, & \text{if } d_k^+ \geq g_k^+ - g_k^- \\ 1 - d_k^+ / (g_k^+ - g_k^-)\,, & \text{if } g_k^+ - g_k^- \geq d_k^+ \geq 0 \\ 1\,, & \text{if } d_k^+ \leq 0 \end{cases} \qquad (6)$$

This modelling of the non-deterministic part in the objective functions consequently permits transforming of the initial fuzzy/stochastic program (1) into a multiobjective equivalent program where the objective functions are deterministic and linear; this equivalent program takes the form:

$$\begin{array}{ll} \text{Max} & P_k(d_k^+) = 1 - d_k^+ /(g_k^+ - g_k^-) \\ \text{s.t.} & d_k^+ = g_k^+ - Z_k^o(x) \text{ , } d_k^+ \geq 0 \text{ , } d_k^+ \leq g_k^+ - g_k^- \text{ ,} \\ & \Sigma_{j=1}^{n} a_{ij}.x_j \leq b_i \text{ , } x_j \geq 0 \text{ , } i=1,...,m \text{ ; } j=1,...,n \text{ ; } k=1,...,K. \end{array} \qquad (7)$$

3.2. Modelling fuzzy/stochastic constraints.

The second step of our method consists of modifying the mathematically ill-defined fuzzy/stochastic constraints to well-defined mathematical entities. To do this, we propose to use the idea of "satisfaction thresholds on the constraints", this approach having the noted advantage of not unduly increasing the size of the initial problem. In order to transform the set of fuzzy/stochastic constraints:

$$a_i.x \leq b_i \qquad (8)$$

where $a_i \in [a_i^-, a_i^+]$ and $b_i \in [b_i^-, b_i^+]$, i=1,..m, each one of these constraints is interpreted as it follows: "the D.M. hopes that $a_i^-.x$ will not be larger than b_i^+ and his satisfaction level will be even higher that $a_i^+.x$ will be closer to b_i^+". Consequently, we introduce the D.M.'s degree of satisfaction relative to the fuzzy/stochastic constraint (8); this degree of satisfaction is designated by μ and is defined by the following linear function (9) which is represented graphically in figure 3.

$$\mu(a_i.x \leq b_i) = \begin{cases} 0 \text{ ,} & \text{if } a_i^-.x \geq b_i^+ \\ 1 \text{ ,} & \text{if } a_i^+.x \leq b_i^- \\ b_i^+ - a_i^-.x \;/\; (b_i^+ - b_i^-) + (a_i^+ - a_i^-).x \text{ ,} & \text{elsewhere} \end{cases} \qquad (9)$$

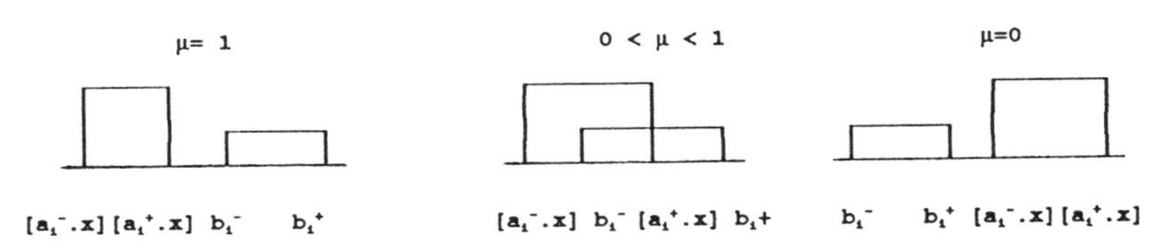

Figure 3 : the degree of satisfaction of the non deterministic constraint

Therefore, in order to transform the fuzzy/stochastic constraints (8) into deterministic constraints, we turn to the idea of satisfaction thresholds on the constraints: the solution is required to individually satisfy each of the fuzzy/stochastic constraints (8) with some satisfaction threshold, a particular satisfaction threshold called "individual satisfaction threshold on constraints" and designated by α_i being defined relatively to each constraint ; this α_i corresponds to the idea of an individual probability threshold in chance-constrained programming. Using this type of threshold, each fuzzy/stochastic constraint (8) is replaced by a deterministic constraint of the form

$$\mu(a_i.x \leq b_i) \geq \alpha_i , \tag{10}$$

where $0 \leq \alpha_i \leq 1$, $i=1,...,n$. From the relation (9), the constraint (10) takes the form of the inequality (11), the values of which are illustrated graphically in Figure 4:

$$(a_i^- + \alpha_i(a_i^+ - a_i^-)).x \leq b_i^+ - \alpha_i(b_i^+ - b_i^-) \tag{11}$$

In this approach, those α_i thresholds would be determined in an interactive manner with the D.M. Through those α_i, the D.M. can modify his requirements relative to his satisfaction levels on the constraints; thus, in the framework of an interactive procedure for solving the obtained deterministic MOLP problem, those α_i thresholds will be set to high values at the beginning of the procedure in order to be slackened in the subsequent steps and, at the same time, to allow for improvement of the obtained values of $Z_k(x)$, $k = 1, ...,K$. Consequently, during these interactions, the D.M. may come to a satisfactory compromise between the quality and the reliability of the obtained compromise solution.

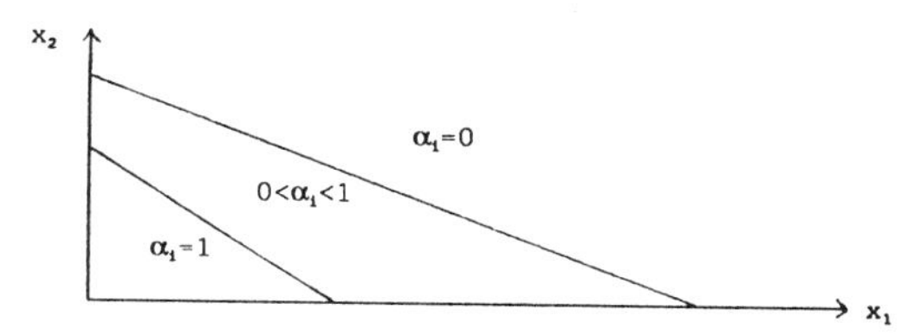

Figure 4:individual satisfaction threshold on constraint

4. The interactive algorithm of solving the equivalent MOLP problem

Without loss of generality, and in the interest of simplicity of exposition, our method is developped in the case where "individual satisfaction thresholds on constraints" are used and where, in addition to the extreme values of the stochastic parameters c_{kj}, the D.M. also knows the central values c^o_{kj} of those parameters. In this case, the deterministic MOLP problem equivalent to the fuzzy/stochastic problem (1) takes the following form :

$$\begin{array}{ll} \text{Max} & P_k(d^+_k) = 1 - d^+_k / (g^+_k - g^-_k) \\ \text{s.t.} & d^+_k = g^+_k - Z^o_k(x) \text{ , } d^+_k \leq g^+_k - g^-_k \text{ , } d^+_k \geq 0 \\ & (a^-_i + \alpha_i(a^+_i - a^-_i)).x \leq b^+_i - \alpha_i(b^+_i - b^-_i) \text{ ; } i=1,...,m \text{ ; } k=1,...,K. \end{array} \tag{12}$$

In what follows, for convenience, we will now use:

$$D_o = \{ \ x \in R^n \ / \ d^+_k = g^+_k - Z^o_k(x), \ d^+_k \geq 0 \text{ and } d^+_k \leq g^+_k - g^-_k, \ k=1,...,K \ \} \tag{13}$$

4.1. *The initial phase*

Defining a target. To solve (12), at first a payoff table must be built. To do this, for each criterion $P_k(d^+_k)$ of (12) and for each combination α, α=0,1, of the least and the more constraining value of the constraints, the optimal value of $P_k(d^+_k)$ is obtained by solving the program:

$$\begin{array}{ll} \max & P_k(d^+_k) \\ \text{s.t.} & x \in D_o \text{ , } (a^-_i + \alpha\ (a^+_i - a^-_i)).x \leq b^+_i - \alpha\ (b^+_i - b^-_i) \text{ , } i=1,...,m. \end{array} \tag{14}$$

This comes down to calculating the best solutions of $P_k(d^+_k)$ for each of the polyhedrons formed by the combinations of the extreme values of the satisfaction thresholds on the non-deterministic constraints. Each of these optimizations gives an optimal solution x^α_k and we define:

$$P^M_k = \max_{\alpha j} \{1 - g^+_k - Z^o_k(x^\alpha_j)/g^+_k - g^-_k\} \text{ ; } \alpha = 0,1 \text{ ; } j = 1,...,K. \tag{15}$$

Therefore, P^M_k represents the ideal point. Similarly, the worst value of $P_k(d^+_k)$ in the payoff table is noted by P^m_k and defined by

$$P^m_k = \min_{\alpha j} \{1 - g^+_k - Z^o_k(x^\alpha_j)/g^+_k - g^-_k\} \text{ ; } \alpha = 0,1 \text{ ; } j = 1,...,K. \tag{16}$$

This information concerning the best and the worst possible value of the objective functions $P_k(d^+_k)$ is translated in M_k and m_k to correspond to the initial problem (1). These parameters are defined by $M_k = g_k^- + P_k^M.(g_k^+-g_k^-)$ and $m_k = g_k^- + P_k^m.(g_k^+-g_k^-)$. With this information, we ask the D.M. to set an aspiration level O_k on each criterion $Z_k(x)$, $k=1,...,K$. Afterwards, the distance, in the space of the criteria, with regard to this target $O=(O_1,..,O_K)$ will be minimized. As the target is set by the D.M., it is possible to come to a solution where all the deviations relative to this target are negative or zero. This situation would reveal the fact that this O objective is dominated, in which case it would be worthwhile to suggest to the D.M. to set a new target point O' and begin the procedure again.

Obtaining a first compromise. As within the STEM method, the weights π_k relative to each criterion $P_k(d_k^+)$ have to be determined. These weights are defined by $\pi_k=b_k/\Sigma_k b_k$ where $b_k=[(P_k^M-P_k^m)/P_k^m].[1/\|c_k^o\|_2]$ and $\|c_k^o\|_2$ is the euclidian norm of the vector c_k^o. A first compromise, denoted by x^1, is then obtained by solving the following program:

$$\begin{aligned} \text{Min} \quad & M.\delta - \Sigma_k \in_k \\ \text{s.t.} \quad & \pi_k (P_k^O - P_k(d_k^+)) \leq \delta-\varepsilon_k \text{ , } x \in D_o \text{ ,} \\ & (a_i^- + \alpha_i(a_i^+-a_i^-)).x \leq b_i^+ - \alpha_i(b_i^+-b_i^-) \text{ , } i=1,...,m \text{ ; } k=1,...,K, \end{aligned} \qquad (17)$$

where $P_k^O=1-(g_k^+-O_k/g_k^+-g_k^-)$. In this program (17), the introduction of parameters M (a larger value than that of all the other coefficients of the program) and $\in_k$ assure us of the efficiency of the first compromixe x^1. This correction is due to Despontin (1984). The subsequent compromises shall be written x^m where m = 2, 3, 4, ... To obtain the first compromise, the values of α_i are at first chosen higher in order to be slackened later if necessary. Generally, the closer levels α_i are to 1.0, the lower the values of the objective functions are.

If the first compromise x^1 is considered satisfactory by the D.M, the interactive algorithm stops there and x^1 is chosen as solution of the initial problem (1). Otherwise, a series of interactive phases begins in order to obtain new compromises until attaining a satisfactory one.

4.2.*The interactive phases*

With each new compromise, the D.M. receives the following information:

1° the obtained value for this compromise x^m (m being the number of the compromise) on each objective function Z_k^o, i.e. the value $Z_k^o(x^m)$. This information is given to the D.M.

simultaneously with the range of variation $[m_k , M_k]$ of the considered criterion; in other words, the D.M. receives the information $[m_k, Z_k^o(x^m), M_k]$ and from there determines O_k;

2° to facilitate the comparison of the obtained values for the different criteria, the relative values of $Z_k^o(x^m)$, noted by $Z_k^{'}(x^m)$, are also calculated; this relative value, generally comprised between 0 and 1, translates the degree of attainment of the aspiration level O_k, i.e. $Z_k^{'}(x^m)=1-[(O_k-Z_k^o(x^m)) / (O_k-m_k)]$. Consequently, the larger $Z_k^{'}(x^m)$ is, the closer the D.M. is to his aspiration level O_k ;

3° a view of the evolution of the compromise solution for levels of satisfaction α_i situated in the proximity of those set of priori by the D.M., is also provided him with. On the basis of this information, the D.M. then chooses a new set of satisfaction thresholds α_i on the constraints.

Choice of a criterion to improve. From the preceding information, the D.M. must state if he considers the compromise satisfactory (in which case the procedure ends), or if he wishes to try to obtain a better one. In the second situation, he must then specify an objective function $Z_{k\bullet}^o$ that he wishes to improve, and, if possible, set the maximum level $\Delta_{k\bullet}$ of this improvement, which means that the maximum value desired by the D.M. for this objective function is $Z_{k\bullet}^o(x^m)+\Delta_{k\bullet}$. If the D.M. is unable to determine $\Delta_{k\bullet}$, this value can then be set at $\Delta_{k\bullet}=O_{k\bullet}-Z_{k\bullet}^o(x^m)$.

The parametric analysis. A parametric analysis allows to explore the consequences of the choices made by the D.M. in the preceding step. It consists of analyzing the repercussions of the possible improvement level $\tau.\Delta_{k\bullet}$ of the different objective functions $Z_{k\bullet}^o$, $\tau \in [0,1]$, $k=1,\ldots,K$. This parametric analysis is done by means of the following program:

$$\begin{aligned} \text{Min} \quad & M.\delta - \Sigma_k\, \varepsilon_k \\ \text{s.t.} \quad & \pi_k\, (P_k^O - P_k(d_k^+)) \leq \delta - \varepsilon_k \,,\; x \in D_o \,, \\ & Z_{k\bullet}(x) = O_{k\bullet} - \tau.\Delta_{k\bullet},\; \tau \leq 1 \\ & (a_i^- + \alpha_i(a_i^+ - a_i^-)).x \leq b_i^+ - \alpha_i(b_i^+ - b_i^-) \,,\; i=1,\ldots,m;\; k=1,\ldots,K \end{aligned} \tag{18}$$

By solving (18), a series of stability intervals of the form $[\tau_i, \tau_{i+1}]$ for the optimal basis of the problem are obtained. If the optimal solution of this parametric problem for a given τ is designated by x^τ, the value $Z_k^o(x^\tau)$ for each objective function $Z_k^o(x)$ can be calculated and presented to the D.M.. From these values , the D.M. can opt for an improvement level τ^*. To facilitate this choice, it may be worthwhile to graphically illustrate the incidence of the improvement level τ on the values $Z_k^o(x^\tau)$. To do this, the curves (linear by parts) $Z_k^o(x^\tau) = f(\tau)$ can be traced. Once τ^* is chosen, the D.M. is faced with a new compromise x^{τ^*} which is simply designated by x^2, and the equations $\pi_{k^*} = 0$ and $Z_{k^*}(x) \leq Z_{k^*}(x^{\tau^*})$ are added to the problem (17). This new compromise x^2 is treated like the preceding one and the procedure ends when the D.M. feels satisfied with the subsequent obtained compromise x^m. In any case, the procedure ends after K steps at the most because the objective function on which an improvement has been made is no longer considered (by adding the constraint $\pi_{k^*}=0$). If that is desired however, a modification to the method allowing one to go back in the procedure is possible; it is just a matter of being able to store the value $Z_{k^*}(x)$ and to re-establish π_{k^*} upon demand.

5. Illustration of the method

5.1. *The fuzzy/stochastic MOLP problem.*

We shall illustrate our method from a didactic example. This example contains two variables x_1 and x_2 and three criteria to maximise. These criteria are expressed as follows:

$$Z_k(x) = \Sigma_{j=1}^{n} c_{kj}.x_j \,, \quad k=1,2,3 \qquad \begin{array}{l} c_{11} \in \{.8,1,1.2\} \,,\ c_{12} \in \{-.5,0,.2\} \\ c_{21} \in \{-.3,0,.2\} \,,\ c_{22} \in \{.7,1,1.2\} \\ c_{31} \in \{.8,1,1.1\} \,,\ c_{32} \in \{.9,1,1.2\} \end{array}$$

The problem implies two constraints (other than those of non-negativity), both non-deterministic, namely:

$$a_{i1}.x_1 + a_{i2}.x_2 \leq b_i, \qquad i=1,2$$

where $a_{11} \in [1.5,2.8]$, $a_{12} \in [.5,1.2]$, $b_1 \in [7,9]$

$a_{21} \in [.5,1.5]$, $a_{22} \in [2,4]$, $b_2 \in [13.5,16]$

5.2. *The equivalent deterministic MOLP problem.*

To transform this fuzzy/stochastic MOLP problem into a deterministic one, it is necessary to determine the goals g_k. To do this, we resolve the following programs:

$$\max Z_k^+(x) \qquad \text{and} \qquad \max Z_k^-(x)$$

$$\text{s.t. } a_i^-.x \leq b_i^+ \qquad\qquad \text{s.t. } a_i^+.x \leq b_i^- \qquad i=1,\ldots,m;k=1,\ldots,K.$$

which correspond to the least (β=0) and the most (β=1) constraining situation. Then we obtain the following optimal solutions x_k^β, β=0,1 : x_1^0=(6,0), x_2^0=(0,8), x_3^0=(3.636,7.091), x_1^1=(2.5,0), x_2^1=(0,3.375), x_3^1=(1.255,2.904). This information is resumed in the goal matrix of Table 1 which gives the values $Z_k^+(x_k^\beta)$ and $Z_k^-(x_k^\beta)$ for β=0,1 , k=1,2,3. From this goal matrix we deduce the values g_k^+ and g_k^- given in Table 2. Consequently the associated deterministic equivalent program of the form (14) can be written.

Table 1

	x_1^0	x_1^1	x_2^0	x_2^1	x_3^0	x_3^1
Z_1^+	7.2	3.0	1.6	.675	5.794	2.087
Z_1^-	4.8	2.0	-4.0	-1.68	-.628	-.448
Z_2^+	1.2	.50	9.6	4.05	9.238	3.736
Z_2^-	-1.8	-.75	5.6	2.362	3.869	1.656
Z_3^+	6.6	2.75	9.6	4.05	12.50	4.866
Z_3^-	4.8	2.0	7.2	3.037	9.299	3.618

Table 2

k	g_k^-	g_k^+	$g_k^+-g_k^-$
1	-4	7.2	11.2
2	-1.8	9.6	11.4
3	2	12.5	10.5

The initial phase. As it has been explained before, the first step of the algorithm consists of building the payoff matrix and identifying a target point O. This payoff matrix is given in the first seven columns of Table 3. This table gives also, for each criterion, the best value M_k and the worst value m_k from which the D.M. determines the goal O_k.

Table 3

	x_1^0	x_1^1	x_2^0	x_2^1	x_3^0	x_3^1	M_k	O_k	m_k
Z_1°	6	2.5	0	0	3.630	1.25	6.0	4	0
Z_2°	0	0	8	3.37	7.091	2.90	8.0	6	0
Z_3°	6	2.5	8	3.37	10.72	4.15	10.7	9	2.5

The following step consists of determining the weights π_k. We obtain here π_1=.34, π_2=.34, π_3=.32. Then the first compromise is obtained by solving the deterministic program of the form (17). To obtain that first compromise, the values chosen for α_i have been, α_1 = α_2 = 0.8 . We obtain x^1 = (x_1=1.3679 , x_2=1.7976). To evaluate x^1, the D.M. receives the information given in Table 4.

Table 4

k	$Z_k^\circ(x^1)$	O_k	$Z_k'(x^1)$
1	1.519	4	.380
2	3.340	6	.560
3	4.859	9	.364

5.3. *The interactive phases.*

Based on the information of Table 4, the D.M. judges the compromise x^1 unsatisfactory and wishes to obtain a better one. To do this, he decides to improve this first compromise in releasing the satisfaction thresholds on constraints to the values $\alpha_1=\alpha_2=.7$. This choice has been taken after consultation of Table 5 showing the evolution of the solution around the first compromise x^1. The values of $Z_k°$ and Z'_k corresponding to this new solution are summed up in Table 6.

Table 5

α_1	α_2	$Z_1°$	Z'_1	$Z_2°$	Z'_2	$Z_3°$	Z'_3
.8	.8	1.52	.38	3.34	.56	4.86	.36
	.7	1.36	.34	3.71	.62	5.07	.40
	.6	1.19	.36	4.12	.69	5.31	.43
.7	.8	1.83	.46	3.3	.54	5.06	.40
	.7	1.67	.42	3.60	.60	5.27	.43
	.6	1.50	.38	4.01	.67	5.51	.47
.6	.8	2.17	.54	3.10	.62	5.27	.43
	.7	2.01	.51	3.48	.58	5.19	.46
	.6	1.85	.46	3.90	.65	5.75	.50

Table 6

k	$Z_k^o(x^1)$	O_k	$Z'_k(x^1)$
1	1.674	4	.420
2	3.600	6	.600
3	5.274	9	.430

The D.M. decides to improve the criterion Z_1^o which is in fact the furthest away from the target 0. A parametric analysis is made to examine the consequence of this choice. From this analysis, for each value of τ, we obtain information on Z_k^o and Z'_k (placed between parentheses) summed up in Table 7. The D.M. chooses the compromise x^2 corresponding to

Table 7

τ	Z_1^o	Z_2^o	Z_3^o
1	1.674	3.600	5.274
	(.42)	(.60)	(.43)
.875	1.965	2.893	4.858
	(.49)	(.48)	(.36)
.378	3.121	.079	3.200
	(.71)	(.01)	(.11)
.364	3.153	0.00	3.153
	(.79)	(.00)	(.10)

$\tau= 0.875$ for which he obtain a table of the form of Table 6 with $Z'_k(x^2)=0.49$, 0.48 and 0.36 for k = 1, 2 and 3. Here again the D.M. does not consider x^2 to be satisfactory. Based on an analysis similar to that of Table 5, the D.M. chooses to release α_i to the values $\alpha_1=0.5$ and

α_2=0.6. To obtain a better compromise, he decides now to improve Z_2^o. By means of a program of the form (18), a parametric analysis is done to determine the desired improvement level τ. The D.M. chooses τ= 1 and the corresponding compromise is $x^3=(x_1=2.233, x_2=3.764)$. The values of $Z_k'(x^3)$ being respectively 0,560, 0.630 and 0.540 for k= 1, 2 and 3, this latter compromise is considered satisfactory by the D.M. and consequently chosen as the best compromise.

6. Conclusion

In the context of MOLP problems, situations where the D.M. knows only the limits of variation of some parameters seem fairly frequent in practice. In such situations, we prefer to avoid the arbitrariness which consists of assigning probability or possibility measures to those indetermined parameters. Our approach which is inspired both by stochastic and fuzzy MOLP methodologies, leads to a fairly simple mathematical formulation and permits a good integration of the D.M. in the decision process.

7. Bibliography

BENAYOUN, R., DE MONTGOLFIER, J., TERGNY, J. and LARICHEV, O., Linear Programming and Multiple Objective Functions: STEP Method (STEM), Mathematical Programming, 1(3), 366-375, 1971.

DESPONTIN, M., Interactive Economic Policy Formulation with Multiregional Econometric Models, in Despontin, Nigkam and Spronk, Macro Planning with Conflicting Goals, Springer Verlag, Britain, 1984.

GOICOECHEA, A., HANSEN, D.R., and DUCKSTEIN, L., Multiobjective Decision Analysis with Engineering and Business Applications, John Wiley, New York, 1982.

ROY, B., Main Sources of Inaccurate Determination, Uncertainty and Imprecision in Decision Models, Mathematical and Computer Modelling, Vol.12, no.10-11, 1245-1254, 1989.

SLOWINSKI, R., A Multicriteria Fuzzy Linear Programming Method for Water Supply System Development Planning, Fuzzy Sets and Systems, 19(1), 1-21, 1986.

TEGHEM, J., DUFRANE, D., THAUVOYE, M., and KUNSCH, P., STRANGE: an Interactive Method for Multiobjective Linear Programming Under Uncertainty, E.J.O.R., 26(1), 65-82, 1986.

THEORETICAL REMARKS ON DYNAMIC TRADE-OFF

Hirotaka Nakayama
Professor
Department of Applied Mathematics
Konan University
8-9-1 Okamoto, Higashinada, Kobe 658, JAPAN

ABSTRACT:

The aspiration level approach is now widely recognized to be effective in many kinds of multi-objective programming problems. Above all, it should be noted that recent developments for rapid and user-friendly interaction between decision makers and decision support systems, such as Pareto Race or VIG by Korhonen-Wallenius, spur the method to many real applications. In such a method, trade-off analysis can be made in some dynamic way. This kind of trade-off analysis will be referred to as dynamic trade-off.

In this paper, several theoretical propeties of dynamic trade-off will be clarified in order to give a foundation of the method.

KEY WORDS:
trade-off analysis, parametric optimization, sensitivity analysis, multi-objective programming, interactive programming.

1. Introduction

Consider the following multi-objective programming:

[MOP] Minimize $F(x) = (f_1(x), f_2(x), \ldots, f_r(x))$
subject to $x \in X \subset R^n$.

The aspiration level approach to multiobjective programming is summarized as follows: The aspiration level at the k-th iteration $\bar{f}^k$ is modified by

$$\bar{f}^{k+1} = T \circ P(\bar{f}^k) \tag{1.1}$$

Here, the operator P selects the Pareto solution nearest in some sense to the given aspiration level $\bar{f}^k$. The operator T is the trade-off operator which changes the k-th aspiration level $\bar{f}^k$ if the decision maker does not compromise with the shown solution $P(\bar{f}^k)$. Of course, since $P(\bar{f}^k)$ is a Pareto solution, there exists no feasible solution which makes all criteria better than $P(\bar{f}^k)$, and thus the decision maker has to trade-off among criteria if he wants to improve some of criteria. Based on this trade-off, a new aspiration level is decided as $T \circ P(\bar{f}^k)$. Similar process is continued until the decision maker obtains an agreeable solution. This idea is implemented in DIDASS (Grauer et.al. 1984) and in the satisficing trade-off method (Nakayama 1984).

Projection on the Pareto Surface

The operation P(.), which gives a Pareto solution nearest to $\bar{f}^k$, is performed by some auxiliary scalar optimizatin. The objective function in this auxiliary optmization is called an achievement function in some literature (Wierzbicki 1986). Let f_i^* be an ideal value which is usually given in such a way that $f_i^* < \text{Min}\,\{f_i(x) \mid x \in X\}$, and let f_{i*} be a nadir value which is usually given by $f_{i*} = \underset{1 \leq j \leq r}{\text{Max}}\ f_i(x_j^*)$ where $x_j^* = \underset{x \in X}{\text{arg Min}}\ f_j(x)$.

Then, typical examples of achievement function based on Tchebyshev norm are given in the following:

$$q = \underset{1 \leq i \leq r}{\text{Max}}\ w_i(f_i(x) - \bar{f}_i^k) + \alpha \sum_{i=1}^{r} w_i f_i(x)$$

where

$$w_i = \frac{1}{f_{*i} - f_i^*}$$

Since the above achievement functions are not smooth, the minimization of it is usually performed by solving the equivalently transformed problem. Instead of minimizing q, for example, we solve the following:

$$[Q] \quad \text{Minimize} \quad z + \alpha \sum_{i=1}^{r} w_i f_i(x)$$

$$\text{subject to} \quad f_i(x) - \bar{f}_i^k \leqq (1/w_i) z$$

$$x \in X.$$

Trade-off

In cases decision makers are not satisfied with the solution for $P(\bar{f}^k)$, they are requested to answer their new aspiration level $\bar{f}^{k+1}$. Since the solutions obtained by the projection $P(\bar{f})$ are Pareto optimal, there is no other feasible solution that improves all objective functions. Therefore, if decision makers want to improve some of objective functions, then they have to agree with some sacrifice of other objective functions. Let x^k denote the Pareto solution obtained by projection $P(\bar{f}^k)$, and classify the objective functions into the following three groups:

(i) the class of criteria which are to be improved more,
(ii) the class of criteria which may be relaxed,
(iii) the class of criteria which are acceptable as they are.

The index set of each class is represented by I_I^k, I_R^k, I_A^k, respectively. Clearly, $\bar{f}_i^{k+1} < f_i(x^k)$ for all $i \in I_I^k$. Usually, for $i \in I_A^k$, we set $\bar{f}_i^{k+1} = f_i(x^k)$. For $i \in I_R^k$, decision makers have to agree to increase the value of $\bar{f}_i^{k+1}$. It should be noted that an appropriate sacrifice of f_j for $j \in I_R^k$ is needed for getting the improvement of f_i for $i \in I_I^k$.

It is of course possible for decision makers to answer new aspiration levels of all objective functions. In practical probelms, however, we often encounter cases with very many objective functions as is seen in the erection management of cable stayed bridge (Ishido et al. 1987). Under this circumstance, decision makers tend to get tired with answering new aspiration levels for all objective functions. Usually, the feeling that decision makers want to improve some of criteria is much stronger than the one that they compromise with some compensatory relaxation of other criteria. Therefore, it is more practical in problems with very many objective functions for decision makers to answer only their improvement rather than both improvement and relaxation.

At this stage, we can use the assignment of sacrifice for

f_j $(j \in I_R)$ which is automatically set in the equal proportion to $\lambda_i+\alpha w_i$, namely, by

$$\Delta f_j = \frac{-1}{N(\lambda_j+\alpha w_j)} \sum_{i \varepsilon I_I} (\lambda_i+\alpha w_i)\Delta f_i \tag{1.2}$$

where N is the number of elements of the set I_R, and λ is the Lagrange multiplier associated with the constraints in Problem [Q].

By doing this, in cases where there are a large number of criteria, the burden of decision makers can be decreased so much. Of course, if decision makers do not agree with this quota Δf_j laid down automatically, they can modify them in a manual way. In general, however, the amount of relaxation given by (1.2) is not sufficient for compensation of improvement of Δf_i $(i \in I_I)$, in particular, in nonlinear and convex cases, because the quota Δf_j is decided in such a way that the new aspiration level is merely on the supporting hyperplane for the Pareto surface. In cases where all functions in [P] are linear and X is a polyhedral set in an n-dimensional Euclidean space, however, a more precise analysis is possible by using parametric optimization techniques. In other words, parametric optimization techniques can give us the exact amount of relaxation so that the new aspiration level may be on the Pareto surface. A method along this line will be suggested in the following sections.

Before proceeding to the next section, consider the following illustrative example:

[EX]

$$f_1=2x_1+ x_2 \rightarrow \text{Max}$$

$$f_2=-x_1+2x_2 \rightarrow \text{Max}$$

subject to

$$-x_1+3x_2 \leqq 21$$

$$x_1+3x_2 \leqq 27$$

$$4x_1+3x_2 \leqq 45$$

$$3x_1+ x_2 \leqq 30$$

$$x_1,\ x_2 \geqq 0$$

The ideal point in this example is $(f_1^*, f_2^*)=(21,14)$ and the nadir point is $(f_{1*}, f_{2*})=(7,-3)$. Suppose that the initial aspiration level is $(\bar{f}_1, \bar{f}_2)=(12,6)$. Then, associated with this aspiration level, we obtain a Pareto solution $(f_1,f_2)=(16.065, 10.935)$ by solving the auxiliary LP [Q]. Now, if DM wants to increase f_1 a little bit more, say, up to 18.0, then he has to relax f_2 up to 9.0 in order to retain the Pareto optimality. Using the automatic trade-off (1.2), we can get the exact amount for f_2 to be relaxed in this case. However, if DM wants to improve f_1 more, say, up to 20.0, then the amount of relaxation of f_2 calculated by (1.2) is not sufficient for retaining the Pareto optimality (see Fig. 1). Even in such a case, as long as the problem is of LP-type, we can obtain the exact amount for f_2 to be relaxed by using parametric optimization technique.

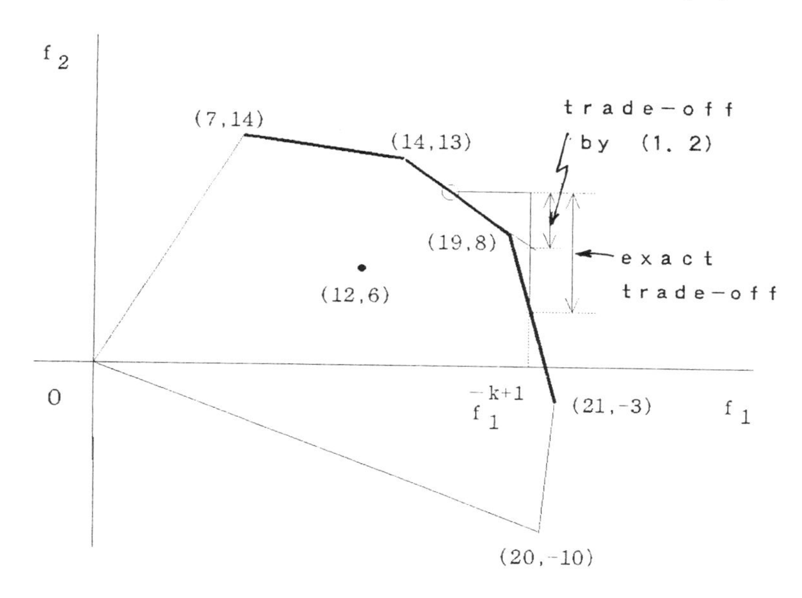

Fig. 1 trade-off relation for the problem [EX]

2. Dynamic Trade-off using Parametric Optimization Techniques in LP

In the following, we consider the following linear multi-objective programming problem:

[MOLP] Minimize $(c_1x,\ldots,c_rx)$

subject to

$$Dx \leqq d$$
$$x \geqq 0.$$

Then the auxiliary min-max problem [Q] becomes

$$\text{[ALP]} \quad \text{Minimize} \quad z + \alpha \sum w_i c_i x$$
$$\text{subject to} \quad c_i x - (1/w_i) z \leqq \bar{f}_i^k, \qquad i=1,\dots,r$$
$$x \in X,$$

where $X := \{x \in R^n \mid Dx \leqq d,\ x \geqq 0\}$.

One way to decrease the burden of decision makers in trade-off is the following:

Step 1. Ask decision makers the new aspiration level $\bar{f}_i^{k+1}$ for the objective functions to be improved, f_i $(i \varepsilon I_I)$.

Step 2. Let r be the index of the objective function to be relaxed most.

Step 3. Let $f^k := f(x^k)$ where x^k is the solution to [ALP]. Then, decide the new aspiration level $\bar{f}_j^{k+1} := f^k + \Delta f_j^k$ $(j \varepsilon I_R)$ for the objective function to be relaxed as follows: Set

$$\Delta f_j^k = \sum_{i \in I_I} (\lambda_i + \alpha w_i) \Delta f_i^k / (N(\lambda_j + \alpha w_j)), \qquad j \varepsilon I_R \setminus \{r\},$$

where

$$\Delta f_i^k = \bar{f}_i^{k+1} - f_i^k, \qquad i \varepsilon I_I.$$

Decide $\bar{f}_r^{k+1}$ by $\bar{f}_r^{k+1} = f_r^k + (1/w_r) z$, where z is the solution to the following linear parametric problem:

$$\text{[PLP]} \quad \text{Minimize} \quad z$$
$$\text{subject to} \quad c_i x \leqq f_i^k - t \Delta f_i^k, \qquad i \varepsilon I_I$$
$$c_j x \leqq f_j^k + t \Delta f_j^k, \qquad j \varepsilon I_R \setminus \{r\}$$
$$c_r x - (1/w_r) z \leqq f_r^k$$
$$x \in X$$

In the parametric linear programming [PLP], at first, the new base is obtained by the sensitivity analysis from the final tableau of [ALP] in which the coefficient vector of objective function is changed from $(1,\alpha c_1',\dots,\alpha c_n')$ into $(1,0,\dots,0)$ where c_i' is the i-th component of the vector $w_1c_1+\dots+w_rc_r$, and the right hand side vector $\bar{f}^k$ into f^k, and finally the column vector associated with z in the coefficient matrix of the constraint is changed from $(1/w_1,\dots,1/w_{r-1},1/w_r)$ into $(0,\dots,0, 1/w_r)$. In many practical cases, the new base in the above process can be obtained in one pivotting as will be shown later (Proposition 3.1).

Secondly, the solution of [PLP] with t=1 is obtained by the right hand side sensitivity analysis. At corner points of the Pareto surface with kink, the solutions are degenerate. As in the usual parametric optimization, the dual simplex method is used for getting the new base at such degenerate solutions.

Note that the obtained new aspiration level is already Pareto optimal. Therefore, since only a few pivotting are usually needed in these techniques, we can obtain the new Pareto solution associated with a new aspiration level very quickly. Moreover, we can make a microjustification around the obtained Pareto solution along the direction $(\Delta f_1^k,\dots,\Delta f_{r-1}^k)$ by modifying the value of t. This operation can be made in some dynamic way, because each solution to [PLP] with a given t is obtained very quickly. Using some computer graphics, this enables us to develop an effective man-machine interface as a decision support system. This idea was originally realized by Korhonen-Wallenius (1987) in a slightly different way.

<u>Remark 2.1</u> The reason why we take f_r, which may be relaxed most, as the objective function in [PLP] is as follows: Since the set $f(X)+R_+^r$ is convex in convex programming such as multi-objective linear programming, the new aspiration point based on (1.2), which is on the supporting hyperplane for $f(X)+R_+^r$, is notalways much enough to compensate for the improvement of f_i ($i\in I_I$). Therefore, we apply the equal distribution (1.2) for objective functions to be relaxed except for f_r, and decide the amount of relaxation of f_r by solving [PLP], because this solution gives usually more relaxation than the one by (1.2).

<u>Remark 2.2</u> In Korhonen-Wallenius (1987), the decision maker is asked to answer his/her new aspiration level for all criteria. Of course, the new aspiration level of criteria which the decision maker does not answer can be set to be the current level. In any case in their approach, Problem ALP (rather than PLP) for the new value of right hand side is solved by some parametric optimization technique. Therefore, the level of improvement can not be neccessarily guaranteed in some cases (in particular, where the relaxation amount is not sufficient). However, this problem can be easily overcome by moving forward or

backward along the direction Δf by modification of the value of t, in other words, by seeing the trade-off along this direction. The main point in their method is that the decision maker can see the tradeoff among criteria in a visual and dynamic way. On the other hand, the point in this paper is to see quickly the exact tradeoff by inputting only the new aspiration level of criteria to be improved, $\bar{f}_i^{k+1}$ $(i \varepsilon I_I)$ even though the problem has very many objective functions.

3. Theoretical Remarks on Dynamic Trade-off

Theorem 3.1 Let $\tilde{x}$ be a solution to (Q). Then, under the assumption of convexity of f_i $(i=1,\dots,r)$ and X, there exists Lagrange multiplier $\tilde{\lambda}=(\tilde{\lambda}_1,\dots,\tilde{\lambda}_r)$ such that

$$\sum_{i=1}^{r} \tilde{\lambda}_i / w_i = 1, \qquad \tilde{\lambda}_i \geqq 0, \qquad i=1,\dots,r$$

$$\sum_{i=1}^{r} (\tilde{\lambda}_i + \alpha w_i)(f_i(x) - f_i(\tilde{x})) \geqq 0 \qquad \text{for all } x \varepsilon X.$$

Lemma 3.1 Under the assumption of convexity of f_i $(i=1,\dots,r)$ and X, the set $f(X)+R_+^r$ is convex. Furthermore, if the set X is convex polyhedral, and if the function f is linear, then the set $f(X)+R_+^r$ is also convex polyhedral.

It is well known that under the convexity of f_i $(i=1,\dots,r)$ and X, each Pareto solution can be characterized by the following problem

$$\begin{aligned} &\text{Min} \quad f_k(x) \\ &\text{subject to} \quad f_i(x) \leqq \varepsilon_i \qquad i=1,\dots,r, \quad i \neq k \\ &\qquad\qquad\qquad x \; \varepsilon \; X. \end{aligned}$$

In case of k=r, setting

$$F(\varepsilon_1,\dots,\varepsilon_{r-1}) := \text{Min}\{f_r(x) \mid f_i(x) \leqq \varepsilon_i, \; i=1,\dots,r-1, \quad x \varepsilon X\},$$

we have the folllowing theorem:

Lemma 3.2 epi F = $f(X) + R_+^r = E(f(X)) + R_+^r$, where E(f(X)) denotes the Pareto solution set.

<u>Remark</u> <u>3.1</u> It should be noted that $(\lambda_1+\alpha w_1,\dots,\lambda_r+\alpha w_r)$ in Theorem 3.1 is a normal vector of the supporting hyperplane of $f(X)+R_+^r$ at $f(x^k)$ and hence the Pareto surface E(f(X)). Therefore, $(-(\lambda_1+\alpha w_1)/(\lambda_r+\alpha w_r),\dots,-(\lambda_{r-1}+\alpha w_{r-1})/(\lambda_r+\alpha w_r))$ is a sub-

gradient of the function F at $(f_1(x^k),\dots,f_{r-1}(x^k))$.

Lemma 3.3 A convex function F(y) is differentiable at $\tilde{y}$ if and only if F(y) has a unique subgradient at $\tilde{y}$. (Rockafellar(1970), Theorem 25.1)

Remark 3.2 In multiobjective linear programming, therefore, if a Pareto optimal solution x^k to [ALP] is nondegenerated, the corresponding simplex multiplier is unique and hence represents the normal vector of the tangent hyperplane of $E(f(X))+R^r_+$ at $f(x^k)$. This implies that the simplex multiplier associated with a nondegenerated primal solution to [ALP] gives us a trade-off information among objective functions.

Next, we shall show the calculation of the exact trade-off can be done very quickly by solving [PLP].

Lemma 3.4 Let (x^k,z^k) be the solution to [ALP]. Then $(x^k,0)$ is a solution to [PLP] with t=0.

Now, we shall show that we can obtain a solution to [PLP] with t=0 in one pivotting for the final tableau of [ALP] in many cases. To this end, the following auxiliary LP plays an important role.

[ALP'] Minimize z

subject to $c_i x - (1/w_i)z \leqq f_i^k, \quad i=1,\dots,r$

$x \in X,$

where $X:=\{x \in R^n \mid Dx \leqq d,\ x \geqq 0\}$.

Lemma 3.5 The optimal basic matrix of [ALP] is also an optimal basic matrix to [ALP'].

Lemma 3.6 Exactly one of z^+ and z^- is basic in [ALP'] and [PLP].

From the above discussion, we can suppose that z^- is basic in [ALP'] without loss of generality. In addition, let a_{z^-} be the s-th column vector of the optimal basic matrix B in [ALP']. Note that [PLP] with t=0 is given by changing the coefficient vector $a_{z^+}=(-1/w_1,\dots,-1/w_r,0,\dots,0)$ in A associated with z^+ in [ALP'] into $a'_{z^+}=(0,\dots,0,-1/w_r,0,\dots,0)$ and by a similar change for a_{z^-}. Let B' be a matrix obtained by replacing s-th column vector of B with

$$a'_{z^-}=(0,\dots 0,1/w_r,0\dots,0).$$

In the following, we shall show that $(B')^{-1}$ is the inverse of an optimal basic matrix in [PLP] with t=0 and can be obtained in one pivotting from
B^{-1} in many cases.

Proposition 3.1

If b_{sr}, the (s,r) component of B^{-1} is nonzero, the inverse of the optimal basic matrix of [PLP] with t=0 can be obtained in one pivotting for that of [ALP].

(Proof) It suffices to show that the inverse of optimal basic matrix of [PLP] can be obtained by one pivotting for B^{-1} of [ALP'], because the optimal basic matrices of [ALP'] and [ALP] are the same due to Lemma 3.5. Define

$$b'_s = B^{-1}a'_{z^-} = (1/w_r)b_r$$

where b_r is the r-th column vector of B^{-1}, and put

$$S = \begin{pmatrix} 1 & & & -b'_{1s}/b'_{ss} & & & \\ & \ddots & & \vdots & & & \\ & & 1 & -b'_{s-1,s}/b'_{ss} & & & \\ & & & 1/b'_{sr} & & & \\ & & & -b'_{s+1,s}/b'_{ss} & 1 & & \\ & & & \vdots & & \ddots & \\ & & & -b'_{qs}/b'_{ss} & & & 1 \end{pmatrix}$$

Then we have

$$(B')^{-1} = SB^{-1}.$$

Since S is a pivotting matrix, it follows that $(B')^{-1}$ can be obtained by one pivotting for B^{-1}. Now, we shall prove that B' is an optimal basic matrix in [PLP] with t=0 as long as B is an optimal basic matrix in [ALP']. To do this, suppose that $(\tilde{x}_B, 0)$ is a solution to [ALP']. Then note that

$$(B')^{-1}(f^k, d) = SB^{-1}(f^k, d) = S\tilde{x}_B. \tag{3.1}$$

Recalling that the s-th component of $\tilde{x}_B$ correspoending to z^- is 0, we have

$$S\tilde{x}_B = \tilde{x}_B.$$

From this and (3.1), it follows that B' is an optimal basic matrix of [PLP] with t=0 because $(\tilde{x}_B, 0)$ is an optimal solution of [PLP] with t=0 according to Lemma 3.4. This completes the proof.

Remark 3.3 If $b_{sr}=0$, the inverse of the optimal basic matrix of [PLP] with t=0 can be obtained in a few pivotting for that of [ALP]. To see this, note that

$$B^{-1}B' = (e_1, \ldots, e_{s-1}, b'_s, e_{s+1}, \ldots, e_q)$$

where $e_i=(0,\ldots,0,1,0,\ldots,0)$ whose i-th component is 1. It follows then that if the s-th component of b'_s is 0, the matrix B' is singular because B^{-1} is nonsingular. This implies that B' obtained by replacing the s-th column vector of B with a'_{z^-} can not be basic in [PLP] with t=0. Since either z^- or z^+ is necessarily basic in [PLP] with t=0 according to lemma 3.6, some of other column vectors of B than a_{z^-} should also be replaced with appropriate vectors so that we might get an optimal basic matrix in [PLP] with t=0. In the final tableau in [ALP'], find an element $b_{sj}\neq 0$ where the j-th simplex criterion $p_j=0$. It is assured that this b_{sj} does exist. For, if there is no p_j with the value 0, then we would have no optmal basic vectors other than B which leads to a contradiction in our situation. In a similar fashion, it leads to a contradiction to suppose that there is no $b_{sj}\neq 0$ for which the j-th simplex criterion is 0. Now take the j-th vector a_j as a basic vector. Namely, sweep out the j-th column vector by the pivot b_{ij}, where the subscript i is decided in the same way as in the usual simplex method, i.e.,

$$\theta_i = \min_k \{b_{k,0}/b_{k,j} \mid b_{k,j}>0\}$$

If we have $b_{sr}\neq 0$ by the above pivotting, then we can apply Lemma 3.6 for the updated basic matrix B. Otherwise we continue the same proceddure until we attain this situation. This situation can be attained necessarily in finite steps. For, unless we get $b_{sr}\neq 0$ for ever, a'_{z^-} can not be basic, as stated above.

4. An Illustrative Example

Consider the same example stated in the section 2. At the stage that we get a Pareto solution (16.065,10.935) associtated with the aspiration level (12.0, 6.0), the inverse of the basic matrix for the auxiliary linear programming problem [ALP] is given by

$$
B^{-1} = \begin{bmatrix}
.3290 & -.2710 & .0000 & .0710 & .0000 & .0000 \\
-.1097 & .0903 & .0000 & .3097 & .0000 & .0000 \\
.6581 & -.5419 & 1.0000 & -.8581 & .0000 & .0000 \\
-.0323 & -.0323 & .0000 & .0323 & .0000 & .0000 \\
-.9871 & .8129 & .0000 & -1.2129 & 1.0000 & .0000 \\
-.8774 & .7226 & .0000 & -.5226 & .0000 & 1.0000
\end{bmatrix}
$$

Since the most relaxable objective is f_2, we have $a_z'-=(0,-17,0,0,0,0)$. Therefore, we have $B^{-1}a_z'-=(4.606,-1.535,9.213,0.548,-13.819,-12.284)$. It is easy to see $(B')^{-1}$, the inverse of the basic matrix of [PLP] with t=0, can be obtained by the following once-pivotting:

$$
\left[\begin{array}{cccccc|c}
.3290 & -.2710 & .0000 & .0710 & .0000 & .0000 & 4.606 \\
-.1097 & .0903 & .0000 & .3097 & .0000 & .0000 & -1.535 \\
.6581 & -.5419 & 1.0000 & -.8581 & .0000 & .0000 & 9.213 \\
-.0323 & -.0323 & .0000 & .0323 & .0000 & .0000 & \boxed{0.548} \\
-.9871 & .8129 & .0000 & -1.2129 & 1.0000 & .0000 & -13.819 \\
-.8774 & .7226 & .0000 & -.5226 & .0000 & 1.0000 & -12.284
\end{array}\right]
$$

Now we have

$$
(B')^{-1} = \begin{bmatrix}
.6000 & .0000 & .0000 & -.2000 & .0000 & .0000 \\
-.2000 & .0000 & .0000 & .4000 & .0000 & .0000 \\
1.2000 & .0000 & 1.0000 & -1.4000 & .0000 & .0000 \\
-.0588 & -.0588 & .0000 & .0588 & .0000 & .0000 \\
-1.8000 & .0000 & .0000 & -.4000 & 1.0000 & .0000 \\
-1.6000 & .0000 & .0000 & .2000 & .0000 & 1.0000
\end{bmatrix}
$$

When DM wants to improve f_1 up to 18.0, the maximum of t, t_{max}, for which the matrix B' remains basic for [PLP], is 1.51666. Therefore, the new aspiration level calculated by the automatic trade-off (1.2) is still Pareto optimal. However, if DM wants to improve f_1 up to 20.0, then we have t_{max}=0.7459. Since this value is less than 1.0, we use the dual simplex method at $t=t_{max}$ (which corresponds to the corner point (19, 8) of Pareto surface) in order to obtain the new basic matrix for $t \geq t_{max}$. This is usually also made by once-pivotting. After such a procedure, we can finally get the exact trade-off $\bar{f}_2^{k+1}$ (=2.5) for compensating the improvement of f_1 up to 20.0. It should be noted that only a few pivotting were made there, and therefore the exact trade-off can be obained very quickly.

REFERENCES

Grauer, M., Lewandowski, A. and Wierzbicki, A.P. (1984), DIDASS-Theory, Implementation and Experiences, in M. Grauer and A.P. Wierzbicki (eds.) Interactive Decision Analysis, Proceeding of an International Workshop on Interactive Decision Analysis and Interpretative Computer Intelligence, Springer: 22-30.

Ishido, K., Nakayama, H., Furukawa, K., Inoue, K. and Tanikawa, K. (1987). Multiobjective Managemnet of Erection for Cablestayed Bridge Using Satisficing Trade-off Mehtod, in Y. Sawaragi, K. Inoue and H. Nakayama (eds.) Toward Interactive and Intelligent Decision Support Systems, Springer: 304-312.

Korhonen, P. and Wallenius, J. (1987). A Pareto Race, Working Paper F-180, Helsinki School of Economics, also Naval Research Logistics (1988)

Korhonen, P. and Laakso, J.: Solving Generalized Goal Programming Problems using a Visual Interactive Approach, European J. of Opertional Res., 26, 355-363 (1986)

Nakayama, H. (1984), Proposal of Satisficing Trade-off Method for Multiobjective Programming, Transact. SICE, 20, 29-35 (in Japanese)

Nakayama, H. (1990), Trade-off Analysis using Parametric Optimization Techniques, Research Paper 90-1, Dept. of Appl. Math., Konan Univ.

Nakayama, H. and Sawaragi, Y. (1984). Satisficing Trade-off Method for Interactive Multiobjective Programming Methods, in M. Grauer and A.P. Wierzbicki (eds.) Interactive Decision Analysis, Proceeding of an International Workshop on Interactive Decision Analysis and Interpretative Computer Intelligence, Springer: 113-122.

Rockafellar, R.T. (1970), Convex Analysis, Princeton Univ. Press

Sawaragi, Y., Nakayama, H. and Tanino, T. (1985), Theory of Multiobjective Optimization, Academic Press, New York

Wierzbicki, A. P. (1981), A Mathematical Basis for Satisficing Decision Making, in J. Morse (ed.), Organizations: Multiple Agents with Multiple Criteria, Springer: 465-485.

Wierzbicki, A. (1986), On the Completeness and Constructiveneess of Parametric, Characterization to Vector Optimization Problems. OR spectrum, 8, 73-87.

ON SOLVING MULTIOBJECTIVE LINEAR PROGRAMS VIA MINIMAX SCALARIZING FUNCTION

Włodzimierz Ogryczak
Marshall University
Computer & Information Sciences
Huntington, WV 25755

Krystian Zorychta
Warsaw University
Institute of Informatics
00-901 Warsaw, Poland

Jamil M. Chaudri
Marshall University
Computer & Information Sciences
Huntington, WV 25755

Abstract

Many approaches to the Multiple Objective Linear Programming (especially the reference point method) need solutions to a sequence of single-objective LP problems with a special minimax scalarizing achievement function. The scalarizing achievement function should not be introduced into the structure of the original problem to retain it in the original form whereas the parameters defining the function are changed during the interactive process. It is extremely important if the original matrix is a specially structured one. In this paper we describe a primal simplex algorithm which handles implicitly the minimax objective function.

1. Introduction

Many approaches to the Multiple Objective Linear Programming (MOLP) need solutions to a sequence (or a single) minimax LP problems with a criterion defined as follows (see eg. [5])

$$minimize\ z = \max_{1 \leq k \leq p} \{f_k(x)\} \qquad (1.1)$$

$$f_k(x) = \frac{1}{w_k}(c^k x - r_k) \qquad (1.2)$$

where: p is the number of objectives,

$\mathbf{x}$ is a column n-vector of variables,

c^k are row n-vectors defining several linear objectives ($k=1,2,\ldots,p$),

w_k and r_k are scalar parameters ($k=1,2,\ldots,p$).

Our interest in such problems is connected with the so-called reference point approach to MOLP which turned out to be very attractive procedure for handling multiple objectives in interactive optimization systems (see [2]). In this approach a scalarizing achievement function is minimized as an artificial criterion. The most typical scalarizing achievement functions use the weighted Chebychev distance to the reference point as a basic measure of the solution quality. It implies a need for an efficient optimization technique taking advantages of the minimax objective function like (1.1). Moreover the reference point and the weights are changed during the interactive process and thereby the scalarizing achievement function should not be introduced into the structure of the original problem.

The minimax LP problem can be transformed into a standard LP one with p additional inequalities

$$f_k(x) \le z \qquad \text{for } k=1,2,\ldots,p \qquad (1.3)$$

It introduces, however, the parameters of the scalarizing achievement function into the problem matrix what leads to significantly complex modifications of the problem during the interactive process. Moreover, if the original matrix is a specially structured one, then by consideration of the inequalities (1.3) explicitly, the structure is destroyed.

As we are interested in interactive systems for structured multiobjective problems, namely for multiobjective transshipment problems, we decided to develop a simplex procedure with implicit handling of the additional inequalities (1.3) what saves the problem structure. In case of the transshipment problems it allows us to take advantages of graph representation of the LP basis (see [3]). Some interesting algorithms (parametric and primal-dual) handling implicitly the inequalities (1.3) were proposed by Ahuja (see [1]). They are applicable, however, only to very specific problems with extremely simple objective functions.

In this paper we describe a primal simplex algorithm which handles implicitly the minimax objective function (1.1). More precisely, it handles implicitly the additional inequalities (1.3) connected with individual achievement functions (1.2). The algorithm is based on the classic decomposition principle. Namely, the inequalities (1.3) are treated as a special kind of constraints and handled outside the LP basis like SUB, GUB or VUB constraints in the respective algorithms (see [4]). It leads, obviously, to more complex formulas for the simplex steps but on the other hand it limits the explicit basis representation to the size of the original problem and thereby it allows us to take advantages of the special basis structure (like graph representation for network problems).

The paper is organized as follows. In Section 2 we show how a basic solution to the expanded problem (with additional inequalities (1.3)) can be represented in terms of the original problem. Three next sections describe in details the main elements of the minimax simplex algorithm. Section 3 and 4 deal with the pricing and pivoting ,respectively, whereas in Section 5 updating rules for the simplex basis are discussed.

2. Basic solution

The problem under consideration is as follows. The numerical data consists of an $m{\times}n$ matrix **A** of rank m, which defines the main linear system. A column m-vector **b** defines the corresponding RHS vector, and p row n-vectors $\mathbf{c}^k$ ($k=1,2,\ldots,p$) represent the individual objective functions. In addition, there are given scalar parameters w_k and r_k ($k=1,2,\ldots,p$) defining the individual achievement functions f_k according to the formula (1.2). One has to minimize the scalarizing achievement function (1.1).

Having introduced additional inequalities of type (1.3) and transformed them into equalities we can form the problem as follows

minimize z

subject to

$$-\mathbf{w}z + \mathbf{C}\mathbf{x} + \mathbf{I}\mathbf{s} = \mathbf{r} \qquad (2.1)$$

$$\mathbf{A}\mathbf{x} = \mathbf{b} \qquad (2.2)$$

$$\mathbf{x} \geq 0 \;,\; \mathbf{s} \geq 0 \qquad (2.3)$$

where:

w denotes a column p-vector of the coefficients w_k,

r denotes a column p-vector of the coefficients r_k,

s denotes a column p-vector of objective slacks,

$C=(c_j^k)_{j=1,2,...,n}^{k=1,2,...,p}$ is a $p\times n$ matrix of objective coefficients consisted of rows c^k ($k=1,2,...,p$).

Consider a basic solution to the problem (2.1)-(2.3). Since z is a free variable one can assume, without loss of generality, that the corresponding column belongs to each basis. So, the set of basic variables consists of the variable z and $m+p-1$ structural variables x_j or objective slacks s_k. Note that any basis cannot include all the objective slacks simultaneously as the set of the corresponding columns considered together with the column z is linearly dependent. Thus the set of basic variables includes at least m structural variables x_j and at most $p-1$ objective slacks s_k.

Let $J=\{1,2,...,n\}$ denote the index set of all the structural variables, and similarly $K=\{1,2,...,p\}$ denote the index set of all the objective slacks. Each basic solution to the problem (2.1)-(2.3) is then characterized by the index sets J_N and K_A which represent the nonbasic variables of the corresponding groups. Further, among the columns of matrix **A** that correspond to the basic variables one can select a set of m linearly independent columns which is in fact a basis of the matrix **A**. Let **B** denote such a kernel basis and J_B denote the corresponding index set. The variables of the set J_B will be referred to as the strict basic structurals whereas the other basic structural variables as the adjoined basic structurals.

Taking into account the above partitioning of the variables we get five index sets defining the basic solution:

K_A – the set of nonbasic objective slacks (it will be also referred to as the set of active objectives);

K_P – the set of basic objective slacks (it will be also referred to as the set of passive objectives);

J_N – the set of nonbasic structurals;

J_B – the set of strict basic structurals;

J_D – the set of adjoined basic structurals;

where

$$|J_B| = m \quad \text{and} \quad |J_B| + |J_D| + |K_P| = m + p - 1$$

Values of the nonbasic variables are defined directly as equal to zero what reduces the linear system (2.1)-(2.3) to some basic linear system which generates values of all the basic variables. Namely, having eliminated the nonbasic variables we get the following expanded basic system:

$$-\mathbf{w}_P z + C^P_D \mathbf{x}_D + C^P_B \mathbf{x}_B + I\mathbf{s}_P = \mathbf{r}_P \qquad (2.4)$$

$$-\mathbf{w}_A z + C^A_D \mathbf{x}_D + C^A_B \mathbf{x}_B = \mathbf{r}_A \qquad (2.5)$$

$$D\mathbf{x}_D + B\mathbf{x}_B = \mathbf{b} \qquad (2.6)$$

where

$B=(\mathbf{A}_j)_{j\in J_B}$ is the kernel basis,

$D=(\mathbf{A}_j)_{j\in J_D}$ is the adjoined basis,

$\mathbf{w}_P=(w_k)_{k\in K_P}$ and $\mathbf{w}_A=(w_k)_{k\in K_A}$ denote the corresponding parts of the vector $\mathbf{w}$,

$\mathbf{r}_P=(r_k)_{k\in K_P}$ and $\mathbf{r}_A=(r_k)_{k\in K_A}$ denote the corresponding parts of the vector $\mathbf{r}$,

$\mathbf{s}_P$ denotes the vector of nonactive objective slacks,

$\mathbf{x}_B$ and $\mathbf{x}_D$ denote the vector of strict basic structurals and adjoined basic structurals, respectively,

$C^P_B=(c^k_j)^{k\in K_P}_{j\in J_B}$, $C^P_D=(c^k_j)^{k\in K_P}_{j\in J_D}$, $C^A_B=(c^k_j)^{k\in K_A}_{j\in J_B}$, $C^A_D=(c^k_j)^{k\in K_A}_{j\in J_D}$

are the corresponding parts of the matrix C.

Making the substitution

$$\tilde{\mathbf{x}}_B = \mathbf{x}_B + B^{-1}D\mathbf{x}_D$$

we transform the system (2.4)-(2.6) into the following one

$$-\mathbf{w}_P z + \bar{C}^P_D \mathbf{x}_D + C^P_B \tilde{\mathbf{x}}_B + I\mathbf{s}_P = \mathbf{r}_P \qquad (2.7)$$

$$-\mathbf{w}_A z + \bar{C}^A_D \mathbf{x}_D + C^A_B \tilde{\mathbf{x}}_B = \mathbf{r}_A \qquad (2.8)$$

$$B\tilde{\mathbf{x}}_B = \mathbf{b} \qquad (2.9)$$

where $\bar{C}_D=C_D-C_B B^{-1}D$ what means that the objective coefficients of variables $\mathbf{x}_D$ are replaced by the corresponding reduced costs generated by the basis B, i.e.,

$$\bar{c}^k_j = c^k_j - c^k_B B^{-1} A_j \tag{2.10}$$

From the system (2.9) we get, directly, $\tilde{x}_B$ as a strict basic solution

$$\tilde{x}_B = B^{-1} b \tag{2.11}$$

Next, due to (2.8) one can find z and x_D (if exists) from the system

$$(-w_A | \bar{C}^A_D) \begin{pmatrix} z \\ x_D \end{pmatrix} = r_A - C^A_B \tilde{x}_B \tag{2.12}$$

The matrix of this system will be referred to as the trade-off matrix for active objectives and it will be denoted thereafter by T. So, the values z and x_D are given by the formula

$$\begin{pmatrix} z \\ x_D \end{pmatrix} = T^{-1}(r_A - C^A_B B^{-1} b) \tag{2.13}$$

Finally, we get the original values of the strict basic variables as

$$x_B = \tilde{x}_B - B^{-1} D x_D = B^{-1} b - B^{-1} D [T^{-1}(r_A - C^A_B B^{-1} b)]_D \tag{2.14}$$

Values of the active objective functions are simply given as

$$c^k x = r_k + w_k z \qquad \text{for } k \in K_A$$

If one is interested in values of the nonactive objective functions then the corresponding objective slacks s_P have to be calculated from the system (2.7) as

$$s_P = r_P + w_P z - \bar{C}^P_D x_D - C^P_B \tilde{x}_B \tag{2.15}$$

Note that the above scheme does not introduce any additional operations if there is only one active objective function. The set J_D is then empty and the vector $x_B = \tilde{x}_B$ is given directly by the formula (2.11) whereas (2.13) takes the form of the simple formula for the value z

$$z = \frac{-1}{w_k}(r_k - c^k_B x_B), \qquad k \in K_A$$

3. Pricing

Consider a basic solution given by a kernel basis B and index sets: J_B, J_D, J_N, K_A, K_P. The edges that emanate from the corresponding vertex of the expanded feasible region (2.1)-(2.3) are defined by relaxing particular active inequalities, i.e., by moving some nonbasic variables from its limits. More precisely, the nonbasic variables from the sets J_N or K_A can be increased, i.e., some nonbasic structural variable can be increased or some active objective function can be decreased below the maximal one. In this section we analyze in details how such changes result in the value z.

Let a nonbasic objective slack s_k $(k \in K_A)$ be increased by a parameter θ. It causes that the right hand side $\mathbf{r}_A$ in inequalities (2.5) and (2.8) is replaced by $\mathbf{r}_A - \mathbf{e}_k\theta$, where $\mathbf{e}_k$ denotes the k-th unit vector of the proper dimension. The formula (2.13) takes then the following form

$$\begin{pmatrix} z(\theta) \\ \mathbf{x}_D(\theta) \end{pmatrix} = \mathbf{T}^{-1}(\mathbf{r}_A - \mathbf{e}_k\theta - \mathbf{C}_B^A\mathbf{B}^{-1}\mathbf{b}) \qquad (3.1)$$

Hence

$$z(\theta) - z = - \mathbf{e}_1^T\mathbf{T}^{-1}\mathbf{e}_k\theta , \qquad k \in K_A$$

Thus we find out that if a nonbasic objective slack s_k $(k \in K_A)$ is increased by a value θ then the objective value z increases proportionally by the value $v_k\theta$ where v_k denotes the corresponding reduced cost given by the formula

$$v_k = - \mathbf{e}_1^T\mathbf{T}^{-1}\mathbf{e}_k$$

So, like in the standard simplex algorithm, a nonbasic objective slack s_k is considered as a potential incoming variable if the corresponding reduced cost v_k is negative. The reduced costs have to be calculated, however, according to specific formulas. The (row) vector of all the reduced costs for active objectives $\mathbf{v}=(v_k)_{k \in K_A}$ can be found as a solution to the linear system

$$\mathbf{v}\mathbf{T} = - \mathbf{e}_1^T \qquad (3.2)$$

Now, let a nonbasic structural variable x_j $(j \in J_N)$ be increased by a parameter θ. Then the right hand side $\mathbf{b}$ in equality (2.9) is replaced by $\mathbf{b} - \mathbf{A}_j\theta$, and the vectors $\mathbf{c}_j^A\theta$ and $\mathbf{c}_j^P\theta$ are subtracted from the corresponding RHS vectors r_A and r_P in (2.7) and (2.8), respectively. It implies changes in the values of $\tilde{x}_B$ as well as z and x_D. Namely

$$\tilde{x}_B(\theta) = \tilde{x}_B - B^{-1}A_j\theta$$

and thereby

$$\begin{pmatrix} z(\theta) \\ x_D(\theta) \end{pmatrix} = T^{-1}(r_A - C_j^A\theta - C_B^A B^{-1}b + C_B^A B^{-1}A_j\theta) \qquad (3.3)$$

Hence

$$z(\theta) - z = -e_1^T T^{-1}(C_j^A - C_B^A B^{-1}A_j)\theta = v(C_j^A - C_B^A B^{-1}A_j)\theta$$

Thus we find out that if a nonbasic structural variable x_j ($j \in J_N$) is increased by a value θ then the objective value z increases proportionally by the value $\bar{d}_j\theta$ where $\bar{d}_j$ denotes the corresponding reduced cost given by a special formula. The reduced costs connected with nonbasic structural variables are given as ordinary reduced costs to an artificial objective defined as a linear combination of all active objective functions

$$\bar{d}_j = \Big(\sum_{k \in K_A} v_k c^k\Big)_j - \Big(\sum_{k \in K_A} v_k c^k\Big)_B B^{-1}A_j \qquad (3.4)$$

So, like in the standard simplex algorithm the term

$$\Big(\sum_{k \in K_A} v_k c^k\Big)_B B^{-1} \qquad (3.5)$$

can be calculated once at the beginning of pricing for structural variables and next multiplied by several columns A_j

The entire pricing procedure in our algorithm requires to perform the following operations:

1° compute the ordinary reduced costs d_j^k for $k \in K_A$ and $j \in J_D$, and build up the trade-off matrix T;

2° compute the reduced cost for active objectives (v_k, $k \in K_A$) as solutions to the linear system (3.2);

3° compute the reduced costs for nonbasic structurals ($\bar{d}_j$, $j \in J_N$) according to the formula (3.4);

4° find a negative reduced cost associated with an

active objective or with a nonbasic structural and select the corresponding variable as the incoming one.

4. Pivoting

As it was discussed in the previous section, two kinds of columns can be selected as the incoming one. It leads, obviously, to the corresponding two kinds of the simplex iterations. One is a typical iteration connected with incoming a structural variable whereas in the second one an objective slack (i.e., a variable which does not exist explicitly in the original problem) is selected as entering the basis.

Let q ($q \in J_N$ or $q \in K_A$) denote the index of a variable (structural or objective slack) chosen to enter the basis. Consider changes of the basic solution while the value of the incoming variable is increased by a nonnegative parameter θ. The value of the incoming variable is then equal to θ. Values of the other nonbasic variables (structural as well objective slacks) remain on the same level, i.e., they are steadily equal to 0. But the values of the basic variables vary proportionally to the parameter θ, i.e.

$$x_j(\theta) = x_j - \xi_j \theta \qquad \text{for } j \in J_B \cup J_D \qquad (4.1)$$

$$s_k(\theta) = s_k - \sigma_k \theta \qquad \text{for } k \in K_P \qquad (4.2)$$

Exact formulas for the coefficients ξ_j and σ_k depend, certainly, on the kind of the incoming variable.

Consider at first the standard case when $q \in J_N$, i.e., the incoming variable is a structural one. Let $\mathbf{y}$ be a representation of the column $\mathbf{A}_q$ in the kernel basis $\mathbf{B}$, i.e.,

$$\mathbf{y} = \mathbf{B}^{-1}\mathbf{A}_q \qquad (4.3)$$

Moreover, let $\mathbf{u}$ be a solution to the linear system

$$\mathbf{T}\mathbf{u} = (\mathbf{C}_q^A - \mathbf{C}_B^A \mathbf{y}) \qquad (4.4)$$

and $\bar{\mathbf{u}}$ denote a vector consisted of u_j corresponding to the index set J_D, i.e., $\bar{\mathbf{u}} = (u_j)_{j \in J_D}$. Due to (3.3), (2.14) and (2.15) the coefficients ξ_j and σ_k are given by the following formulas:

$$\xi_j = u_j \qquad \text{for } j \in J_D \qquad (4.5)$$

$$\xi_j = y_j - e_j^T B^{-1} D\bar{u} \qquad \text{for } j \in J_B \qquad (4.6)$$

$$\sigma_k = (w_k | -\bar{c}_D^k) u - c_B^k y \qquad \text{for } k \in K_P \qquad (4.7)$$

In the case when $q \in K_A$, i.e., the incoming variable is an objective slack, the coefficients ξ_j and σ_k are given by the same formulas (4.5)-(4.7) provided that the vector $\mathbf{y}$ is omitted and the vector $\mathbf{u}$ is a solution to the system with another RHS vector. More precisely in this case one must put $\mathbf{y} = \mathbf{0}$ and take $\mathbf{u}$ as a solution to the system

$$Tu = e_q \qquad (4.8)$$

Taking into consideration the constraints (2.3) and the formulas (4.5)-(4.7) for changes of the basic variables we get the following upper bound on θ:

$$\theta \le \bar{\theta} = \min \{\theta_1, \theta_2, \theta_3\} \qquad (4.9)$$

where

$$\theta_1 = \begin{cases} \min\limits_{j \in J_B} \left\{ \dfrac{x_j}{\xi_j} : \xi_j > 0 \right\} & \text{if exists} \\ +\infty & \text{otherwise} \end{cases}$$

$$\theta_2 = \begin{cases} \min\limits_{j \in J_D} \left\{ \dfrac{x_j}{\xi_j} : \xi_j > 0 \right\} & \text{if exists} \\ +\infty & \text{otherwise} \end{cases}$$

$$\theta_3 = \begin{cases} \min\limits_{k \in K_P} \left\{ \dfrac{s_k}{\sigma_k} : \sigma_k > 0 \right\} & \text{if exists} \\ +\infty & \text{otherwise} \end{cases}$$

If such a defined $\bar{\theta}$ is equal to the infinity then the problem under consideration is unbounded. Otherwise, the corresponding $\theta_t = \bar{\theta}$ determines a new active constraint which defines a new basic solution. The argument of the corresponding minimum (denoted in the following by $r \in J_B \cup J_D \cup K_P$) points out the outgoing variable. So, we have three types of the simplex iteration corresponding to several quantities θ_t.

Note that the incoming variable not always can be put into the same subset of basic indices (J_B, J_D or

K_P) which the outgoing variable is eliminated from. It is connected with two different kinds of variables: the structurals and the objective slacks. Apart from one case it does not introduce any doubt. However, if the incoming variable is not a structural one ($q \in K_A$) and simultaneously a strict basic structural turns out to be selected as the outgoing one ($r \in J_B$), then the direct modification of the index sets leads us to too small set J_B. In this case one must additionally move one variable (say r') from the set J_D to the set J_B. Due to the considerations from Section 2 such a transformation is possible.

Changes of the basic solution structure depending on the iteration type are summarized in Table 1. Rows of the table correspond to several types of the simplex iterations (θ_t) whereas columns are connected with two kinds of the incoming variable. Each cell provide us with the result index sets of basic variables. It is assumed that the index sets before the iteration are given as J_B, J_D, K_P, respectively.

Table 1. Exchange rules

	$q \in J_N$	$q \in K_A$
θ_1	$J_B \cup \{q\} \setminus \{r\}$ J_D K_P	$J_B \setminus \{r\} \cup \{r'\}$ $J_D \setminus \{r'\}$ $K_P \cup \{q\}$
θ_2	J_B $J_D \cup \{q\} \setminus \{r\}$ K_P	J_B $J_D \setminus \{r\}$ $K_P \cup \{q\}$
θ_3	J_B $J_D \cup \{q\}$ $K_P \setminus \{r\}$	J_B J_D $K_P \cup \{q\} \setminus \{r\}$

Recapitulating, the pivoting procedure requires to perform the following operations:

1° in case of $q \in J_N$, find a representation of the column $\mathbf{A}_q$ in the kernel basis $\mathbf{B}$ (see (4.3));

2° find the vector of coefficients ξ_j for adjoined basic variables (see (4.5)) as a solution to the linear system (4.4) or (4.8), depending on the kind of the incoming variable ($q \in J_N$ or $q \in K_A$, respectively);

3° calculate the vector of coefficients ξ_j for strict basic variables (see (4.6)) as the linear combination of the adjoined basic structural columns represented in the kernel basis **B** with added vector **y** (if exists), i.e.,

$$\mathbf{y} - \sum_{j \in J_D} u_j \mathbf{B}^{-1} \mathbf{A}_j \tag{4.10}$$

4° calculate the vector of coefficients σ_k (see (4.7)) as the linear combination of columns of the nonactive objective coefficients defined as follows

$$u_1 \mathbf{w}_p - \sum_{j \in J_D} u_j \bar{C}^p_j - \sum_{j \in J_B} y_j C^p_j$$

5° compute the quotients defining the quantities θ_t and select the outgoing variable r.

5. Basis transformations

The exchange rules summarized in Table 1 provide us with the structure of a new basic solution. The nonbasic structurals are given directly as equal to zero whereas the strict and adjoined basic structurals are defined according to the formulas (2.11)-(2.14). The new sets of basic and nonbasic indices (i.e., J_B, J_D, J_N, K_P and K_A) define, obviously, a new kernel basis **B** and a new adjoined basis **D**. However, the new basic matrices **B** and **D** are very close to the previous ones and therefore we are interested in transformations performed on them in connection with several exchange rules of Table 1.

Analyzing carefully the formulas for pricing and pivoting one can easily notice that in both these steps we need to solve (similarly as in the standard simplex algorithm) some linear system with the kernel basis **B** (see (3.5) and (4.3), respectively) and additionally a linear system with the matrix T (see (3.1) and (4.4), respectively). Moreover, during the pivoting a linear combination of the adjoined basic structural columns represented in the kernel basis **B** is computed as well as these columns can be used while computing the reduced costs $\bar{c}_j^k$ for the adjoined basic variables to build the matrix **T**. Therefore the algorithm should use (keep and update) some representation (factorization) of the inverse kernel basis $\mathbf{B}^{-1}$ and the inverse matrix $\mathbf{T}^{-1}$ as well as a column form of the adjoined basis represented in the kernel basis, i.e., $\mathbf{B}^{-1}\mathbf{D}$. Furthermore the basic solution vector $\mathbf{x}$ (including both $\mathbf{x}_B$ and $\mathbf{x}_D$) can be, similarly as in the standard simplex algorithm, updated at each iteration instead of direct calculations according to the formulas from Section 2.

In the standard simplex algorithm there is only one kind of basis transformation: replacing a basis column by another one. In the algorithm under consideration such a transformation of the kernel basis **B** appears in two cases (among six, see Tab.1). In both cases it is exactly the same simple transformation provided that the matrix $\mathbf{B}^{-1}\mathbf{D}$ is stored by columns, i.e., $\mathbf{B}^{-1}\mathbf{A}_{r'}$ is directly available like $\mathbf{B}^{-1}\mathbf{A}_q$ which is known from the pivoting. So, the kernel basis **B** is changed by the standard simplex basis transformation and thereby any factorization scheme like Bartels-Golub, Forrest-Tomlin or Fletcher-Mathews as well as some special techniques like graph representation for network problems can be used for representation of its inverse.

The adjoined basis **D** can be changed in various manners. In two cases: $q \in J_N$ and $r \in J_B$ or $q \in K_A$ and $r \in K_P$ it remains without any changes. In one case $q \in J_N$ and $r \in J_D$ the adjoined basis **D** saves its dimension and the transformation depends only on replacing one column by another one. Further, in two cases: $q \in K_A$ and $r \in J_B$ or $q \in K_A$ and $r \in J_D$ dimension of the matrix is decreased since one column is eliminated from the matrix whereas in one case $q \in J_N$ and $r \in K_P$ it is increased as one column is added to the matrix **D**.

Instead of the original matrix **D** the algorithm should rather handle its representation in the kernel basis, i.e., $\mathbf{B}^{-1}\mathbf{D}$. In such an approach the transformations of the kernel basis **B** are composed with the transformations of matrix **D**. Fortunately, these two kinds of transformations are quite independent of each other. Thus while updating $\mathbf{B}^{-1}\mathbf{D}$ the following operations can be needed:

(a) one column is deleted;

(b) the elimination process equivalent to the kernel basis transformation is performed on all the columns;

(c) a representation of one column in the kernel basis is added.

The operation (a) is, obviously, a trivial one. Similarly the operation (c) is extremely simple since a representation of the entering column in the kernel basis is explicitly known from pivoting. The most complex operation (b) is performed only when the kernel basis is changed (two cases among six) and it requires only to extend the kernel basis elimination process on all the columns of matrix $\mathbf{B}^{-1}\mathbf{D}$.

6. Concluding remarks

The presented minimax simplex algorithm has been implemnted on microcomputer IBM-PC/AT as a solver for a pilot version of a new decision support system for multiple criteria transportation problems. The implementation has confirmed that the algorithm allows to joint the efficient interactive scheme of the aspiration/reservation approach with preserving the original structure of the problem. Namely, it allows to use the graph representation technique for handling the kernel basis $\mathbf{B}$. Due to using these techniques, while analyzing a problem with a few hundreds of arcs, the system needs only a several seconds (on a regular IBM-PC/AT with 8MHz processor) to find out a new efficient solution that meets the decision maker's aspiration and reservation levels.

References

[1] Ahuja R. K.: Minimax linear programming problem. *Operations Research Letters* **4** (1985), 131-134.

[2] *Aspiration Based Decision Support Systems - Theory, Software and Applications.* A. Lewandowski and A. P. Wierzbicki (eds.), Springer, Berlin 1989.

[3] Grigiriadis M. D.: An Efficient Implementation of the Network Simplex Method. *Mathematical Programming Study* 26 (1986).

[4] Nazareth L.: *Computer Solution of Linear Programs.* Oxford Univ. Press 1988.

[5] Steuer R. E.: *Multiple Criteria Optimization - Theory, Computation & Applications.* J. Wiley 1986.

DEVELOPING AN EXPERT SYSTEM PROTOTYPE FOR INTELLIGENT DECISION SUPPORT

Vladimir M. Ozernoy
School of Business and Economics
California State University, Hayward
Hayward, California 94542 U.S.A.

ABSTRACT

The paper is concerned with the use of expert system technology to formalize knowledge about discrete alternative MCDM methods and the usage of these methods. In this paper, a prototype advisory MCDM knowledge-based system is described and discussed. A sample consultation sessions is presented. The system uses the information about the decision problem and available MCDM methods to guide the user through an analysis of the decision situation. Thus, the system helps the decision maker or analyst both select and justify the selection of the most appropriate MCDM method.

1. INTRODUCTION

This paper presents a formal procedure to assist the user in systematically selecting the most appropriate method in a particular application. By integrating the procedure with a large amount of MCDM knowledge and the powerful reasoning and explanation capabilities of a microcomputer based advisory expert system, a decision-aiding tool providing new input into the decision-making process could be developed.

Based on a typology suggested by Zionts (1985), MCDM methods can be classified into those for solving discrete alternative MCDM problems and those for solving multiobjective mathematical programming (MOMP) problems. While much of the MCDM research is devoted to developing and/or investigating MOMP methods, the focus of this paper is on discrete alternative MCDM methods for solving discrete alternative MCDM problems.

The differences that substantiate the need for the consideration of discrete alternative MCDM methods as a distinct area of research and applications are discussed in detail in Zionts (1985), Hwang and Yoon (1981), and many other publications. Briefly put, MOMP methods address problems whose constraints have explicit mathematical expressions and in which alternatives are implicit and possibly infinite in number. In contrast, the distinguishing feature of discrete alternative MCDM problems is that there is a finite number of alternatives (although it could be a very large number), the constraints are implicit, and the required ordering of a set of feasible alternatives should be obtained and justified.

In (Ozernoy, 1989), a conceptual framework was developed for a rule-based

expert system that would assist the decision maker or analyst to justify the selection of the most appropriate discrete alternative MCDM method in a given decision situation. The purpose, structure, and possible applications of the expert system were also discussed.

There are three major steps in the development of an MCDM expert system: (1) MCDM knowledge identification, (2) MCDM knowledge acquisition, and (3) MCDM knowledge representation. This paper describes one of these major steps: MCDM knowledge representation. An expert system prototype developed is characterized and some screens of a sample consultation are demonstrated. A small prototype called "MCDM Advisor" was developed in order to (1) investigate the possibility of using the expert system technology to formalize knowledge about discrete alternative MCDM methods, and (2) test the basic structure and concept of an MCDM expert system before committing substantial resources for its development.

2. "MCDM ADVISOR": AN OVERVIEW

"MCDM Advisor" was developed using a microcomputer based expert system shell called VP-Expert (Hicks and Lee, 1988). The knowledge base contains nine MCDM methods, including both compensatory and noncompensatory methods. These are: disjunctive method (MacCrimmon, 1973), conjunctive method (MacCrimmon, 1973), dominance (MacCrimmon, 1973), lexicographic method (Fishburn, 1974), weighted-additive evaluation function with partial information (Kirkwood and Sarin, 1985), simple multiattribute rating technique (Edwards, 1977), multiattribute utility functions (Keeney and Raiffa, 1976), aspiration level interactive model (Lotfi et al., 1988), and ELECTRE-1 (Roy, 1971).

In the MCDM knowledge base, backward chaining is used to find the appropriate MCDM method. With this approach, the system starts with the hypothesis that a particular method is appropriate for a given decision situation, then reasons backward, looking for facts and rules to support it. If the first hypothesis fails, the system switches to another.

The MCDM knowledge base is composed of three parts or blocks (see Hicks and Lee, 1988 for more detail). The first block, the Actions Block, controls the user consultation session. The only actions are to display an opening message, find the value of the goal variable Advice and then display that value. The Find clause instructs the inference engine to seek the value of Advice, which becomes the goal variable for the consultation. When all the instructions in this section of the knowledge base have been performed, the consultation will be complete.

The Rules Block contains rules written in an IF/THEN format. The inference engine first looks for a rule having the final goal, Advice, in its conclusion. Next, it tests the conditions for the rule as specified in the IF part of the rule. These constitute sub-goals for the search. Some of these sub-goals are satisfied directly by asking the user questions. Other sub-goals are determined through other rules. This process continues until all the conditions of the original goal, Advice, have been satisfied. If at any point in

this search process a sub-goal cannot be satisfied, the inference engine "backs up" and looks for another rule to satisfy the goal. If no other rules can be found, it makes attempts to obtain the value from the user. If this approach does not work, the search fails and no value is found.

The last block, the Statements Block, contains messages that direct the interaction with the user to elicit the information about the decision problem as well as the decision maker's preference information. This block contains the ASK statements and any special conditions associated with them. The presence of a CHOICES statement indicates that the user will be presented with a menu of available values, from which he or she selects an answer by positioning the cursor. If an ASK statement does not have an accompanying CHOICES statement, the user will be expected to enter a value.

3. A SAMPLE CONSULTATION SESSION

When the user starts the dialogue, the inference engine begins by displaying the welcome message in the DISPLAY cause (Figure 1) and then processing the FIND clause. The inference engine uses backward chaining to find the value of the goal.

Figure 1 shows the structure of the consultation screen. The upper window is the Consultation Window. The program will display messages to the user in this location, such as requests for information and displays of results.

The left window is the Rules Window. It displays the value that is now sought and the rule that is being evaluated. The right window is the Results Window. It tells the user which variables have obtained values, what those values are, and what confidence factor is given to each value.

The first question to the user in the consultation is about the type of the user's decision problem (Figure 2). The question displayed in the Consultation Window can be answered by moving the lightbar to the correct answer. In the sample consultation session, the choice is 1, "To find the preferred alternative".

The Rules Window provides the user with a trace of the rules presently being used in the consultation. In this window, the user can see which rule is being evaluated. The Results Window shows the current status of the reasoning.

The second question in the sample consultation is "Are you willing to make tradeoffs between attributes?" (Figure 3). The answer is NO. At this point, all of the windows are active, and we can observe the progress of the consultation.

The third question is "Can you rank-order all the attributes in decreasing order of importance?" The answer is YES. The fourth question in the sample consultation is "Is an alternative acceptable if the attribute levels of the chosen alternative exceed the cutoff values ('levels of aspiration') of each of the attributes?" The answer is YES.

The preceding questions remain on the screen until they are scrolled off by subsequent questions. The last question (Fig. 4) is "Would you like to explore the nondominated frontier by allowing the user to establish and adjust levels of aspiration?" The answer is YES.

```
Welcome to the MCDM Advisor!
            This is a prototype advisory MCDM knowledge-based
            expert system.

            The following consultation will help you to determine
            the most appropriate discrete alternative MCDM method
            for your decision problem.

            Press 1 to begin the consultation.

1Help    2Go     3WhatIf  4Variable 5Rule   6Set    7Edit    8Quit
1Help 2How? 3Why? 4Slow 5Fast 6Quit
```

Figure 1. Welcome message

```
            Press 1 to begin the consultation.
Please identify the type of your decision problem
            1 - To find the preferred alternative
            2 - To rank-order feasible alternatives
            3 - To find all acceptable alternatives
            4 - To identify all nondominated alternatives
            5 - To partially order nondominated alternatives
 1 ◀                  2                    3
 4                    5

Testing 1                               Consultation = 1 CNF 100
RULE 1 IF                               Consultation = 1 CNF 100
Problem = 3 AND
Tradeoffs = No AND
One_cutoff_value = Yes
THEN
Advice = Disjunctive_Method CNF 100
Finding Problem

    Enter to select   END to complete   /Q to Quit   ? for Unknown
```

Figure 2. First question in the sample consultation

```
Are you willing to make tradeoffs between attributes?
 Yes                    No ◀

Can you rank-order all the attributes in
decreasing order of importance?
 Yes ◀                  No

Is an alternative acceptable if  the attribute levels of the chosen
alternative exceed the cutoff values for each of the attribute?
 Yes ◀                  No
```

```
Problem = 1 AND                          Pause = 1 CNF 100
Tradeoffs = No AND                       Pause = 1 CNF 100
All_cutoff_values = Yes AND              Problem = 1 CNF 100
Change_cutoff_values = Yes               Tradeoffs = No CNF 100
THEN                                     Rank_attributes = No CNF 100
Advice = AIM-Aspiration_Level_Interacti
ve_Model CNF 100
Finding All_cutoff_values
```

```
Enter to select   END to complete   /Q to Quit   ? for Unknown
```

Figure 3. Second, third, and fourth questions in the sample consultation

```
 Yes ◀                  No

Would you like to explore the nondominated frontier by allowing the
user to establish and adjust levels of aspirations?
 Yes ◀                  No

The best advice we have for you is as follows:
        the recommended MCDM method is AIM-Aspiration Level Interactive Model
CNF 100.
```

```
Tradeoffs = No AND                       Pause = 1 CNF 100
All_cutoff_values = Yes AND              Problem = 1 CNF 100
Change_cutoff_values = Yes               Tradeoffs = No CNF 100
THEN                                     Rank_attributes = No CNF 100
Advice = AIM-Aspiration_Level_Interacti  All_cutoff_values = Yes CNF 100
ve_Model CNF 100                         Change_cutoff_values = Yes CNF 100
Finding All_cutoff_values                Advice = AIM-Aspiration_Level CNF 100
Finding Change_cutoff_values
```

```
1Help   2Go   3WhatIf   4Variable 5Rule   6Set   7Edit   8Quit
1Help 2How? 3Why? 4Slow 5Fast 6Quit
```

Figure 4. Last question and advice

If the answer is YES, there will be no request for additional information. Instead, the message states that the goal has been reached. The best advice that this MCDM knowledge base has for the user is AIM - the Aspiration Level Interactive Model.

If the answer to the last question were NO, the goal would not have been reached and the Advice would have been "The recommended method is not found in the MCDM knowledge base" (Figure 5).

Once the rules have been entered, a convenient way of monitoring their execution is a decision trace. The system records the steps taken in a consultation and can display a trace after the consultation has been completed. Two forms of traces are used to show the consultation process: the Graphics display or the Text display. Figure 6 shows the Text display corresponding to the selection of the AIM method. The display clearly shows the path of the inference engine and the values that it obtains as it tests various rules.

Several options are available during a consultation. For example, the WHY? option shows why a question is asked or why a specific recommendation has been made.

In an additional sample consultation session, the first question to the user is about the type of the user's decision problem. The choice is 3 - "To find all acceptable alternatives". The second question is: "Are you willing to make tradeoffs between attributes?" The answer is NO (Figure 7). The third question is: "Is an alternative acceptable if at least one attribute level of the chosen alternative exceeds a desirable level?" The answer is NO (Figure 7).

And the fourth question is: "Is an alternative acceptable if the attribute levels of the chosen alternative exceed the cutoff values of each of the attributes?" If the user asks WHY? at this point, the display will be as shown in Figure 8. Thus, the WHY? option shows why a question was asked.

4. SUMMARY

A small demonstration advisory MCDM expert system called "MCDM Advisor" was developed using a microcomputer based expert system shell called VP-Expert. Experiments with "MCDM Advisor" indicate that the basic structure of the system and the interrelationship of its components will permit the development of a prototype MCDM expert system. Experimentation has already shown that an increase in the number of MCDM methods in the knowledge base did not result in a significant increase of the consultation time.

As VP-Expert interfaces with external programs (such as dBaseII and Lotus 1-2-3) that may offer greater possibilities to the knowledge base designer, these capabilities can be used both in consultation and in computer-aided instruction. As the user may require additional information, such as definitions of various terms (e.g., "preference independence" or "value function"), clarifying examples, the description of various methods, summaries of case studies and published analyses of the strengths and weaknesses of various methods, information about existing software, and further

```
Yes ◀                    No

Would you like to explore the nondominated frontier by allowing the
user to establish and adjust levels of aspirations?
Yes                      No ◀

The best advice we have for you is as follows:
        the recommended MCDM method is not found in the MCDM knowledge base C
F 100.
```

```
Advice <> Kirkwood_&_Sarin_Method AND      Consultation = 1 CNF 100
Advice <> SMART AND                        Problem = 1 CNF 100
Advice <> AIM-Aspiration_Level_Interact    Tradeoffs = No CNF 100
ive_Model AND                              Rank_attributes = No CNF 100
Advice <> Electre_I                        All_cutoff_values = Yes CNF 100
THEN                                       Change_cutoff_values = No CNF 100
Advice = not_found_in_the_MCDM_knowledg    Advice = not_found_in_the_MCD CNF 100
e_base CNF 100
```

```
1Help    2Go     3WhatIf   4Variable 5Rule    6Set     7Edit    8Quit
1Help 2How? 3Why? 4Slow 5Fast 6Quit
```

Figure 5. Advice when the goal has not been reached

```
Testing b:\MCDM.kbs
(= yes CNF 0 )
(= 1 CNF 100 )
!   Advice
!   !   Testing 1
!   !   !   Problem
!   !   !   !   (= 1 CNF 100 )
!   !   Testing 2
!   !   Testing 3
!   !   Testing 4
!   !   !   Tradeoffs
!   !   !   !   (= No CNF 100 )
!   !   !   Rank_attributes
!   !   !   !   (= No CNF 100 )
!   !   Testing 5
!   !   Testing 6
!   !   Testing 7
!   !   !   All_cutoff_values
!   !   !   !   (= Yes CNF 100 )
!   !   !   Change_cutoff_values
!   !   !   !   (= Yes CNF 100 )
!   !   (= AIM-Aspiration_Level_Interactive_Model CNF 100 )
```

Figure 6. Example of a decision trace

```
Are you willing to make tradeoffs between attributes?
 Yes                      No ◀

Is an alternative acceptable if at least one attribute level
of the chosen alternative exceeds a desirable level?
 Yes                      No ◀

Is an alternative acceptable if  the attribute levels of the chosen
alternative exceed the cutoff values for each of the attribute?
 Yes                      No

Testing 2                                 Consultation = 1 CNF 100
RULE 2 IF                                 Consultation = 1 CNF 100
Problem = 3 AND                           Problem = 3 CNF 100
Tradeoffs = No AND                        Tradeoffs = No CNF 100
All_cutoff_values = Yes                   One_cutoff_value = No CNF 100
THEN
Advice = Conjunctive_Method CNF 100
Finding All_cutoff_values

 Enter to select   END to complete    /Q to Quit   ? for Unknown
```

Figure 7. Additional sample consultation session

```
Are you willing to make tradeoffs between attributes?
 Yes                      No ◀

Is an alternative acceptable if at least one attribute level
of the chosen alternative exceeds a desirable level?
 Yes                      No ◀

Is an alternative acceptable if  the attribute levels of the chosen
                              [ WHY ]
  The question is being asked because:
  If you want to classify the decision alternatives into acceptable/
  not acceptable categories AND an alternative is acceptable for you
  if attribute levels of the chosen alternative exceed the cutoff values
  for each of the attributes AND tradeoffs between attributes are not
  permitted, then the appropriate MCDM method is CONJUNCTIVE METHOD.
  (Press Any Key To Continue)

Finding All_cutoff_values

  1Help     2How?     3Why?     4Slow     5Fast     6Quit
  Ask why a question was asked
```

Figure 8. WHY? question

references, this additional information, to the degree it is available for different methods, will be provided by a menu driven sequence of screens.

Subsequently, a fifth-generation MCDM decision support system can be developed based on a stand-alone advisory system. It would allow not only recommending but also executing the recommended MCDM method.

REFERENCES

Edwards, W., "Use of Multiattribute Utility Measurement for Social Decision Making." In D.E. Bell, R.L. Keeney and H. Raiffa (eds.), Conflicting Objectives in Decisions. Wiley, New York, pp. 247-276, 1977.

Fishburn, P.C., "Lexicographic Order, Utilities and Decision Rules: A Survey", Management Science, Vol. 20, pp. 1442-1471, 1974.

Hicks, R. and Lee, R., VP-Expert for Business Applications. Holden- Day, Inc., Oakland, 1988.

Hwang, C.L. and Yoon, K., Multiple Attribute Decision Making -- Methods and Applications: A State-of-the Art Survey. New York:Springer- Verlag, 1981.

Keeney, R.L. and Raiffa, H., Decisions with Multiple Objectives. Wiley, New York, 1976.

Kirkwood,C.W. and Sarin, R.K., "Ranking with Partial Information: A Method and an Application", Operations Research, Vol. 33, pp. 38-48, 1985.

Lotfi, V., Stewart, T.J., and Zionts, S., "An Aspiration - Level Interactive Model for Multiple Criteria Decision Making," Working Paper No. 701, Department of Management Science and Systems, School of Management, State University of New York at Buffalo, Buffalo, New York 14260, 1988.

MacCrimmon, K.R., "An Overview of Multiple Objective Decision Making". In: J.L. Cochrane and M. Zeleny (eds.), Multiple Criteria Decision Making, pp. 18-44. Columbia, South Carolina, The University of South Carolina Press, 1973.

Ozernoy, V.M. "Some Issues in Designing an Expert System for Multiple Criteria Decision Making". In: B. Rohrman, L.R. Beach, C. Vlek, and S.R. Watson (eds.), Advances in Decision Research, pp. 237- 254. North Holland, Amsterdam, 1989.

Roy, B., "Problems and Methods with Multiple Objective Functions", Mathematical Programming, Vol. 1, pp. 239-266, 1971.

Zionts, S., "Multiple Criteria Mathematical Programming: An Overview and Several Approaches. In: G. Fandel and J. Spronk (eds.), Multiple Criteria Decision Methods and Applications, pp. 85-128. Heidelberg: Springer-Verlag, 1985.

WEEKLY OPERATION OF A WATER RESOURCE SYSTEM WITH RELIABILITY CRITERIA

E. PARENT, Applied Maths Dpt, ENGREF, 19 av du Maine, 75015 Paris, France.
F. LEBDI, Rural Engineering Dpt, INAT, 43 av Charles Nicolle, 1002 Tunis, Tunisia.

1 INTRODUCTION

The paper presents a case study of the weekly operation of a French water resource system where two conflicting objectives have to be satisfied, namely irrigation and water quality demand. The technique for solution is multi-objective dynamic programming (BELLMAN, 1957).
Since the works of MASSE (1959), dynamic programming methods have been widely used in water resource operation and planning (YAKOWITZ ,1982);YEH ,1985). In most systems possessing an agricultural component, a multicriterion decision making approach should be considered. For instance, SZIDAROVSZKY and DUCKSTEIN (1986) used sequential multicriterion decision making techniques by combining dynamic programming and multiobjective decision making. A second difficulty comes from the stochastic aspects of hydrology and the unpredictable attitude of water users. One can then apply dynamic programming on hydrological scenarii (GOUSSEBAILLE and ROCHE 1988) or check the adequacy of a restricted class of operating rules (BHASKAR and WHITLACH, 1987). Performing complete stochastic dynamic programming algorithms increases computational burden. It must also take into account both risk and reliability analysis .
In this paper, the two objective problem involving irrigation and water quality is modeled by two parameters and solved by stochastic dynamic programming under reliability constraints. It can be shown that each set of parameters is associated with a compromise equilibrium and can be adjusted by simulation. The model can also be used to derive the impact of a non-continuous (Boolean) decision such as a "crisis" situation enforced by law when all farmers have to suspend irrigation.

2 THE NESTE WATER RESOURCE SYSTEM

The Lannemezan Plateau in the Gascogne basin is bordered by the Pyrenees Mountains in the South and the Gascogne and Adour rivers in the East and West . Consequently, the Gascogne basin rivers are perhaps the only French "oueds" that dry out in summer when no precipitation occurs. As early as 1850, a 25 km long canal (the Neste canal) was built in order to convey water from the mountain river Neste to the top of the Gascogne watershed. From now on, the "Neste system" will refer to the so-called Association of the Canal and the Gascogne Watershed Basin including its irrigation zones and domestic water users. For more than a century, this abundant new facility has allowed the development of 30,000 ha of intensive irrigated corn agriculture.
A second purpose of the Neste canal is to supply continuous minimum levels in the Gascogne rivers so as to permit water quality control and domestic supply. In order to fulfill increasing irrigation demand, new dams [42 hm^3] have been built in the hills of the Gascogne basin in addition to the 48hm^3 reservoir in the Pyrenees Mountains that already supplies the river Neste.
Nowadays, water supply and irrigation demand compete for the same limited resource. The aim of our study is to develop a multiple-purpose model so that the decision maker can reach a rational allocation of the Neste water resource.

3 SYSTEM APPROACH

In a first step, we will consider the weekly operation of the Neste system from a global point of view as follows:

- a single reservoir S(t) at the beginning of week t including both mountains and Gascogne storages. t is expressed in weeks.

- a lumped release L(t) that is to be selected before knowing what the inflows or demands will be.

- The river Neste inflow N(t). In a former hydrological study based on 15 years data, it was found that Log N(t) could be represented as a Markov chain with lag 1.

- The water is diverted into two valleys:
  * the natural Neste river valley,
  * the Gascogne watershed via the Neste canal.

- D(t) is the total irrigation demand that occurs in week t in the Gascogne plateau.
- G(t) is the natural inflow of the lumped Gascogne river. D(t) and G(t) are assumed to be statistically independent variables.

Figure 1 represents the Neste system with identification of the above variables. r_n and r_g are the outputs of the system; here target output requirements for domestic use and quality control must be achieved. The study period begins on the 1st of June when all reservoirs are full and lasts 20 weeks.

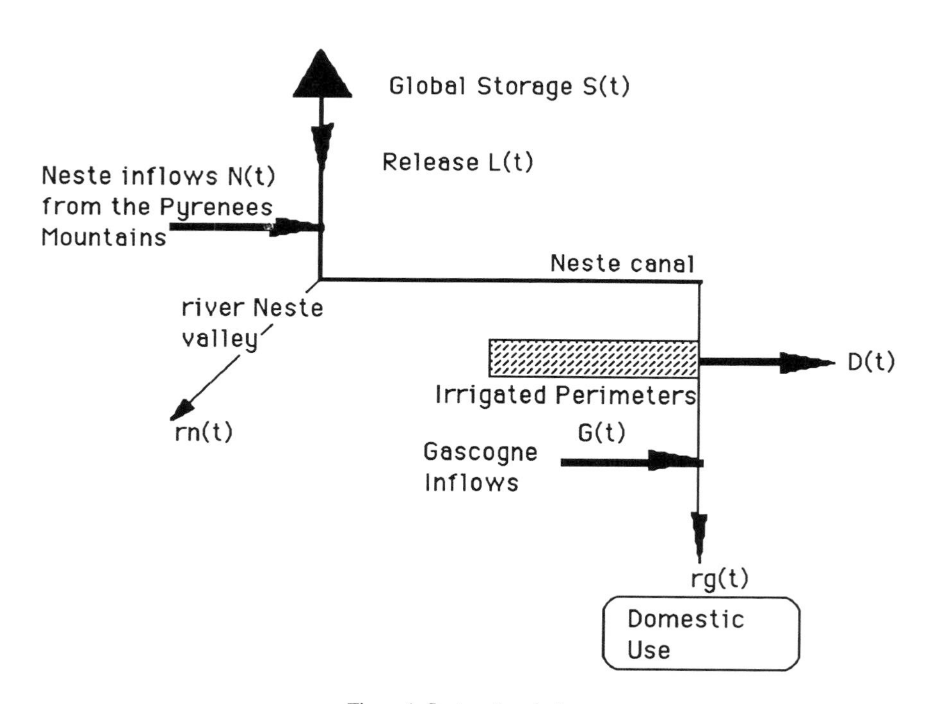

Figure 1: System description

The Neste system operates in three possible modes:

- Normal mode

In normal periods, the reservoirs release enough water to satisfy irrigation and maintain a level $r_n(t) \geq qobjN = 4m^3/s$ and a certain level $r_g(t)$ belonging to the interval [qminG,qobjG] with qminG= $5.4m^3$ and qobjG = $12.4m^3/s$.

- Precrisis mode

When the system manager feels that there could be difficulties to satisfy the irrigation purpose, the output level of the river Neste $r_n(t)$ can be dropped to qminN = $3m^3/s$. The farmers' union has often asked the system manager to do so during the past twenty years.

- Crisis mode

The next step consists in stopping irrigation (crisis mode). This action, which can only be enforced by law , has occurred only very rarely in the past. In such a case of shortage, domestic use and water quality are higher priority objectives than irrigation. When the system enters this crisis mode, it practically cannot switch back to either normal or precrisis modes because of administrative inertia and because restarting irrigation once the crop has received no water for a week or two would be useless.

Which criteria should be used?

While we could have used a multiattribute utility approach which has already been applied to such agro-hydraulic engineering studies (see for instance KRZYSZTOFOWICZ, 1982), the present approach is a decision making aid designed to help the system manager to formulate which incidents should be avoided . Indeed, it is often easier to think in terms of incidents and failures than in performance indices to be maximized. We will focus on deficit incidents that are events that occur beyond a prespecified range of bounds. In addition to the three modes (normal, precrisis, and crisis) formerly described, two other incidents can happen in the Neste system:

- quality failure, i.e. reaching r_g < qminG which could be caused by either an inadequate release, unexpected low levels of natural inflows or high levels of irrigation demands.
- storage failure which occurs when the reservoirs are empty before the 20th week due to poor operation rules or poor hydrological conditions.

Various performance indices which can be associated with each incident once an operation rule has been set, include:

- risk, that is, the probability that the incident will occur at time t;
- resilience, that is, the average recovery time of the system needs once an incident has occurred;
- vulnerability, which is an indicator related to the severity of the incident and its consequences for the resource users.

According to these concepts developed in Plate and Duckstein (1988) we could define as many as 15 criteria from which the ones given below have been selected.

Performances indices :

* Vfinal : average final volume at T = 20 weeks.

* VulnerabilityP : vulnerability index estimating the average discrepancy between the real output r_g and the minimum level qminG in case of a quality failure: it can be computed, using the expectation operator E over the stochastic variables N(t), D(t) and G(t), as the quantity:

VulnerabilityP = E(QminG-r_g) when $r_g \leq$ QminG.

* Vcrisis : frequency of operating in crisis mode.
It is also related to the length of the irrigation period. Let a binary function g being such that

g(mode(t)) = 0 if mode(t) = crisis
g(mode(t)) = 1 if mode(t) belongs to (precrisis, normal)
The performance index Vcrisis can be written as

$$(T-t_0)*(1-\text{ Vcrisis}) = E\left\{ \sum_{t=t_0}^{T} g(\text{mode}(t)) \right\}$$

where t_0 is the initial time and T the horizon of operation.

*Vprecrisis : frequency of operating in a precrisis mode.
In the same way, defining f_n such that f_n(normal) = 1, f_n(crisis) = 0 and f_n(precrisis) = 0, one can write

$$(T-t_0)*(1-\text{ Vcrisis- Vprécrisis}) = E\left\{ \sum_{t=t_0}^{T} f_n(\text{mode}(t)) \right\}$$

* Vquality : a vulnerability criterion related to quality of Gascogne rivers. Let $f_g(r_g)$ be a function between 0 and 1 defined in the following way :

$$f_g(r_g) = \left(-\left(\frac{r_g - \text{qminG}}{\text{qobjG} - \text{qminG}} \right)^2 + 2\left(\frac{r_g - \text{qminG}}{\text{qobjG} - \text{qminG}} \right) \right)$$

if QminG < r_g < QobjG and $f_g(r_g)$ = 0 otherwise. The following equation holds :

$$(T-t_0)*(\text{VqualitéG}) = E\left\{ \sum_{t=t_0}^{T} f_g(r_g(t)) \right\}$$

4 COMPROMISE BETWEEN IRRIGATION AND WATER QUALITY

Let α and β be two coefficients between 0 and 1. We are interested in the two parameter family of objectives to be maximized over the strategy of releases :

$$\beta E \left\{ \sum_{t=t_0}^{T} ((1-\alpha) f_g(r_g(t)) + \alpha f_n(\text{mode}(t))) \right\} + (1-\beta) E \left\{ \sum_{t=t_0}^{T} ((1-\alpha) g(\text{mode}(t))) \right\}$$

β is a coefficient trading off quality objectives and the expected length of the irrigation period .
α refects the relative importance of quality criteria for Gascogne versus Neste rivers.
At time t, according to the mode of operation (normal, precrisis or crisis) the compromise function is the sum of three parts that are given in Table 1. The goal is to find relative weights for irrigation and water quality. One can use simulated or historical series to adjust the coefficients (α, β) so as to reflect the system manager's preference function.

Figure 8 illustrates that acceptance of the crisis mode may help increasing the quality performance index. By considering these trade-off curves the system manager can select a pair (α, β) that would fit his own attitude towards risk.

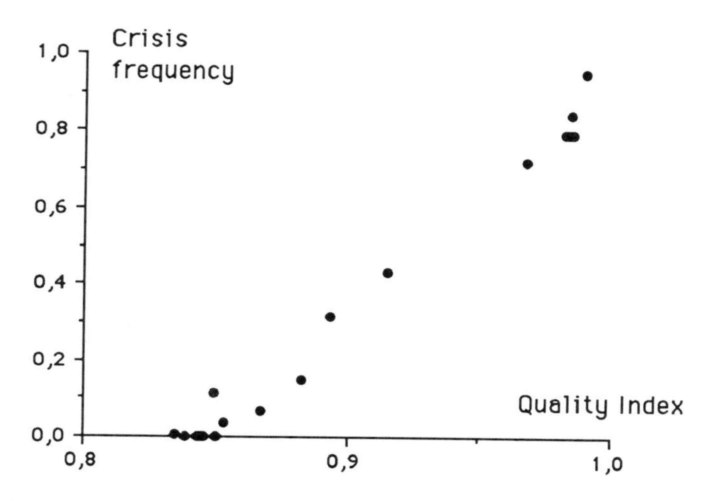

Figure 8: Trade-off between crisis mode and quality performance index

7 CONCLUDING COMMENTS

The operating rules of a hydro-system with irrigation and quality objectives have been defined within the framework of a classical stochastic dynamic programming model with a small number of parameters to be adjusted by stochastic simulation.

From the present application to the Neste system, our conclusions are :

1) this procedure allows the system manager to formulate and quantify objectives in a rational way. An operating rule can be derived to allocate water so as to meet a combination of the various objectives. Of course this optimal allocation should be carefully considered because of model uncertainties influencing both the system response and the hydrological behavior.

2) Stochastic dynamic programming can be efficiently used to derive optimal feedback rules of operations and can routinely deal with a discontinuous decision such as stopping irrigation when crisis occurs.

3) By simulating the rules obtained with different sets of parameters, multi-objective trade-off curves can help the system manager to choose his own set of parameters.

4) Such models are designed to serve only as multicriterion decision making aids. In very dry days such as occured in summer 1976 or 1989 in France they cannot create additional water resources...still they can help the system manager by providing constant updated multidimensional estimates of the risks and trade-offs that may be encountered when following different operation rules.

REFERENCES

BELLMAN R.E., DYNAMIC PROGRAMMING , PRINCETON UNIVERSITY PRESS, 1957.

BHASKAR N.R. and WHITLATCH, COMPARISON OF RESERVOIR LINEAR OPERATION RULES USING LINEAR AND DYNAMIC PROGRAMMING. WATER RESOURCES BULLETIN 23,6,p 1027-1036, 1987.

COLLETER P., DELEBECQUE F., FALGARONE F. and QUADRAT J.P., APPLICATION DU CONTROLE STOCHASTIQUE A LA GESTION DES MOYENS DE PRODUCTION D'ENERGIE EN NOUVELLE-CALEDONIE. In "E.D.F. BULLETIN DE LA DIRECTION DES ETUDES ET RECHERCHE, SERIE C. MATHEMATIQUES-INFORMATIQUE"(E.D.F.), p 1-28, 1978.

GOUSSEBAILLE J. and ROCHE P. A., MONTE-CARLO SIMULATION FOR STOCHASTIC DYNAMIC PROGRAMMING . A CASE STUDY: THE SEINE RESERVOIR DAILY OPERATION. Document interne CERGRENE, 8 p , 1988.

KLEMES V., DISCRETE REPRESENTATION OF STORAGE FOR STOCHASTIC RESERVOIR OPTIMIZATION. WATER RESOURCES RESEARCH 13,1,p 149-158, 1977.

KRZYSZTOFOWICZ R., UTILITY CRITERION FOR WATER SUPPLY : QUANTIFYING VALUE OF WATER AND RISK ATTITUDE. In "DECISION -MAKING FOR HYDROSYSTEMS - FORECASTING AND OPERATION"(T. UNNY), 43-62, 1982.

MASSE P.: LE CHOIX DES INVESTISSEMENTS . DUNOD 1959.

PLATE E. J. and DUCKSTEIN L., RELIABILITY-BASED DESIGN CONCEPTS IN HYDRAULIC ENGINEERING. WATER RESOURCES BULLETIN 24,2,p 235-245, 1988.

SZIDAROVSZKY F. and DUCKSTEIN L., DYNAMIC MULTIOBJECTIVE OPTIMIZATION : A FRAMEWORK WITH APPLICATION TO REGIONAL WATER AND MINING MANAGEMENT. E J O R 24,p 305-317, 1986.

TURGEON A., OPTIMAL OPERATION OF MULTIRESERVOIR POWER SYSTEMS WITH STOCHASTIC INFLOWS. WATER RESOURCES RESEARCH 16,2,p 275-283, 1981.

YAKOWITZ S., DYNAMIC PROGRAMMING APPLICATIONS IN WATER RESOURCES. WATER RESOURCES RESEARCH 18,4,p 673-696, 1982.

YEH W.W.G., RESERVOIR MANAGEMENT AND OPERATIONS MODELS : A STATE-OF-THE-ART REVIEW. WATER RESOURCES RESEARCH 21,12,p 1797-1818, 1985.

PRESIDENTIAL ELECTIONS, THE SUPERCONDUCTING SUPERCOLLIDER, AND ORGAN TRANSPLANT DECISIONS

Thomas L. Saaty
University of Pittsburgh
Pittsburgh, PA 15260

ABSTRACT

Here we briefly describe some applications of the Analytic Hierarchy Process in the private and public sectors. Who will be the next president? Where should the superconducting supercollider be located? Finally, who should get the next liver transplant at Pittsburgh's Presbyterian University Hospital, and on the basis of what criteria?

1. Introduction [3,6,7]

A decision is defined by a structure which represents the elements of the problem: a goal, criteria, subcriteria and alternatives (options) and a set of judgments to establish relationships among them. The aim is to derive a scale of relative importance for the alternatives. How do we formalize human judgments or perceptions, and what conditions do we impose to develop ideas needed to deal with the question of deriving a ratio scale of importance from numerical judgments expressed by using a unique fundamental scale? The Analytic Hierarchy Process (AHP) uses hierarchic structures, matrices and linear algebra to formalize the decision process. To a person unfamiliar with the subject, there may be some concern about what to include and where to include it in the structure. When constructing hierarchies, one must include enough relevant detail to: represent the problem as thoroughly as possible, but not so thoroughly as to lose sensitivity to change in the elements; consider the environment surrounding the problem; identify the issues or attributes that contribute to the solution; and identify the participants associated with the problem. Arranging the goals, attributes, issues, and stakeholders in a hierarchy serves two purposes: (1) it provides an overall view of the complex relationships inherent in the situation; and (2) helps the decision maker assess whether the issues in each level are of the same order of magnitude, so he can compare such homogeneous elements accurately.

One certainly cannot compare according to size a football with Mt. Everest and have any hope of getting a meaningful answer. The football and Mt. Everest must be compared in sets of objects of their class. Because people have an ability to compare objects according to length, weight, etc., and this ability is independent of the presence of any scales, we can use such comparisons to generate approximations to readings on underlying scales and also to express our preferences for such readings. When making comparisons between two alternatives falling in one level of a hierarchy with respect to an attribute falling in the next higher level of the hierarchy, one identifies the larger of the two and asks, (according to one's perception), how many times more of the attribute does the larger alternative have than the smaller one? The smaller one would have the reciprocal value when compared with the larger one. The response to such a question must take the form of a ratio, as there are no numbers yet available to form the difference of two measurements. The smaller of the two alternatives is assumed to have a unit amount of the attribute and the larger one is measured in terms of multiples of that unit. We are in the process of compiling a dictionary of hierarchies pertaining to all sorts of problems, from personal to corporate to public.

The fundamental scale used in making comparisons (among elements of a comparable set) consists of verbal judgments ranging from equal to extreme (equal, moderately more, strongly more, very strongly more, extremely more) corresponding to the verbal judgments are the absolute numerical judgments (1, 3, 5, 7, 9) and compromises (2, 4, 6, 8) between these values. The reciprocal value is used in comparing the smaller object with the larger one. When two elements being compared are nearly the same, we enlarge the interval 1 to 2 and approximately use the scale 1.3 for moderate, 1.5 for strong and so on. Thus moderately more important gives a ratio of 1.3/1. The total for the two elements is 2.3 and 1/2.3 = .43. Also 1.3/2.3 = .57. If the difference is slight it would be 1.2 to 1 and 1.2/2.2 = .55. Also 1/2.2 = .45. In this manner we can use our perception to distinguish between two close elements. The derived ratio scale of relative values has been proven to be insensitive to small changes in the numerical judgments. A useful characteristic of the process is to allow for inconsistency and then offer a way to improve the judgments if the individual or group can modify some of their estimates. Still there is a tolerable level above which inconsistency would be detrimental to the making of a sound decision. The software package, Expert Choice [3], allows one to carry out sensitivity analyses to test the effect of the uncertainty in the criteria on the choice of a best alternative. We now turn to some applications of this process.

2. Predicting the Presidential Elections [8]

Since 1976 we have used the AHP after the summer political conventions, held prior to the presidential election every four years in the United states, to predict which candidate would become president. The idea was to simulate the judgments of the U.S. public at large. In each of the four elections, 1976 through 1988, the predicted outcome coincided with the actual election results. Usually a group of two dozen or more business leaders participated in the discussion, and either achieved concurrence or we used the AHP method of synthesizing individual judgments after the discussion is completed, to arrive at a group judgment. We take the geometric mean of the judgments because it is the only way to combine the paired comparison judgments on an alternative so that the reciprocal of the result is equal to what one obtains by combining the reciprocals of the judgments. Here is the hierarchy and judgments used in the 1980 elections that resulted in the winning of Ronald Reagan. In constructing a hierarchy one invites everyone to participate. The hierarchy of Figure 1 descends from the general to the more specific in an effort to determine the net

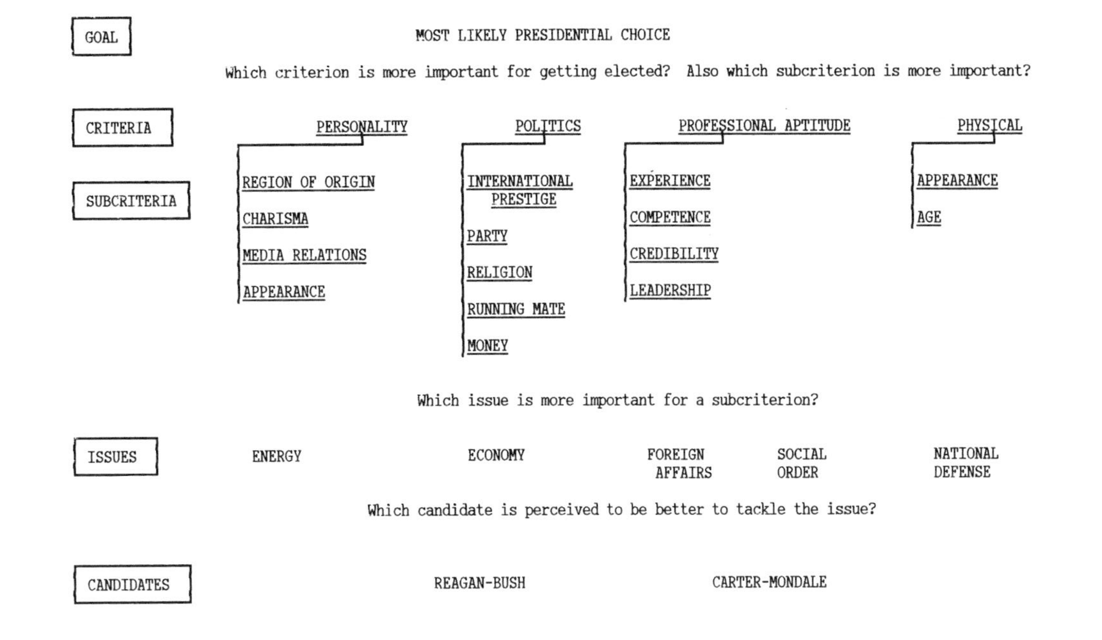

Figure 1
The 1980 Presidential Race

Table 1
CARTER vs REAGAN 1980

Priorities of Criteria

Election 1980	Personality	Politics	Professional Aptitude	Physical	Priorities
Personality	1	1/5	1/7	3	.08
Politics	5	1	5	9	.61
Professional Aptitude	7	1/5	1	7	.27
Physical	1/3	1/9	1/7	1	.04

Priorities of the factors which define the criteria

Personality	Region of Origin	Charisma	Media Relations	Appearance	Priorities
Region of Origin	1	1/7	1/5	1/2	.06
Charisma	7	1	1/2	3	.32
Media Relations	5	2	1	7	.52
Appearance	2	1/3	1/7	1	.10

Politics	International Prestige	Party	Religion	Running Mate	Money	Priorities
International Prestige	1	1/4	7	1/4	1/3	.11
Party	4	1	9	2	1	.34
Religion	1/7	1/9	1	1/7	1/6	.03
Running Mate	4	1/2	7	1	1/2	.22
Money	3	1	6	2	1	.30

Professional Aptitude	Experience	Competence	Credibility	Leadership	Priorities
Experience	1	1/3	1/5	1/6	.06
Competence	3	1	1/5	1/3	.12
Credibility	5	5	1	3	.54
Leadership	6	3	1/3	1	.28

Physical	Age	Appearance	Priorities
Age	1	1/5	.17
Appearance	5	1	.83

Overall priorities of the factors

(1) Region of Origin	(2) Charisma	(3) Media Relations	(4) Appearance	(5) International Prestige	(6) Party	(7) Religion
.005	.025	.041	.008	.065	.206	.019

(8) Running Mate	(9) Money	(10) Experience	(11) Competence	(12) Credibility	(13) Leadership	(14) Age	(15) Appearance
.136	.183	.016	.032	.148	.076	.006	.033

We group these under the following four headings as follows:

Cluster	Party	Credibility	Media	International Standing
Members	(6)(8)(9)	(12)(13)	(3)(4)(15)	(5)(11)
Total Priorities Normalized for factors included	.57	.25	.09	.10

Table 1 - continued

Priorities of the Issues with respect to the four clusters of factors

Party	A	B	C	D	E	Priorities
Energy	1	1/5	6	3	3	.21
Economy	5	1	6	9	5	.54
Foreign Affairs	1/6	1/6	1	2	1/5	.06
Social Order	1/3	1/9	1/2	1	2	.08
National Defense	1/3	1/5	5	1/2	1	.11

Credibility	A	B	C	D	E	Priorities
A	1	1	1	1	1	.20
B	1	1	1	1	1	.20
C	1	1	1	1	1	.20
D	1	1	1	1	1	.20
E	1	1	1	1	1	.20

Media	A	B	C	D	E	Priorities
A	1	1/2	1/4	1/3	3	.12
B	2	1	5	3	5	.39
C	4	1/5	1	1/4	1/4	.11
D	3	1/3	4	1	5	.27
E	1/3	1/5	4	1/5	1	.11

Inter.Stand.	A	B	C	D	E	Priorities
A	1	6	1/4	5	1/7	.17
B	1/6	1	1/3	5	1/3	.09
C	4	3	1	5	1	.31
D	1/5	1/5	1/5	1	1/5	.04
E	7	3	1	5	1	.39

Priorities of Carter & Reagan with respect to the issues

Energy	Carter	Reagan	Priorities
Carter	1	1/3	.25
Reagan	3	1	.75

Economy	Carter	Reagan	Priorities
Carter	1	1/3	.25
Reagan	3	1	.75

Foreign Affairs	Carter	Reagan	Priorities
Carter	1	2	.67
Reagan	1/2	1	.33

Social Order	Carter	Reagan	Priorities
Carter	1	3	.75
Reagan	1/3	1	.25

National Defense	Carter	Reagan	Priorities
Carter	1	1/5	.17
Reagan	5	1	.83

Composite Weights

Carter	.35
Reagan	.65

outcome of the political forces in the country. It has the criteria, subcriteria, clusters, issues and candidates. The matrices following the hierarchy, given in Table 1, show the paired comparisons, and should be easy to follow, by descending from the top and comparing the elements in each level, or grouping in a level with respect to their parent node in the next level above. This is followed by a descending process of weighting and adding the relative scales generated from the matrices to obtain the priorities of the candidates. It was predicted then that Reagan-Bush would win by a .65 priority as compared with Carter-Mondale with .35 priority. If the exercise is done with care, it could reflect the relative strength of how the population vote would split.

3. The Superconducting Supercollider [4,5]

In 1987, at the suggestion of a colleague in the Department of Physics at the University of Pittsburgh, I sent a detailed proposal to the American Institute of Physics suggesting that the AHP be applied to assist in choosing the new site for the forthcoming superconducting supercollider (SSC). A long time went by without hearing from them. In 1989, after the decision was announced by the government, we obtained information from one of the people who worked on the problem in Washington, to test the decision within the framework of the AHP and verify the outcome. The analysis and the results given below were communicated to that individual. One notes that the results of the AHP were close to the outcome obtained in Washington by the working committee. Here is a brief account of the study.

In January of 1987, the U.S. government administration proposed the construction of a superconducting supercollider. Thirty-five sites were initially proposed and it was the job of the Department of Energy with assistance from the National Academies of Sciences and Engineering to select the best site. A report was later issued which described how the committee doing the ranking chose the best site from among the eight finalists selected. Later, I worked with Nicholas Rudenko and Christopher P. Sparta to apply the absolute mode of the Analytic Hierarchy Process and the "Expert Choice" software package using information obtained directly from the "GAO Report to Congressional Requesters" [4] to validate the ranking of the site finalists. Some of the information had to be summarized or interpreted because it was not given in explicit form, perhaps to preserve confidentiality. The criteria were:

A. GEOLOGY AND TUNNELING: (1) suitability of the topography, geology and associated geohydrology for efficient and timely construction of the proposed SSC underground structures; (2) stability of the proposed geology against settlement and seismicity and other features that could adversely affect SSC operations; (3) installation and operational efficiency resulting from minimal depths for the accelerator complex and experimental halls; and (4) risk of encountering major problems during construction.

B. REGIONAL RESOURCES: (1) proximity of communities within commuting distance of the proposed SSC facilities capable of supporting the SSC staff, their families, and visitors. Adequacy of community resources - e.g., housing, medical services, community services, educational and research activities, employment opportunities for family members, recreation, and cultural resources, all available on a nondiscriminatory basis; (2) accessibility to the site, e.g., major airport(s), railroad(s), and highway system(s) serving the vicinity and site; (3) availability of a regional industrial base and skilled labor pool to support construction and operation of the facility; (4) extent and type of state, regional and local administrative and institutional support that will be provided, e.g., assistance in obtaining permits and unifying codes and standards.

C. ENVIRONMENT: (1) significance of environmental impacts, constructing, operating, and decommissioning the SSC; (2) projected ability to comply with all applicable, relevant, and appropriate federal, state and local environmental/ safety requirements within reasonable bounds of time, cost, and litigation risk; (3) ability of the proposer, DOE, or both to reasonably mitigate adverse environmental impacts to minimal levels.

Note: the above subcriteria were developed from the following 15 environmental factors: (1) Earth resources; (2) Water resources; (3) Air resources; (4) Noise/ vibration matters; (5) Ecological resources; (6) Health and safety matters; (7) Land use; (8) Socioeconomics; (9) Scenic/visual resources; (10) Cultural (historical, archaeological, and paleontological) resources; (11) Compliance with federal laws and regulations; (12) Compliance with state laws and regulations; (13) Compliance with local laws and regulations; (14) Alternative mitigative measures available; and (15) Cost effectiveness of mitigative measures.

D. SETTING: (1) ability of the proposed to deliver defendable title, in accordance with the schedule for land estates in land that will adequately protect the government's interest and the integrity of the SSC during construction and operation; (2) flexibility to adjust the position of the SSC in the nearby vicinity of the proposed location; and (3) presence of natural and man-made features of the region that could adversely affect the siting, construction and operations of the SSC.

E. REGIONAL CONDITIONS: (1) presence of man-made disturbances, such as vibration and noise, that could adversely affect the operation of the SSC; and (2) presence of climatic conditions that could adversely affect the construction and operation of the SSC.

F. UTILITIES: (1) reliability and stability of the electric-power-generating and transmission grid system. Flexibility for future expansion; and (2) reliability, quality, and quantity of water to meet the needs of the facility.

Paired comparisons and ratings were inferred from the condensed information available in the GAO report, to simulate the committee's thought processes and utilize the information provided in the report, the criteria listed in the report were condensed as follows: Geology and Tunneling, Setting, and Regional Conditions - are given without subcriteria.

Regional resources - treated in the model in the detailed way described above.

Environmental - the three subcriteria were aggregated into one main criterion. 15 environmental factors were used to develop the ratings for this main criterion. The available information was used to indicate the number of factors that were rated Good and the number that were rated Satisfactory. If the site had 12 or more factors which were rated Good, we gave it a "Good" rating. If the site had less than 12 factors rated Good, we gave it a "Good/Satisfactory" rating.

Utilities - were not included in the final group of criteria because there was little discrimination among the sites as far as utilities were concerned.

The hierarchy is shown in Figure 2 followed by the ranking of alternatives according to the absolute mode of measurement of the AHP [6] applied to the sites by rating them with respect to the intensities indicated at the bottom level of the hierarchy. The intensities for each criterion are divided by the largest intensity and then weighted by the priority of its corresponding criterion or subcriterion. Thus an alternative that is best on all criteria receives the value 1. All others receive a proportionate value. The outcome shown in Table 2 coincided with the committee ranking of the eight sites except that sites 4 and 6 are interchanged. What is important is that the best site, Texas, is the same. That is where the SSC will be located.

4. Organ Transplants [1,2]

The City of Pittsburgh has become a leader in the world in performing organ transplant operations. Because there are more patients who need livers, hearts and kidneys than there are organs available, it has become essential to assign priorities to the patients. The two charts given here were a result of a several month study by Alison R. Casciato and John P. O'Keefe, working with the author in coordination with doctors and research scientists at the Presbyterian University Hospital (University of Pittsburgh).

One value of this research was to attract the attention of doctors to the potential use of the AHP for this purpose. A result of that was the subsequent work by Drs. Cook, Staschak and Green on the equitable allocation of livers to patients [2]. The goal of the hierarchy of Figure 3 is divided into: Family, Medical History, Social Factors and Funding. Family is subdivided into emotionally dependent, cohesion and financially dependent and the first and third are in turn divided into single, married, and divorced with and without dependents. Then each of them is further represented by intensities shown in Figure 3B. Cohesion is simply represented by the intensities: weak, average and strong.

Most of the bottom level factors in Figure 3A are subdivided into corresponding

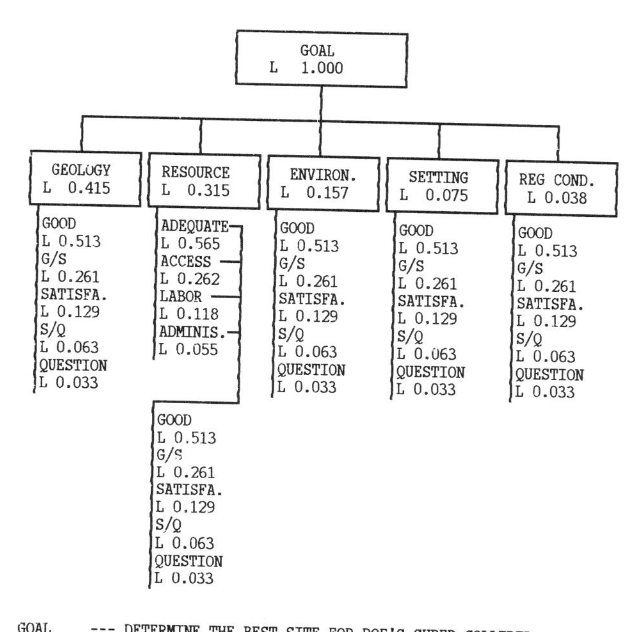

GOAL --- DETERMINE THE BEST SITE FOR DOE'S SUPER COLLIDER
ACCESS --- ACCESSIBILITY TO THE SITE, E.G. MAJOR AIRPORTS,RAILROADS
ADEQUATE --- ADEQUACY OF COMMUNITY RESOURCES
ADMINIS. --- EXTENT OF STATE, REGIONAL AND LOCAL ADMINISTRATIVE AND INSTITUTIONAL SUPPORT
ENVIRON. --- ENVIRONMENT
G/S --- GOOD/SATISFACTORY
GEOLOGY --- GEOLOGY AND TUNNELING
GOOD --- GOOD
LABOR --- AVAILABILITY OF REGIONAL INDUSTRIAL BASE AND SKILLED LABOR POOL
QUESTION --- QUESTIONABLE
REG COND.--- REGIONAL CONDITIONS
RESOURCE --- REGIONAL RESOURCES
S/Q --- SATISFACTORY/QUESTIONABLE
SATISFA. --- SATISFACTORY
SETTING --- SETTING

L --- LOCAL PRIORITY : PRIORITY RELATIVE TO PARENT NODE

Note that only Resources had subcriteria and then intensities of subcriteria. All others are criteria directly linked to intensities.

Figure 2
Criteria and Intensities for Rating the Eight Sites

Table 2
Rating of Sites on the Intensities

ALTERNATIVES	GEOLOGY .4153	RESOURCE ADEQUATE .1778	RESOURCE ACCESS .0825	RESOURCE LABOR .0370	RESOURCE ADMINIS. .0174	ENVIRON. .1570	SETTING .0747	REG COND. .0383	TOTAL
1 ONE	GOOD	GOOD	GOOD	GOOD	GOOD	GOOD	GOOD	G/S	.981
2 TWO	GOOD	GOOD	GOOD	GOOD	GOOD	G/S	G/S	SATISFA.	.857
3 THREE	GOOD	GOOD	SATISFA.	GOOD	GOOD	G/S	G/S	GOOD	.824
4 FOUR	GOOD	S/Q	GOOD	SATISFA.	GOOD	GOOD	GOOD	G/S	.798
5 FIVE	GOOD	SATISFA.	SATISFA.	GOOD	SATISFA.	GOOD	SATISFA.	SATISFA.	.708
6 SIX	GOOD	SATISFA.	GOOD	SATISFA.	SATISFA.	G/S	SATISFA.	G/S	.674
7 SEVEN	SATISFA.	SATISFA.	GOOD	GOOD	SATISFA.	G/S	GOOD	SATISFA.	.438
8 EIGHT	SATISFA.	SATISFA.	SATISFA.	GOOD	GOOD	GOOD	G/S	SATISFA.	.429

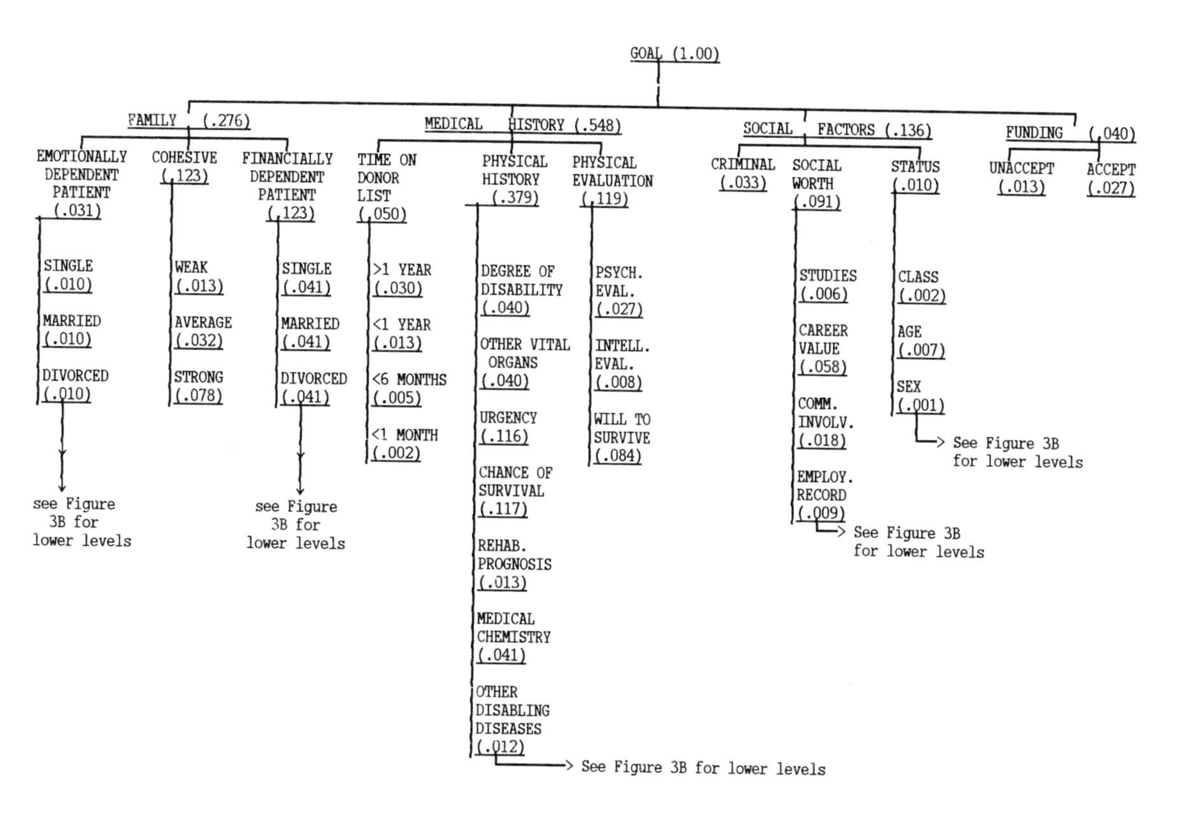

Figure 3A
Hierarchy For Rating Organ Recipients (upper levels)

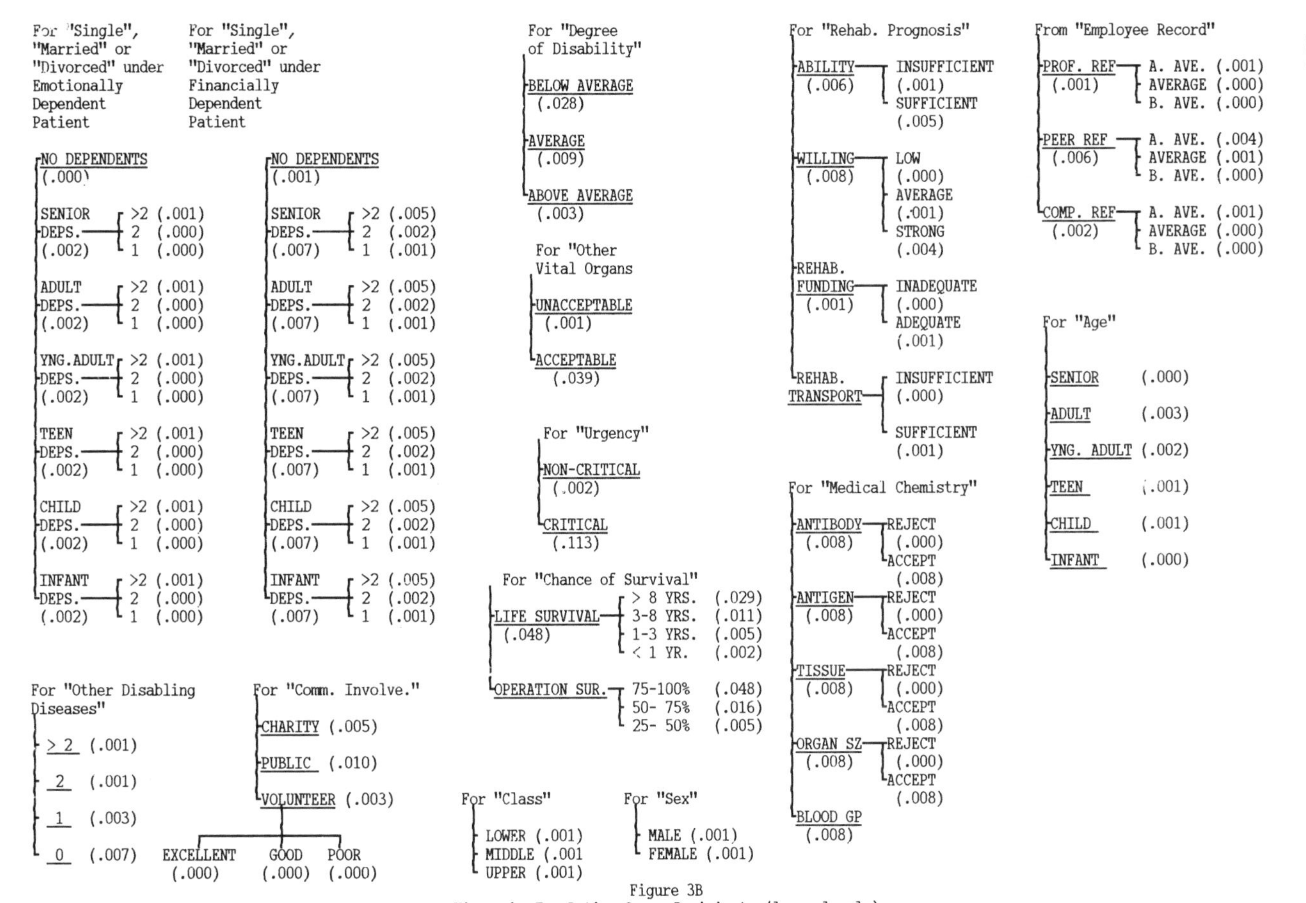

Figure 3B
Hierarchy For Rating Organ Recipients (lower levels)

0.000 value means the significance is in the fourth or greater place.

intensities indicated in Figure 3B. The priorities (derived from judgments obtained from expert physicians in the area of organ transplants and other people in the field) are indicated next to each factor and sum to one for each level. A patient is ranked according to the priorities of the factors when no intensities are involved for these factors, or according to the intensities on the other factors by rating that patient on the intensities. The higher the total score the better the chance to receive a transplant. We are hopeful that by getting more doctors to participate in this type of study, more people in the medical profession will learn to use the multicriteria approach in the field of organ transplants.

5. Conclusion

Group participation in the AHP provides an opportunity to construct a rich but balanced representation of the hierarchy. The judgment process starts out by seeking consensus, but as the participants become familiar with each other, judgments are synthesized according to the geometric mean. The reason for it is that the geometric mean is the only way to combine individual judgments in comparing two alternatives, so that the reciprocal of the result is equal to what one obtains by combining the geometric mean of the reciprocals of the judgments. The software package Expert Choice, the computer arm of the AHP makes it possible to combine group judgments in this way.

REFERENCES

1. Casciato, A.R. and J.P. O'Keefe, Organ Transplant Priorities : Who Lives ?, University of Pittsburgh, August 1987.

2. Cook, D.R., and S. Staschak and W.T. Green, "Equitable allocation of livers for orthotopic transplantation: An application of the Analytic Hierarchy Process", European Journal of Operational Research, 48, 1, 49-56, 1990.

3. Forman, E. and T.L. Saaty, Expert Choice: Software Package for Personal Computers. Expert Choice, Inc. 4922 Ellsworth Avenue, Pittsburgh, PA 15213, 1989.

4. Government Accounting Office (GAO) Report to Congressional Requesters", titled "Determination of the Best Qualified Sites for DOE's Supercollider", January 1989.

5. Rudenko, N. and C.P. Sparta, "Determination of the Best Qualified Sites for DOE's Supercollider Using Expert Choice", University of Pittsburgh, Pittsburgh, PA, July 1989.

6. Saaty, Thomas L., Decision Making: The Analytic Hierarchy Process. McGraw-Hill Publishers, 1980. RWS Publications, 4922 Ellsworth Avenue, Pittsburgh, PA 15213, 1990.

7. Saaty, Thomas L. and K. Kearns, Analytical Planning. International Series in Modern Applied Mathematics and Computer Science, Vol. 7, Pergamon Press, 1985.

8. Saaty, Thomas L. and L. Vargas, Prediction, Projection and Forecasting. Kluwer Academic Publishers, Boston, 1991.

PROCESSING INTERVAL JUDGMENTS IN THE ANALYTIC HIERARCHY PROCESS

Ahti Salo
Research Assistant

Raimo P. Hämäläinen
Professor

Systems Analysis Laboratory
Helsinki University of Technology
Otakaari 1 M, 02150 Espoo, Finland

Abstract - This paper investigates how approximate preference statements can be incorporated into the analytic hierarchy process (AHP) by means of interval judgments. Earlier work is extended by developing a computationally feasible method for synthesizing interval judgments in the entire hierarchy to obtain weight intervals for the decision alternatives.

As the interval judgments gradually describe the decision maker's (DM) preferences more and more precisely the alternatives' weight intervals become narrower. Thus the method leads to an iterative process where the DM can start by articulating his preferences about those aspects he feels most certain about. Later on he can proceed to refine earlier judgments and examine the corresponding changes in the weight intervals.

A numerical example is given to illustrate how the method can help the DM to establish a preference order for the decision alternatives.

1. INTRODUCTION

The analytic hierarchy process (AHP), developed by Saaty (1977,1980), is a methodology for problem structuring and analysis which in recent years has become increasingly popular. For instance, Vargas (1990) reports a large number of applications to economic, social, political and technological problems.

In the AHP the decision maker (DM) constructs a hierarchy which relates the relevant issues in the decision problem under an overall objective. The hierarchy serves as a framework for addressing both qualitative and quantitative factors.

In order to establish a preference order for the alternatives the DM carries out a series of pairwise comparisons. For each pairwise comparison the DM specifies the intensity of preference of one factor over another with respect to a given criterion. The pairwise comparisons are processed so that each decision alternative receives a weight which reflects its desirability in view of the DM's preferences.

This paper extends the AHP methodology by allowing the DM to enter ranges of numerical values, i.e. intervals, in addition to point estimates when making the pairwise comparisons. By using interval judgments the DM can enter approximate preference statements when he is either unwilling or unable to be explicit about his preferences.

A method of synthesizing interval judgments into weight intervals for the alternatives is developed. In this method the alternatives' weight intervals shrink as the interval judgments capture the DM's preferences more and more closely. After each new preference statement the weight intervals are recomputed and displayed graphically to the DM. When the weight intervals no longer overlap a preference order for the alternatives has been found.

From the user's point of view the proposed method has considerable practical potential. Since the weight intervals for the alternatives can be recomputed at any point of the process the method leads to an improved interactive decision support process. An important feature is that a preference order for the alternatives may be established even before the DM has supplied interval judgments for all the pairwise comparisons. In this way the method can substantially reduce the amount of comparison work.

The paper is organized as follows. Section 2 gives a brief summary of the AHP. Section 3 relates interval judgments to local priorities, which in section 4 are synthesized into weight intervals. Section 5 illustrates the proposed method in the context of an energy production problem.

2. THE ANALYTIC HIERARCHY PROCESS

Here only the key features of the AHP are summarized. For a full exposition, see e.g. Saaty (1980).

The decision support process in the AHP can be divided into three phases, which are problem structuring, preference elicitation and synthesis.

In problem structuring the relevant criteria in the decision problem are organized into a hierarchy. The topmost element of the hierarchy stands for the overall goal in the decision problem. The level immediately below it consists of subgoals, which contribute to the attainment of the overall goals. Each subgoal is decomposed further until a sufficiently detailed representation of the decision problem is obtained. The decision alternatives are placed on the lowest level of the hierarchy. For example, figure 1 shows a hierarchy constructed to help Parliament members compare different forms of energy production (Hämäläinen, 1988; see also Hämäläinen, 1990, 1991).

Preference elicitation consists of a series of pairwise comparisons where the DM considers the relative importance of two subelelements at a time with respect to a given criterion. To indicate the relative importance the DM makes a verbal statement, which is then cast into an integer in the one to nine range.

Assuming that a given criterion has n subelements, the pairwise comparisons lead to the following *comparison matrix*

$$A = \begin{pmatrix} 1 & a_{12} & \dots & a_{1n} \\ a_{21} & 1 & \dots & a_{2n} \\ \vdots & \vdots & \ddots & \vdots \\ a_{n1} & a_{n2} & \dots & 1 \end{pmatrix} \tag{1}$$

This matrix conveys, for example, that the ith subelement is considered a_{ij} times more important than the jth subelement. The matrix in (1) is a reciprocal one, i.e. $a_{ij} = \frac{1}{a_{ji}}$.

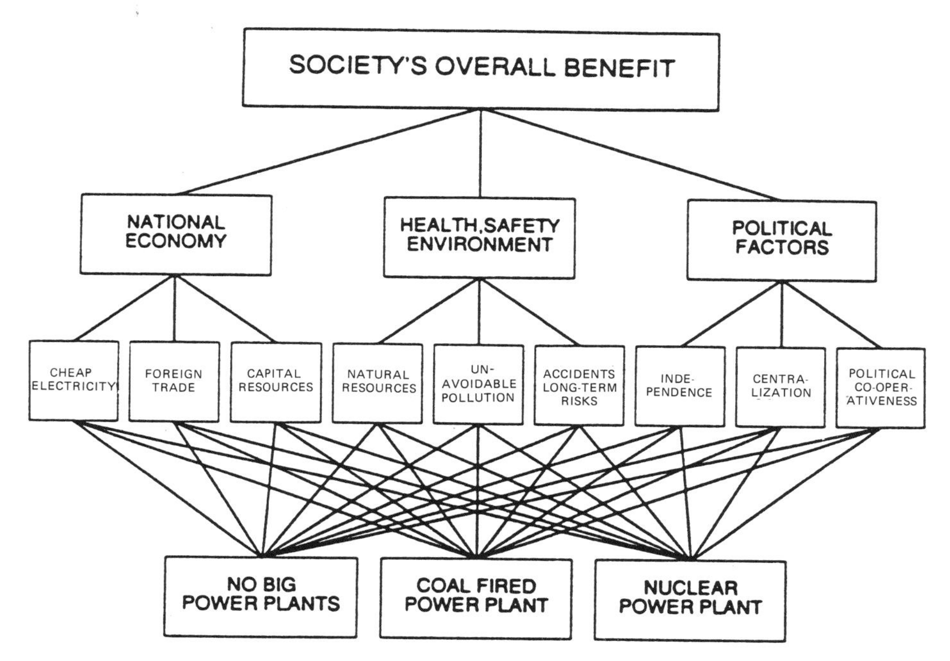

Figure 1: A hierarchy for comparing different forms of energy production

A *local priority vector* $w = (w_1, \ldots, w_n)$ is derived from the comparison matrix (1) as the unique solution to $\sum_{i=1}^{n} w_i = 1$ and

$$Aw = \lambda_{\max} w \tag{2}$$

where $\lambda_{\max}$ is the largest eigenvalue of A.

The local priorities are processed to assign a (global) weight to each element of the hierarchy. By definition, the weight of the topmost element is one. This weight is propagated downward so that the weight an upper level element, say e_1, gives to one of its subelements, e_2, is the product of the weight of e_1 and the component corresponding to e_2 in the local priority vector of e_1. In this way each decision alternative finally receives a unique weight.

3. INTERVAL JUDGMENTS AND LOCAL PRIORITIES

The DM can find it difficult to specify a point estimate when making a pairwise comparison. Instead, he may be more confident in stating a range of values. Such an interval of numerical values is called an *interval judgment* (Saaty and Vargas, 1987; Arbel, 1989).

Interval judgments allow the DM to be loose in his preferences statements. For instance, instead of asserting that the ith subelement is three times as important as the jth subelement with respect to the given criterion, he can state that the ith subelement is at least two but no more than four times as important as the jth subelement. From this statement the interval $I_{ij} = [l_{ij}, u_{ij}] = [2, 4]$ is obtained.

In terms of interval judgments the comparison matrix can be written as

$$\begin{pmatrix} 1 & [l_{12}, u_{12}] & \dots & [l_{1n}, u_{1n}] \\ [l_{21}, u_{21}] & 1 & \dots & [l_{2n}, u_{2n}] \\ \vdots & \vdots & \ddots & \vdots \\ [l_{n1}, u_{n1}] & [l_{n2}, u_{n2}] & \dots & 1 \end{pmatrix} \tag{3}$$

If ith subelement is at least two times more important than the jth subelement ($l_{ij} = 2$) then the jth subelement can have no more than half of the importance of the ith subelement, i.e. $u_{ji} = \frac{1}{2}$. In general, from the reciprocal nature of the pairwise comparisons it follows that $l_{ij}u_{ji} = 1, i \neq j$.

Saaty and Vargas (1987) discuss ways of deriving local priorities from (3). They point out that determining all the right eigenvectors corresponding to reciprocal matrices such that $a_{ij} \in [l_{ij}, u_{ij}]$ is an intractable task from the computational point of view. The eigenvector in (2) is a nonlinear function of the elements of the comparison matrix, and no simple methods for determining bounds for the components of the local priority vectors exist. Even when the numbers a_{ij} are restricted to the first nine integers and their reciprocals the amount of computation is formidable.

Yoon (1988) studies the sensitivity of the local priority vector to errors in the comparison matrix using the propagation of errors technique. However, in order to avoid complicated algebraic calculations he discards the right principal eigenvector in favor of the normalized row sum of the comparison matrix.

Arbel (1989) interprets the interval judgment $I_{ij} = [l_{ij}, u_{ij}]$ in terms of linear constraints on the local priorities. By definition, the judgment $I_{ij} = [l_{ij}, u_{ij}]$ is equivalent to stating that the ith subelement is at least l_{ij} but no more than u_{ij} times as important as the jth subelement. Thus any local priority vector $w = (w_1, \dots, w_n)$ consistent with this statement must satisfy the constraints $l_{ij}w_j \leq w_i \leq u_{ij}w_j$.

In Arbel's paper *loose articulation of preference* refers to a set of constraints such that some the bounds l_{ij}, u_{ij} in (3) are missing. *Approximate articulation of preferences* refers to the case where all the entries in (3) have been specified. *Feasible region*, denoted by S, is the largest subset of $Q^n = \{w = (w_1, \dots, w_n) | w_i \geq 0, \sum_{i=1}^n w_i = 1\}$ such that the vectors in S satisfy all the constraints resulting from the user-specified interval judgments. Arbel discusses properties of the feasible region and presents a numerical example where the feasible region is determined through linear programming. However, he deals with local priorities only.

To clarify the way in which interval judgments define the feasible region, consider the case of three subelements among which the DM has specified the interval judgments $I_{12} = [1, 2], I_{13} = [3, 5]$. These judgments are equivalent to stating that the first subelement is one to two times more important than the second one, and three to five times more important than the third one. The feasible region, shown in Figure 2, therefore consists of those local priorities in Q^3 which satisfy the constraints $w_2 \leq w_1 \leq 2w_2$ and $3w_3 \leq w_1 \leq 5w_3$.

4. SYNTHESIZING INTERVAL JUDGMENTS

This section synthesizes the interval judgments in the hierarchy to derive a weight interval for each decision alternative. Each weight interval consists of those weights which are generated by some set of local priorities in the feasible regions.

We assume that some, or even all, of the entries in (3) may be missing. Moreover, the DM is allowed to enter the bounds of an interval one at a time. That is, instead of specifying both bounds he may enter either the upper or the lower bound of an interval.

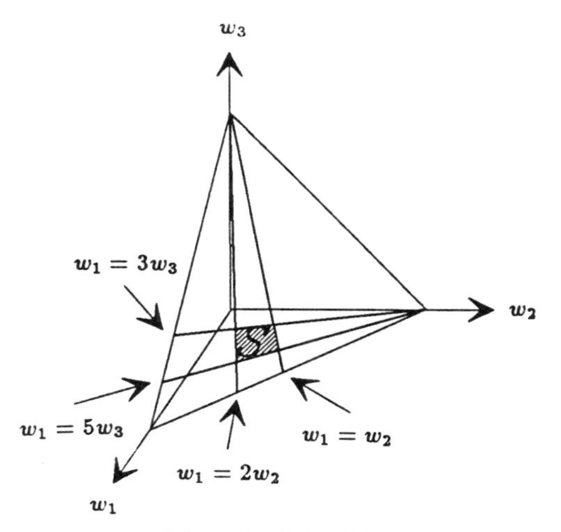

Figure 2: A feasible region

The feasible region corresponding to the interval judgments is

$$S = Q^n \cap \{w | l_{ij} w_j \leq w_i \leq u_{ij} w_j\} \tag{4}$$

where the constraints are implied by those bounds l_{ij}, u_{ij} in (3) which the DM has specified. These bounds can be restricted to the integers 1 through 9 and their reciprocals. However, the subsequent results hold even if the bounds are allowed to assume values in $(0, \infty)$ or if the DM is allowed to cancel earlier interval judgments.

Given a non-empty feasible region at each criterion of the hierarchy, the problem is to synthesize the preferences contained in these regions so that the related weight ranges for the alternatives can be found.

The first approach that comes to mind is to compute lower and upper bounds for the components of the local priority vector, i.e. $\underline{w}_i = \min_{w \in S} w_i$, $\overline{w}_i = \max_{w \in S} w_i$ and to use interval arithmetic (see e.g. Moore, 1966) to establish the weight intervals for the decision alternatives. However, direct application of interval arithmetic leads to meaningless results; for example, the upper bound of the weight of an alternative may be greater than one.

Instead, we suggest that the upper bound for the weight of an alternative is obtained as a solution to an optimization problem, where the objective function is the weight of the alternative and the variables are the local priority vectors constrained to the feasible regions. Similarly the lower bound is obtained by minimizing the weight of the alternative subject to the same constraints.

This approach for determining the weight intervals for the alternatives involves $2n_m$ optimization problems where n_m is the number of alternatives. Fortunately due to the hierarchical structure these problems decompose into a series of linear optimization tasks over the feasible regions.

In order to present theorem 1, the main result of our paper, the following notation is introduced. Assume that there are $m + 1$ levels in the hierarchy numbered 0 through m so that the topmost element is on level 0 and the alternatives are on level m. The number of elements on level k is n_k, and the lth element on level k is $e_{k,l}$. The index set of the subelements of element $e_{k,l}$ is $D_{k,l} \subset \{1, \ldots, n_{k+1}\}$. That is, $i \in D_{k,l}$ only if $e_{k+1,i}$ has been structured under $e_{k,l}$.

The feasible region at element $e_{k,l}$ is denoted by $S_{k,l}$ so that any $w \in S_{k,l}$ satisfies the constraints implied by the interval judgments made among the elements $e_{k+1,i}, i \in D_{k,l}$ with respect to $e_{k,l}$. If $w \in S_{k,l}$ and $i \in D_{k,l}$ then $w[i]$ is the share of its weight that $e_{k,l}$ gives to $e_{k+1,i}$. The (global) weight of the element $e_{k,l}$ is $v_{k,l}$.

For any fixed set of local priorities $w_{i,j} \in S_{i,j}$ the global weights of the elements $v_{k,l}$ are obtained from the formulas

$$v_{0,1} = 1 \tag{5}$$

$$v_{1,l} = w_{0,1}[l] \tag{6}$$

$$v_{2,l} = \sum_{\{i | l \in D_{1,i}\}} v_{1,i} w_{1,i}[l] \tag{7}$$

$$\vdots$$

$$v_{m,l} = \sum_{i=1}^{n_{m-1}} v_{m-1,i} w_{m-1,i}[l] \tag{8}$$

Note that $v_{k,l}$ does not depend on the priorities $w_{i,j}, i \geq k$. Straightforward calculation shows that $\sum_{i=1}^{n_k} v_{k,i} = 1$ for any $k \in \{0, \ldots, m\}$.

For any fixed choice of local priorities in the feasible regions the equations (5)-(8) assign a unique weight to each element of the hierarchy. As the local priorities vary over the feasible regions the element $e_{k,l}$ receives a set of weights denoted by $V_{k,l}$. In other words, $r \in V_{k,l}$ if only if there exists some feasible combination of local priorities which assigns the weight r to $e_{k,l}$.

Lemma 1 *$V_{k,l}$ is a closed convex subset of* $[0, 1]$.

Proof. In the appendix. □

Let $a \in \{1, \ldots, n_m\}$ so that $e_{m,a}$ is a decision alternative. Theorem 1 shows how the interval $V_{m,a}$ can be computed by solving a series of linear programming problems.

Theorem 1 *At level $m - 1$ define for $l = 1, \ldots, n_{m-1}$*

$$\overline{\nu}^a_{m-1,l} = \max_{w \in S_{m-1,l}} w[a] \tag{9}$$

$$\underline{\nu}^a_{m-1,l} = \min_{w \in S_{m-1,l}} w[a] \tag{10}$$

For $k = m - 2, \ldots, 0$, $l = 1, \ldots, n_k$ define recursively the functions $\overline{\mu}^a_{k,l}, \underline{\mu}^a_{k,l} : S_{k,l} \mapsto [0, 1]$ by

$$\overline{\mu}^a_{k,l}(w) = \sum_{\{j | j \in D_{k,l}\}} \overline{\nu}^a_{k+1,j} w[j] \tag{11}$$

$$\underline{\mu}^a_{k,l}(w) = \sum_{\{j | j \in D_{k,l}\}} \underline{\nu}^a_{k+1,j} w[j] \tag{12}$$

and put

$$\overline{\nu}^a_{k,l} = \max_{w \in S_{k,l}} \overline{\mu}^a_{k,l}(w) \tag{13}$$

$$\underline{\nu}^a_{k,l} = \min_{w \in S_{k,l}} \underline{\mu}^a_{k,l}(w) \tag{14}$$

Then $V_{m,a} = [\underline{\nu}^a_{0,1}, \overline{\nu}^a_{0,1}]$.

Proof. Choose any feasible local priorities $w_{i,j} \in S_{i,j}$ and compute the weights for the elements as in (5)-(8). Then for an arbitrary $k, 0 \leq k < m-1$

$$\begin{aligned}
\sum_{i=1}^{n_{k+1}} v_{k+1,i}\overline{\nu}^a_{k+1,i} &= \sum_{i=1}^{n_{k+1}} \Big(\sum_{\{j|i \in D_{k,j}\}} v_{k,j} w_{k,j}[i] \Big) \overline{\nu}^a_{k+1,i} \\
&= \sum_{j=1}^{n_k} v_{k,j} \Big(\sum_{\{i|i \in D_{k,j}\}} \overline{\nu}^a_{k+1,i} w_{k,j}[i] \Big) \\
&= \sum_{j=1}^{n_k} v_{k,j} \overline{\mu}^a_{k,j}(w_{k,j}) \leq \sum_{j=1}^{n_k} v_{k,j} \overline{\nu}^a_{k,j}
\end{aligned}$$

Applying the above inequality repeatedly gives

$$\begin{aligned}
v_{m,a} &= \sum_{i=1}^{n_{m-1}} v_{m-1,i} w_{m-1,i}[a] \\
&\leq \sum_{i=1}^{n_{m-1}} v_{m-1,i} \overline{\nu}^a_{m-1,i} \\
&\leq \sum_{i=1}^{n_0} v_{0,i} \overline{\nu}^a_{0,i} = \overline{\nu}^a_{0,1}
\end{aligned}$$

Thus $\max_{r \in V_{k,l}} r \leq \overline{\nu}^a_{0,1}$. By choosing $\overline{w}_{k,l} \in S_{k,l}$ such that the linear programs in (9) and (13) are maximized the above inequalities become equalities. The case for the lower bound is proved similarly. □

Theorem 1 immediately suggests an algorithm for computing the weight intervals for the alternatives. First compute the numbers $\overline{\nu}^a_{m-1,l}, \underline{\nu}^a_{m-1,l}$ as solutions to the linear programs (9)-(10). Then form the objective functions $\overline{\mu}^a_{m-2,l}, \underline{\mu}^a_{m-2,l}$ in (11)-(12) and solve the linear programs in (13)-(14) to obtain the numbers $\overline{\nu}_{m-2,l}, \underline{\nu}_{m-2,l}$. Proceed to higher levels of the hierarchy until the topmost element of the hierarchy has been reached.

If the feasible region is changed at level k then the scalars $\overline{\nu}^a_{i,j}, \underline{\nu}^a_{i,j}$ remain unchanged on levels $i > k$. Thus only levels $0, \ldots, k$ of the hierarchy need be recomputed if the scalars $\overline{\nu}^a_{i,j}, \underline{\nu}^a_{i,j}$ are stored for later use. This means that changes to upper parts of the hierarchy typically require less computation than changes on the lower levels of the hierarchy.

The optimum of a linear program is attained at an extreme point. Thus the linear programs (9)-(10),(13)-(14) can be solved by enumeration if *ext* $S_{k,l}$, the set of extreme points of the feasible region $S_{k,l}$, is known. Since $2n_m$ linear programs must be solved at each criterion, it can be computationally advantageous to determine *ext* $S_{k,l}$ first.

The proposed method can be used even when the feasible region is derived from a more general description of the DM's preferences. If in addition to pairwise comparisons the DM is allowed to put constraints on the components of the local priority vector (such as $0.25 \leq w_1 \leq 0.80$) the feasible regions are still compact and convex sets, and the weight intervals can be computed by applying theorem 1.

Belton and Gear (1983) point out that in the AHP the ranking of some alternatives may become reversed when a new alternative is taken into consideration. This phenomenon,

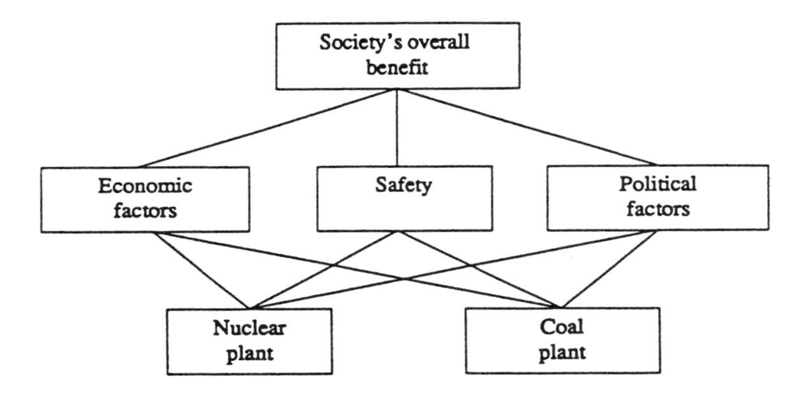

Figure 3: The hierarchy in the example

called rank reversal, has been extensively discussed and is often seen as a major problem in the AHP (see e.g. Dyer, 1990).

Working with interval judgments does not eliminate rank reversal as such. The reason for this is that the DM can enter interval judgments such that any combination of local priorities from the feasible regions causes rank reversal to take place. However, in practice the interval approach may reduce the number of rank reversals since the ranking of the alternatives becomes complete only in the later stages of the analysis. This topic is currently under further investigation.

5. AN EXAMPLE

This section demonstrates the results of section 4 in the context the hierarchy shown in Figure 3, which is a simplified version of the hierarchy for comparing alternative forms of energy production reported in Hämäläinen (1988, 1990, 1991) (see Figure 1). In the present example the overall benefits of the society are decomposed into three subcriteria, i.e. the economic, safety related and political dimensions of the decision. There are two alternatives, a nuclear power plant and a coal power plant.

In the indeces B refers to the topmost element of the hierarchy, E, S, P refer its three subcriteria and N, C refer to the two alternatives. Thus, for example, S_B is the feasible region at the topmost element. The components of the local priorities are indexed in the above order: if $w \in S_B$ then the first component of w is the weight given to economic factors etc. We emphasize that the preferences in this example are hypothetical and have been designed to illustrate the interval approach to the AHP.

Assume that the DM starts the analysis by making the following preference statements

- with respect to economic factors, a nuclear power plant is at least four but no more than five times more profitable than a coal power plant
- with respect to safety, neither alternative is more than two times safer than the other
- with respect to political factors, a nuclear power plant is more controversial than a coal power plant, but not more than by a ratio of two

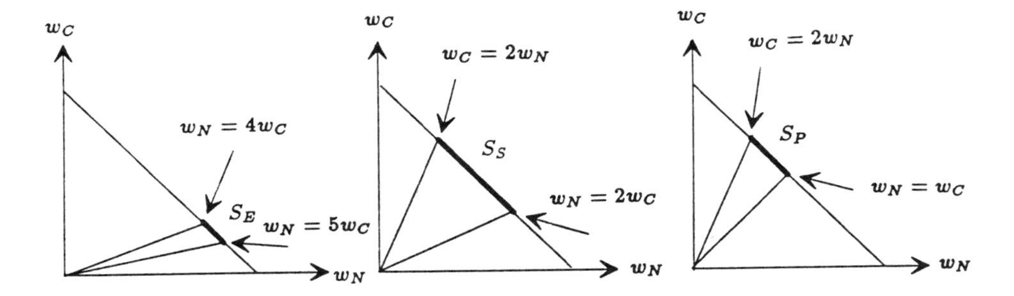

Figure 4: The feasible regions at the subcriteria

In view of the first statement, at the element economic factors the feasible region S_E consists of those vectors $w = (w_N, w_C)$ in the positive orthant which, in addition to the requirement $w_N + w_C = 1$, satisfy the constraints $4w_C \leq w_N \leq 5w_C$. This feasible region, characterized by its extreme points (0.80,0.20) and (0.83,0.17), is shown in Figure 4.

As discussed in section 3, the inequalities $4w_C \leq w_N \leq 5w_C$ correspond to the interval judgment $I_{NC} = [4, 5]$. Alternatively by the reciprocality of the bounds of the interval judgments these inequalities can be expressed as $I_{CN} = [\frac{1}{5}, \frac{1}{4}]$.

Likewise, the second statement implies that the feasible region S_S, also shown in Figure 4, consists of local priorities which satisfy the constraints $\frac{1}{2}w_N \leq w_C \leq 2w_N$. These inequalities correspond to the interval judgment $I_{NC} = I_{CN} = [\frac{1}{2}, 2]$.

The third statement leads to the constraints $w_N \leq w_C \leq 2w_N$; therefore by his statement the DM has specified the interval judgment $I_{CN} = [1, 2]$.

To summarize, after the above three statements the extreme points of the feasible regions at the three subcriteria are

$$ext\ S_E = \{(0.80, 0.20), (0.83, 0.17)\} \tag{15}$$
$$ext\ S_S = \{(0.33, 0.67), (0.67, 0.33)\} \tag{16}$$
$$ext\ S_P = \{(0.33, 0.67), (0.50, 0.50)\} \tag{17}$$

In order to compute the upper bound for the weight of the nuclear plant using theorem 1 we first compute upper bounds for how much of their weight the three criteria can give to the nuclear plant. These bounds are obtained from the linear programming problems in (9), whose solutions, found by inspecting the extreme points in (15)-(17), are

$$\overline{\nu}_E^N = \max_{w \in S_E} w_N = 0.83 \tag{18}$$
$$\overline{\nu}_S^N = \max_{w \in S_S} w_N = 0.67 \tag{19}$$
$$\overline{\nu}_P^N = \max_{w \in S_P} w_N = 0.50 \tag{20}$$

Since no judgments have been made at the topmost element the feasible region S_B contains all the vectors $w = (w_E, w_S, w_P)$ in the positive orthant such that $w_E + w_S + w_P = 1$. Combining (11) with (18)-(20) gives $\overline{\mu}_B^N(w) = 0.83w_E + 0.67w_S + 0.50w_P$ where $w \in S_B$. In view of (13), the upper bound for the weight of the nuclear plant is the maximum of this function over the feasible region at the topmost element , i.e.

$$\overline{\nu}_B^N = \max_{w \in S_B} \overline{\mu}_B^N(w) = 0.83 \times 1.0 + 0.67 \times 0.0 + 0.50 \times 0.0 = 0.83 \tag{21}$$

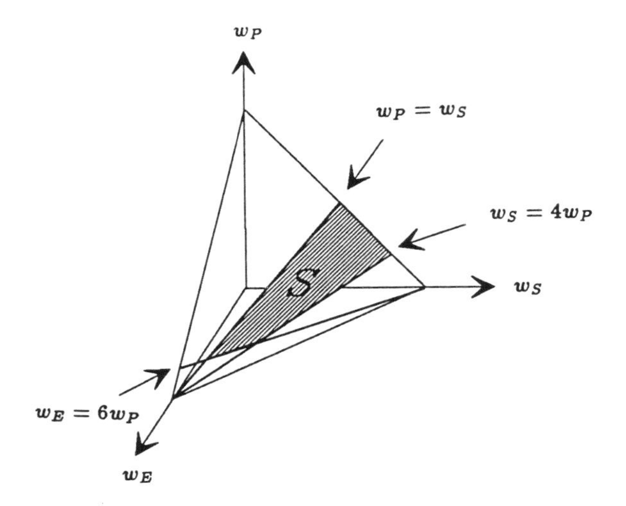

Figure 5: The feasible region at the topmost element

Likewise, the lower bound for the weight of the nuclear alternative is found by first solving the linear programs in (10) by inspecting the extreme points in (15)-(17)

$$\underline{\nu}_E^N = \min_{w \in S_B} w_N = 0.80 \tag{22}$$
$$\underline{\nu}_S^N = \min_{w \in S_S} w_N = 0.33 \tag{23}$$
$$\underline{\nu}_P^N = \min_{w \in S_P} w_N = 0.33 \tag{24}$$

Inserting these into (12) gives $\underline{\mu}_B^N(w) = 0.80w_E + 0.33w_S + 0.33w_P$, and consequently the minimum in (14) at the topmost element is $\underline{\nu}_B^N = \min_{w \in S_B} \underline{\mu}_B^N(w) = 0.33$. Combining this result with (21) shows that the weight interval of the nuclear alternative is $V_N = [0.33, 0.83]$.

A similar series of computations shows that the weight interval for the coal plant, in view of the statements the DM has entered so far, is $V_C = [0.17, 0.67]$. Note that at this point of analysis the intervals V_N and V_C overlap.

Assume that the DM next makes the following statements at the topmost element

- economic factors are at most six times and safety is at most four times as important as political factors
- political factors are less important than safety

The first of these statements implies the inequalities $w_E \leq 6w_P$ and $w_S \leq 4w_P$, and the second implies that $w_P \leq w_S$. The feasible region S_B constrained by these three inequalities is shown in Figure 5, and its extreme points are found to be

$$ext\ S_B = \{(0.00, 0.50, 0.50), (0.00, 0.80, 0.20), (0.55, 0.36, 0.09), (0.75, 0.13, 0.13)\} \tag{25}$$

The interval matrix (3) corresponding to the DM's statements at the topmost element is

$$\begin{array}{c} \\ E \\ S \\ P \end{array} \begin{array}{c} \begin{array}{ccc} E & S & P \end{array} \\ \left(\begin{array}{ccc} 1 & [-,-] & [-,6] \\ [-,-] & 1 & [1,4] \\ [\frac{1}{6},-] & [\frac{1}{4},1] & 1 \end{array} \right) \end{array}$$

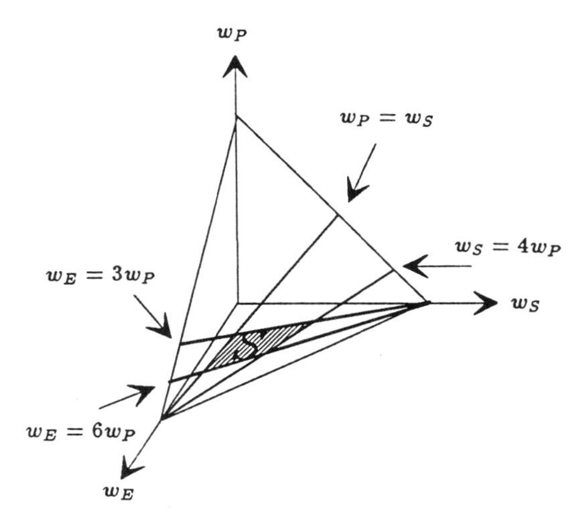

Figure 6: The feasible region in the last phase of the analysis

where the dashes in the judgments I_{ES}, I_{SE} indicate that nothing has been said about the relative importance of economic and safety related factors. Moreover, the lower bound of the interval I_{EP} is still unspecified, indicating that the DM can enter bounds of the interval judgments one at a time.

After the new preference information the alternatives weight intervals are updated as follows. Since no changes have been made to the feasible regions S_E, S_S, S_P the numbers (18)-(20), (22)-(24) and consequently the functions $\overline{\mu}_B^N(\cdot), \underline{\mu}_B^N(\cdot)$ remain the same as in the first phase. To compute the revised bounds for the weight of the nuclear plant it therefore suffices to solve the problems (13) and (14) over the modifed feasible region at the topmost element. Inspecting the extreme points in (25) gives

$$\overline{\nu}_B^N(w) = \max_{w \in S_B} \overline{\mu}_B^N(w) = 0.83 \times 0.75 + 0.67 \times 0.13 + 0.50 \times 0.13 = 0.77 \quad (26)$$
$$\underline{\nu}_B^N(w) = \min_{w \in S_B} \underline{\mu}_B^N(w) = 0.80 \times 0.00 + 0.33 \times 0.50 + 0.33 \times 0.50 = 0.33 \quad (27)$$

Thus $V_N = [0.33, 0.77]$. In addition to $(0.00, 0.50, 0.50)$ S_B contains also other local priorities like $(0.00, 0.80, 0.20)$ which also give the minimum weight to the nuclear alternative.

Updating the bounds for the weight of the coal plant gives that its weight interval is $V_C = [0.23, 0.67]$. Since the intervals V_N, V_C still overlap it is necessary to continue the elicitation of the DM's preferences.

Finally, assume that the DM states that economic factors are at least three times more important than political ones. This statement imposes the constraint $w_E \geq 3w_P$ on the feasible region, which now becomes much smaller (see Figure 6). Its extreme points are

$$ext\ S_B = \{(0.38, 0.50, 0.13), (0.55, 0.36, 0.09), (0.60, 0.20, 0.20), (0.75, 0.13, 0.13)\}$$

Since the point $(0.75, 0.13, 0.13)$, which gave the maximum weight to the nuclear alternative in (26), still belongs to S_B the upper bound $\overline{\nu}_B^N$ remains the same. However, the lower bound increases since

$$\underline{\nu}_B^N(w) = \min_{w \in S_B} \underline{\mu}_B^N(w) = 0.80 \times 0.38 + 0.33 \times 0.50 + 0.33 \times 0.13 = 0.51$$

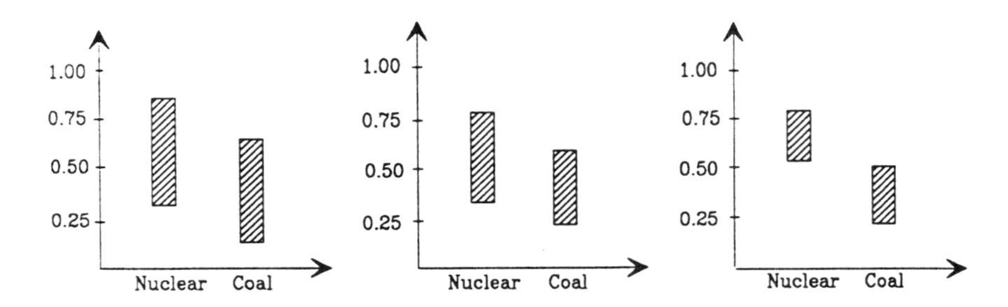

Figure 7: Changes in weight intervals due to refined interval judgments

Moreover, the upper bound for the weight of the coal alternative decreases from 0.67 to 0.49. Since the weight intervals $V_N = [0.51, 0.77]$ and $V_C = [0.23, 0.49]$ no longer overlap the nuclear energy alternative receives the higher weight for any set of feasible local priorities. Thus it can be concluded that the nuclear plant is the more preferred alternative. The final as well as the earlier weight intervals are shown in Figure 7.

Observe that it was not necessary to elicit any information about the relative importance of the criteria safety and economic factors in order to identify the most preferred alternative. It is in this sense that the interval approach may, at least in some cases, help to reduce the amount of comparison effort in the analysis, especially when the number of subelements is large.

6. CONCLUSIONS

This paper has described a method for integrating interval judgments into the AHP. The method associates a weight interval with each decision alternative by processing interval judgments in the hierarchy. These weight intervals consist of weights generated by those local priority vectors which are consistent with the DM's judgments, and they become narrower as the DM gradually refines his preferences.

The interval approach leads to a more interactive decision support process than the conventional AHP. This is because after each new interval judgment the weight intervals can be recomputed and graphically displayed to the DM. In our view this is a significant improvement to the methodology.

A complete ranking for the alternatives is found when the alternatives' weight intervals no longer overlap. Such an order may be established before assessing all the interval judgments from the DM, and thus the method helps to reduce the amount of comparison effort required by the analysis.

Based on the proposed method we have developed a decision support tool called INPRE. Testing the method in real-life applications will help to determine how the DMs can in practice make use of the potential benefits of the interval approach to the AHP.

ACKNOWLEDGMENTS

This work has been supported by the Research Council for Technology of the Academy of Finland.

BIBLIOGRAPHY

Arbel, A., "Approximate Articulation of Preference and Priority Derivation," *European Journal of Operational Research*, vol. 43, 317-326, 1989.

Belton, V. and T. Gear, "On a Short-coming of Saaty's Method of Analytic Hierarchies," *Omega*, vol. 11, no. 3, 228-230, 1983.

Dyer, J. S., "Remarks on the Analytic Hierarchy Process," *Management Science*, vol. 36, no. 3, 249-258, 1990.

Hämäläinen, R. P., "Computer Assisted Energy Policy Analysis in the Parliament of Finland," *Interfaces*, vol. 18, no. 4, 12-23, 1988.

Hämäläinen, R. P., "A Decision Aid in the Public Debate on Nuclear Power," *European Journal of Operational Research*, vol. 48, 66-76, 1990.

Hämäläinen, R. P., "Facts or Values - How do Parlamentarians and Experts See Nuclear Power," to appear in *Energy Policy*, 1991.

Moore, R. E., *Interval Analysis*, Prentice-Hall, Englewood Cliffs, N.J., 1966.

Saaty, T. L., "A Scaling Method for Priorities in Hierarchical Structures," *Journal of Mathematical Psychology*, vol. 15, no. 3, 234-281, 1977.

Saaty, T. L., *The Analytic Hierarchy Process*, McGraw-Hill, New York, 1980.

Saaty, T. L. and L. G. Vargas, "Uncertainty and Rank Order in the Analytic Hierarchy Process," *European Journal of Operational Research*, vol. 32, 107-117, 1987.

Salo, A. and R. P. Hämäläinen, "Decision Support under Ambiguous Preference Statements." In *Proc of the Annual Conference of the Operational Research Society of Italy : Models and Methods for Decision Support*, Sorrento, October 1990, 229-243.

Vargas, L. G., "An Overview of the Analytic Hierarchy Process and Its Applications," *European Journal of Operational Research*, vol. 48, 2-8, 1990.

Yoon, K., "The Analytic Hierarchy Process (AHP) with Bounded Interval Input," *Preprints of the International Symposium on the Analytic Hierarchy Process*, Tianjin, China, September 1988, 149-156.

APPENDIX

Proof of lemma 1. By (5)-(8) $V_{k,l} \subset [0,1]$. Fix $k > 0$ and l and choose $\overline{w}_{i,j} \in S_{i,j}, i = 1,\ldots,k-1, j = 1,\ldots,n_i$ such that $v_{k,l}$ is maximized, and denote this maximum by $\overline{v}_{k,l}$. Such vectors $\overline{w}_{i,j}$ exist, because the sets $S_{i,j}$ are compact and $v_{k,l}$ is a continuous function of the local priorities. Similarly choose $\underline{w}_{i,j} \in S_{i,j}$ which minimize $v_{k,l}$, and let $\underline{v}_{k,l}$ be this minimum. It remains to show that there are vectors in $S_{i,j}$ such that every value in

$[\underline{v}_{k,l}, \overline{v}_{k,l}]$ is attained.
Define functions $f_{i,j} : [0,1] \mapsto [0,1]$, $(i = 1,\ldots,k, j = 1,\ldots,n_i)$ recursively by

$$\begin{aligned} f_{1,j}(t) &= t\overline{w}_{0,1}[j] + (1-t)\underline{w}_{0,1}[j] \\ f_{2,j}(t) &= \sum_{\{i|j\in D_{1,i}\}} (t\overline{w}_{1,i}[j] + (1-t)\underline{w}_{1,i}[j])f_{1,i}(t) \\ &\vdots \\ f_{k,j}(t) &= \sum_{\{i|j\in D_{k-1,i}\}} (t\overline{w}_{k-1,i}[j] + (1-t)\underline{w}_{k-1,i}[j])f_{k-1,i}(t) \end{aligned}$$

Then $f_{k,l}(0) = \underline{v}_{k,l}$ and $f_{k,l}(1) = \overline{v}_{k,l}$. Since $f_{k,l}$ is a continuous function for any $r \in [\underline{v}_{k,l}, \overline{v}_{k,l}]$ there exists a $t^r \in [0,1]$ such that $f_{k,l}(t^r) = r$. Define vectors $w^r_{i,j} =^{def} t^r\overline{w}_{i,j} + (1-t^r)\underline{w}_{i,j}$, which by the convexity of $S_{i,j}$ are feasible. These vectors give the weight r to $e_{k,l}$. □

WEAKLY-EFFICIENT SOLUTIONS OF LIMITING MULTICRITERIA OPTIMIZATION PROBLEMS

M.E.Salukvadze
Professor
A.L.Topchishvili
Leading Researcher
Institute of Control Systems
Georgian Academy of Sciences
Tbilisi, Georgia, 380060, USSR

Abstract

This paper concerns the connection among efficient, weakly-efficient and proper efficient sets of multicriteria optimization problems solutions. For the family of bicriteria optimization problems the limiting properties of the sets of weakly-efficient solutions are determined. In particular, based on the properties and structure of the limiting matrix of the family of stochastic matrices it is shown that the infinite intersection of the compact weakly-efficient solution sets, each corresponding to the specially constructed bicriteria optimization problem, coincides with the set of all weakly-efficient solutions of bicriteria optimization problem of the special form.

1. Introduction

A whole series of production processes, economic systems of different types and technical objects is described by mathematical models which are multicriteria optimization problems , see, for example, Norenkov (1980) and Brahman (1984). This situation is quite usual, because frequently it is necessary to take simultaneously into account the influence of a number of contradictory extremal factors on the system.

The most intensive development of the theory and the methods of multicriteria optimization took place in the 1970's. A number of papers, a detailed bibliographic description of which is given in Salukvadze (1979) and Steuer (1986), are devoted to the investigation and the analysis of modern methods for solving linear and nonlinear multicriteria optimization problems. Some classifications of the methods of this type, oriented to the specific user, are given in Salukvadze, Kilasonia and Topchishvili (1990), and the multicriteria optimization problems with contradictory constraints were explored in Salukvadze and Topchishvili (1989). Very interesting results generalized into general domination cone for different classes of solutions of multicriteria problems are given in Yu (1985).

Now one of the widely developing fields in multicriteria optimization is its qualitative theory; the most important results are given in Podinovskey and Nogin (1982). The well-known algorithms can be modified and new ones can be created based on new theoretical results.

The object of this article is to examine some properties of different classes of multicriteria optimization problem solutions. Also, our aim is to construct the limiting set of a family of converging sets of weakly-efficient solutions for multicriteria optimization problems.

2. Connection among Different Classes of Solutions

Consider the following model of a multicriteria optimization problem:

$$\Gamma = < X , \{ f_i(x) \}_{i\in N} >. \quad (1)$$

Here X is a nonempty set of all feasible solutions, $X\subset R^m$, x is an element of X, $f_i(x)$ is a scalar utility function which is assumed to be continuous on X, $N=\{1,2,\dots,n\}$. We define the utility vector-function by $f(x)=(f_1(x),\dots,f_n(x))^T$. A transpose is designated by the superscript T.

Throughout this article we shall assume that X is a fixed, nonempty, convex, compact set.

Stated briefly, a multicriteria optimization problem consists in the choice of a particular solution $x^*\in X$, for which all of the utility functions $f_i(x)$, $i\in N$, simultaneously approach bigger values or at least not decrease.

Let us recall some concepts of multicriteria optimization problem solutions; see, for example, Podinovskey and Nogin (1982).

Definition 2.1. Solution $x^P\in X$ is called Pareto-optimal (or efficient) for the problem (1), iff for every $x\in X$ the system of inequalities

$$f_i(x) \geq f_i(x^P), \ i\in N,$$

where, at least one inequality is strict, is inconsistent.

Definition 2.2. Solution $x^S\in X$ is called Slater-optimal (or weakly-efficient) for the problem (1), iff for every $x\in X$ the following system of strict inequalities

$$f_i(x) > f_i(x^S), \ i\in N,$$

is inconsistent.

Definition 2.3. Solution $x^G \in X$ is called Geoffrion-optimal (or proper efficient) for the problem (1), iff it is a Pareto-optimal solution for the problem (1) and there exists a positive number $\theta>0$ such that for each $i \in N$ we have

$$\frac{f_i(x) - f_i(x^G)}{f_j(x^G) - f_j(x)} \le \theta$$

for some j such that $f_j(x) < f_j(x^G)$ whenever $x \in X$ and $f_i(x) > f_i(x^G)$.

Let $S_f(X)$, $P_f(X)$, $G_f(X)$ denote the sets of weakly-efficient, efficient, proper efficient solutions, respectively, for the problem (1).

It is obvious that $G_f(X) \subset P_f(X) \subset S_f(X)$.

Let A be a square matrix of the n-th order with all positive elements: $a_{ij}>0$ $\forall i,j \in N$.

Consider the family of problems

$$\Gamma_{A^k} = < X , \{f_i^{(k)}(x)\}_{i \in N} > \qquad (2)$$

corresponding to problem (1), where

$$f_i^{(k)}(x) = (A^k f(x))_i, \; i \in N,$$

and $k=1,2,\ldots,\bar{k}$, where $\bar{k}$ is an arbitrary big natural number.

The sets of weakly-efficient, efficient and proper efficient solutions for the problem (2) are correspondingly denoted by $S_{f^{(k)}}(X)$, $P_{f^{(k)}}(X)$, $G_{f^{(k)}}(X)$.

In Salukvadze and Topchishvili (1990) the validity of the following two facts was obtained.

Theorem 2.1. If $x^* \in S_{f^{(k+1)}}(X)$, then $x^* \in S_{f^{(k)}}(X)$.

Theorem 2.2. If $x^* \in S_{f^{(k+1)}}(X)$, then $x^* \in G_{f^{(k)}}(X)$.

Corollary 2.1. The chain of the following inclusions

$$S_f(X) \supset P_f(X) \supset G_f(X) \supset S_{f^{(1)}}(X) \supset P_{f^{(1)}}(X) \supset$$
$$\supset G_{f^{(1)}}(X) \supset \ldots \supset S_{f^{(k)}}(X) \supset P_{f^{(k)}}(X) \supset G_{f^{(k)}}(X) \supset \ldots$$

is valid.

It is well known (Zhukovskii and Salukvadze (1991)) that for arbitrary natural number k the set $S_{f^{(k)}}(X)$ is nonempty compact set. Then, by a well known theorem of functional analysis, the family of nested sets $\{ S_{f^{(k)}}(X) \}_{k=1}^{\infty}$ has a nonempty intersection: this means that there exists a set X^* such that

$$X^* = \bigcap_{k=1}^{\infty} S_{f^{(k)}}(X) = \lim_{k\to\infty} S_{f^{(k)}}(X).$$

Now let us suppose that $n=2$, so that we have a bicriterion problem (1) and, connected with it, a bicriterion series of problems (2).

Suppose that the matrix A, given above, is a second order stochastic matrix of the type

$$A = H = \begin{Vmatrix} a, & 1-a \\ 1-b, & b \end{Vmatrix},$$

where $0<a<1$, $0<b<1$.

3. Some Properties of Matrices

Let $\lambda_s^{(k)}$, $s=1,2$, denote the eigenvalues of matrix

$$H^k = \underbrace{H\cdot H\cdot \ldots \cdot H}_{k};$$

that is, they satisfy

$$\det (H^k - \lambda I) = 0,$$

where I is the identity matrix of second order.

Proposition 3.1. $\lambda_1^{(k)}=1$, $\lambda_2^{(k)}=(a+b-1)^k$ for an arbitrary natural number k.

The truth of this proposition follows immediately from the more general fact which is proved in Theorem 5.3.4 of Lankaster (1978).

Corollary 3.1. $\lambda_1^{(k)} \neq \lambda_2^{(k)}$ for an arbitrary natural number k.

Corollary 3.2. H^k is a prime matrix for an arbitrary natural number k.

Theorem 3.1. Let B be a prime matrix of the n-th order with real elements and eigenvalues λ_1, λ_2, ..., λ_n, such that $|\lambda_1|>|\lambda_j|$, $j=2,3,\dots,n$. Then

$$\lim_{k\to\infty} (\lambda_1^{-1}B)^k = G_1,$$

where G_1 is accompanying matrix corresponding to λ_1; this means (see, for example, Lankaster [2]) that

$$G_1 = v^{(1)}u^{(1)^T},$$

where $v^{(1)}$, $u^{(1)}$ are the right and the left eigenvectors of matrix B corresponding to λ_1.

One can find the proof of it in Salukvadze and Topchishvili [8].

Corollary 3.3. $\lim_{k\to\infty} H^k=H^*$, where H^* is the second order accompanying matrix corresponding to the eigenvalue $\lambda_1^{(1)}=1$ of the stochastic matrix H.

Let us define the structure of the matrix H^*.

Construct the right and the left eigenvector systems for the matrix H corresponding to $\lambda_1^{(1)}= 1$, $\lambda_2^{(1)}= a+b-1$.

Correspondingly we consider two cases.

I. $\lambda_1^{(1)} = 1$.

Then the right eigenvector can be defined by the following vector equation:

$$Hv = v.$$

The solution of this system is vector $v^{(1)}= (\alpha,\alpha)^T$, where α is some constant.

The left eigenvector can be defined by the vector equation

$$H^Tu = u.$$

Then the solution of this system is vector

$u^{(1)} = (\beta,\frac{a-1}{b-1}\beta)^T$, where β is some constant.

II. $\lambda_2^{(1)} = a+b-1$.

Analogously we obtain the presentation of the right and the left eigenvectors of matrix H corresponding to $\lambda_2^{(1)}$. They are of the following type:

$$v^{(2)} = (\gamma, \frac{b-1}{1-a}\gamma)^T, \quad u^{(2)} = (\delta, -\delta)^T,$$

where γ, δ are some constants.

Define the following square matrices:

$$V = (v^{(1)}, v^{(2)}), \quad U = (u^{(1)}, u^{(2)}).$$

The two systems of vectors $\{v^{(1)}, v^{(2)}\}$, $\{u^{(1)}, u^{(2)}\}$ must be quasi-biorthogonal systems. It follows that $U^T V = I$, and

$$\left\| \begin{matrix} \beta, & \frac{a-1}{b-1}\beta \\ \delta, & -\delta \end{matrix} \right\| \left\| \begin{matrix} \alpha, & \gamma \\ \alpha, & \frac{b-1}{1-a}\gamma \end{matrix} \right\| = I. \tag{3}$$

From the equation (3) we get

$$\beta = \frac{b-1}{a+b-2} \cdot \frac{1}{\alpha}, \quad \delta = \frac{a-1}{a+b-2} \cdot \frac{1}{\gamma}.$$

Hence

$$H^* = v^{(1)} u^{(1)T} = \left\| \begin{matrix} \frac{b-1}{a+b-2}, & \frac{a-1}{a+b-2} \\ \frac{b-1}{a+b-2}, & \frac{a-1}{a+b-2} \end{matrix} \right\| \tag{4}$$

Thus the following fact is proved.

Theorem 3.2. The accompanying matrix of the eigenvalue 1 of the second order stochastic matrix H is determined by the equality (4).

4. Main Result

Theorem 4.1. The infinite intersection of the compact weakly-efficient solution sets $S_{f^{(k)}}(X)$, each corresponding to the multicriteria optimization problem Γ_{A^k}, $k=1,2,3,\ldots$, coincides with the set of all weakly-efficient solutions $S_{f^*}(X)$ of the multicriteria optimization problem

$$\Gamma^* = \langle X, \{f_i^*(x)\}_{i=1,2} \rangle,$$

where $f_i^*(x)$ $(i=1,2)$ is the i-th component of the vector-function $H^*f(x)$, and the second-order matrix H^* is the accompanying matrix, corresponding to the maximal eigenvalue $\lambda_1=1$ of the stochastic second-order matrix H.

Proof. We must obtain that the equality

$$X^* = S_{f^*}(X) \tag{5}$$

is valid. Then for the proof of (5) we must show that the inclusions

$$X^* \subset S_{f^*}(X) \quad \text{and} \quad S_{f^*}(X) \subset X^*$$

are fulfilled.

First of all we shall prove the validity of the first inclusion.

Suppose $x^* \in X^*$; then we want to show that $x^* \in S_{f^*}(X)$.

With $X^* = \lim_{k\to\infty} S_{f^{(k)}}(X)$ and $x^* \in X^*$, there exists a sequence of points $x^{(k)} \in S_{f^{(k)}}(X)$, $k=1,2,3,\ldots$, such that

$$\lim_{k\to\infty} x^{(k)} = x^*.$$

While $x^{(k)} \in S_{f^{(k)}}(X)$, there does not exist $x \in X$ such that the inequality system

$$(H^k f(x))_i > (H^k f(x^{(k)}))_i, \; i=1,2,$$

is consistent. This means that for every $x \in X$ there exists an index $i(x)$ such that the inequality

$$(H^k(f(x) - f(x^{(k)})))_{i(x)} \le 0 \tag{6}$$

is fulfilled.

We fix an arbitrary vector $x \in X$.

Define the following two sets:

$$K_x^1 = \{ k \in \{1,2,3,\ldots\}: (H^k f(x))_1 \le (H^k f(x^{(k)}))_1 \},$$

$$K_x^2 = \{ k \in \{1,2,3,\ldots\}: (H^k f(x))_2 \le (H^k f(x^{(k)}))_2 \}.$$

Using (6) we have that $K_x^1 \cup K_x^2 = \{1,2,3,\ldots\}$.

It is obviuos that at least one of the sets k_x^1 or $K_{x_1}^2$ is infinite; suppose, for instance, that the set K_x^1 is infinite.

While $\lim_{k\to\infty} H^k = H^*$, then we have $\lim_{\substack{k\to\infty \\ k\in K_x^1}} H^k = H^*$. So we get

$$\lim_{\substack{k\to\infty \\ k\in K_x^1}} (H^k f(x) - H^k f(x^{(k)}))_1 = (H^* f(x) - H^* f(x^*))_1. \quad (7)$$

Thus, from the equality (7) and the structure of the set K_x^1 we obtain the validity of the inequality

$$(H^* f(x))_1 \le (H^* f(x^*))_1.$$

While x was the arbitrary vector in the set X, then analogously to the mentioned above it follows for every $x\in X$ that there exists an index $i(x)\in\{1,2\}$ such that the inequality

$$(H^* f(x))_{i(x)} \le (H^* f(x^*))_{i(x)}$$

is fulfilled.

So we have $x^*\in S_{f^*}(X)$, and thus $X^* \subset S_{f^*}(X)$, and the first inclusion is proved.

Now we shall prove the validity of the inverse inclusion, that is

$$S_{f^*}(X) \subset X^*.$$

Suppose $x^*\in S_{f^*}(X)$. Then we want to prove that

$$x^*\in X^* = \lim_{k\to\infty} S_{f^{(k)}}(X).$$

Suppose that $x^*\notin X^*$. Then there exists a natural number $N(x^*)=n_o$ such that $x^*\notin S_{f^{(k)}}(X)$ for $k\ge N(x^*)$. This means that there exists a sequence of points $x^{(k)}\neq x^*$ such that the system of strict inequalities

$$(H^k f(x^{(k)}))_i > (H^k f(x^*))_i, \quad i=1,2, \tag{8}$$

is fulfilled.

We shall show for the system (8) that for every $k \geq n_o$ we can take the same point $x^{(n_o)}$. In the case of $k=n_o$, it follows from the system (8) that

$$(H^{n_o}(f(x^{(n_o)}) - f(x^*)))_i > 0, \quad i=1,2. \tag{9}$$

From the system (9) it follows that the system of inequalities

$$(H^{n_o+1}(f(x^{(n_o)}) - f(x^*)))_i > 0, \quad i=1,2,$$

is valid, because H is a matrix with all positive elements.

Continuing this process, we get

$$(H^{n_o+l}(f(x^{(n_o)}) - f(x^*)))_i > 0, \quad i=1,2, \tag{10}$$

$l=0,1,2,3,\ldots$

Let us implement the limiting process in the system (10). We get

$$(H^*(f(x^{(n_o)}) - f(x^*)))_i \geq 0, \quad i=1,2, \tag{11}$$

because $\lim_{l\to\infty} H^{n_o+l} = H^*$ and $f(\cdot)$ is a continuous vector-function.

By our supposition $x^* \in S_{f^*}(X)$, and so we obtain the system of equalities

$$(H^*(f(x^{(n_o)}) - f(x^*)))_i = 0, \quad i=1,2, \tag{12}$$

from the system (11).

From the system of an equalities (12), using the structure of matrix H^* (Theorem 3.2), we get

$$(1-b)(f_1(x^{(n_o)})-f_1(x^*))+(1-a)(f_2(x^{(n_o)})-f_2(x^*)) = 0,$$

or

$$\frac{f_2(x^{(n_o)}) - f_2(x^*)}{f_1(x^{(n_o)}) - f_1(x^*)} = -\frac{1-b}{1-a}. \tag{13}$$

But we must note that $0<a<1$, $0<b<1$,. Thus, the

value

$$\alpha = \frac{1-b}{1-a}$$

is positive, and then from the equality (13) we have that the nominator and the denominator of the left side must have the different signs. For instance, suppose that

$$f_1(x^{(n_o)}) > f_1(x^*),$$
$$f_2(x^{(n_o)}) < f_2(x^*). \tag{14}$$

From the equality (13) we have that

$$f_2(x^{(n_o)}) - f_2(x^*) = -\alpha(f_1(x^{(n_o)}) - f_1(x^*)). \tag{15}$$

It is well-known (Lankaster [2]) that the power of a stochastic matrix is a stochastic matrix. Then the matrix H^k can be written as the follows:

$$H^k = \begin{Vmatrix} a_k , & 1-a_k \\ 1-b_k , & b_k \end{Vmatrix}. \tag{16}$$

Let us transform the left part of inequality system (10), using the formulas (15) and (16). We get

$$\begin{Vmatrix} a_k , & 1-a_k \\ 1-b_k , & b_k \end{Vmatrix} \begin{Vmatrix} f_1(x^{(n_o)}) - f_1(x^*) \\ f_2(x^{(n_o)}) - f_2(x^*) \end{Vmatrix} =$$

$$= \begin{Vmatrix} (a_k - \alpha(1-a_k))\cdot(f_1(x^{(n_o)}) - f_1(x^*)) \\ ((1-b_k) - \alpha b_k)\cdot(f_1(x^{(n_o)}) - f_1(x^*)) \end{Vmatrix}. \tag{17}$$

Using (14) and (17), it follows from the system (10) that for every $k \geq n_o$ the inequalities

$$a_k - \alpha(1-a_k) > 0,$$
$$(1-b_k) - \alpha b_k > 0$$

are fulfilled, or

$$a_k > \frac{\alpha}{1+\alpha}, \quad b_k < \frac{1}{1+\alpha} \tag{18}$$

for $k \geq n_o$.

From the fact that $H^{k+1} = H^k \cdot H$, we have

$$\begin{aligned} a_{k+1} &= a_k(a+b-1) + (1-b), \\ b_{k+1} &= b_k(a+b-1) + (1-a). \end{aligned} \tag{19}$$

Then, in view of (19) we get

$$\begin{aligned} a_{k+1} - a_k &= a_k(a+b-2) + (1-b), \\ b_{k+1} - b_k &= b_k(a+b-2) + (1-a). \end{aligned} \tag{20}$$

Using the inequalities $-2 < a+b-2 < 0$ and (18) it follows from (20) that

$$a_{k+1}-a_k<\frac{\alpha}{\alpha+1}(a+b-2)+(1-b)=\frac{1-b}{2-b-a}(a+b-2)+(1-b)= 0, \tag{21}$$

$$b_{k+1}-b_k >\frac{1}{\alpha+1}(a+b-2)+(1-a)=\frac{1-a}{2-b-a}(a+b-2)+(1-a)=0. \tag{22}$$

Thus, for every $k \geq n_o$, from (21) and (22) we have the strict inequalities

$$a_{k+1} < a_k, \tag{23}$$

$$b_{k+1} > b_k, \tag{24}$$

and this means that, for a particular k, the inequalities of the opposite type are fulfilled. We shall show that this is impossible.

For every $s=1,2,3...$ we find

$$b_{s+1}-b_s = (a+b-1)(b_s-b_{s-1})=\ldots=(a+b-1)^s(b-1); \qquad (25)$$

$$a_{s+1}-a_s = (a+b-1)(a_s-a_{s-1})=\ldots=(a+b-1)^s(a-1). \qquad (26)$$

Using the inequalities $0<a<1$, $0<b<1$, it follows from (25) and (26) that for every fixed $s=1, 2,3,\ldots,$ either the inequalities

$$b_{s+1} - b_s > 0, \quad a_{s+1} - a_s > 0$$

are fulfilled together or the inequalities

$$b_{s+1} - b_s < 0, \quad a_{s+1} - a_s < 0$$

are fulfilled together.

The inequalities (23) and (24) contradict this fact.

So, the system of strict inequalities

$$(H^k f(x))_i > (H^k f(x^*))_i, \ i=1,2,$$

is inconsistent for every $x \in X$, $k \geq N(x^*)$, or $x^* \in S_{f^{(k)}}(X)$.

The last inclusion contradicts the supposed one. This contradiction implies that $x^* \in X^*$ and $S_{f^*}(X) \subset \subset X^*$. This finishes the proof of the theorem.

5. Illustrative Example

Consider the following bicriteria optimization problem:

$$\Gamma = < X, f_1(x), f_2(x) >,$$

where $X = \{ (x_1,x_2): 0 \leq x_1 \leq 1, \ 0 \leq x_2 \leq 1 \}$ is the set of all feasible solutions; $f_1(x)=x_1-x_2$, $f_2(x)=x_1+x_2$ are criteria functions. The problem consists in the choice of a particular solution $x^* \in X$ for which both objective functions $f_1(x)$ and $f_2(x)$ simultaneously approach possibly bigger values.

It is obvious that $S_f(X) = \bigcup_{x_2^* \in [0,1]} (1,x_2^*)$.

Denote by H the secondorder stochastic matrix of the type

$$H = \begin{Vmatrix} 1/3 & ; & 2/3 \\ 1/2 & ; & 1/2 \end{Vmatrix}.$$

Then the matrix H has the eigenvalues $\lambda_1=1$; $\lambda_2= -1/6$.

According to Theorem 3.2 the accompanying matrix of the eigenvalue 1 of the matrix H has the following presentation:

$$H^* = \begin{Vmatrix} 3/7 & ; & 4/7 \\ 3/7 & ; & 4/7 \end{Vmatrix}.$$

Now we consider the bicriteria optimization problem

$$\Gamma^* = < X,\ f_1^*(x),\ f_2^*(x) >,$$

where $f_1^*(x)=f_2^*(x)=(H^*f(x))_1=(H^*f(x))_2= x_1+(1/7)x_2$.

Then we have

$$S_{f^*}(X) = \{(1,1)\}.$$

Thus according to Theorem 4.1 we get that

$$\bigcap_{k=1}^{\infty} S_{f^{(k)}}(X) = \{(1,1)\}.$$

6. Conclusions

One can see from Theorem 4.1 that the problem Γ^* is a singlecriteria optimization problem. Then the set $S_{f^*}(X)$ is simply a set of it's optimal solutions , and thus the problem is to find all extremal points of the singlecriteria optimization problem of the following type:

$$\max_{x\in X} \left(\frac{b-1}{a+b-2} f_1(x) + \frac{a-1}{a+b-2} f_2(x) \right).$$

The Theorem 4.1 allows one in the limit to define the structure of the set of Slater-optimal solutions for the family of multicriteria optimization problems (2).

References

1. Brahman, T.R., "Multicriteriality and Choice of the Alternative in Technical Problems", Radio i Sviaz, Moscow, USSR, 1984 (in Russian).
2. Lankaster, P., "Theory of Matrices", Nauka, Moscow, USSR, 1978 (in Russian).
3. Norenkov, I.P., "Introduction to the Automatized Planning of Technical Devices and Systems", Visshaia Shkola, Moscow, USSR, 1980 (in Russian).
4. Podinovskey, V.V., and Nogin, V.D., "Pareto-Optimal Solutions of Multicriteria Problems", Nauka, Moscow, USSR, 1982 (in Russian).
5. Salukvadze, M.E., "Vector-Valued Optimization Problems in Control Theory", Academic Press, New York, New York, 1979.
6. Salukvadze, M.E., Kilasonia, N.A., and Topchishvili, A.L., "On One Classification Approach to the Multicriteria Optimization Methods", Institute of Control Systems of Georgian Academy of Sciences, Tbilisi, USSR, 1990 (in Russian).
7. Salukvadze, M.E., and Topchishvili, A.L.,"Insoluble Multicriteria Linear Programming Problems", JOTA, Volume 61, Number 3, pp. 487-491, June 1989.
8. Salukvadze, M.E., and Topchishvili, A.L., "Properties of Weakly-Efficient Solutions of Sequence of Multicriteria Problems", Institute of Control Systems of Georgian Academy of Sciences, Tbilisi, USSR, 1990 (in Russian).
9. Steuer, R.E., "Multiple Criteria Optimization: Theory, Computation, and Application", John Wiley and Sons, New York, New York, 1986.
10. Yu, P.L., "Multiple-Criteria Decision Making: Concepts, Techniques and Extensions", Plenum, New York, New York, 1985.
11. Zhukovskii, V.I, and Salukvadze, M.E., "Multicriteria Control Problems under the Uncertainty Conditions", Metsniereba, Tbilisi, USSR, 1991 (in Russian).

APPLICABILITY OF IDEAL POINTS IN MULTICRITERIA DECISION MAKING

Andrzej M. J. Skulimowski*,
Institute of Computer Science,
The University of St. Gallen, St. Gallen, Switzerland

Abstract. An ideal point for the vector optimization problem $(F : U \to E) \to \min(\Theta)$, where Θ is a convex cone introducing the partial order in the criteria space E, is defined as a maximal element of $\{y \in E : F(U) \subset y+\Theta\}$. If $E=\mathbb{R}^N$, and Θ is the positive orthant then the coordinates of ideal points are equal to the global infima of F_i on U, for i=1,..N. In this paper we investigate the properties of ideal points from the point of view of their applicability in multicriteria decision aid. We examine their existence and uniqueness, as well as the properties of totally dominating points and here defined local ideal points. It turns out that the uniqueness of ideal points for all Θ-bounded subsets can only be guaranteed in the criteria space ordered by a cone isomorphic to $\mathbb{R}^N_+$. Based on the properties of local ideal points, we propose a class of scalarization methods for non-convex vector optimization problems. Finally, we point out the consequences of the above results for the MCDM methodology.

1. PRELIMINARIES

The notion of an ideal point, sometimes called also the utopia point, is one of those most frequently used in multicriteria optimization, cf. e.g. [3], [4], [7], or [8]. For multicriteria problems with the natural (coordinatewise) partial order in the criteria

---

*/ Permanent Address :

Institute of Automatic Control,
University of Mining & Metallurgy, Krakow, Poland

space $\mathbb{R}^N$ the coordinates of the ideal point x^* are defined as the global infima of the criterion functions F_i, for $i=1,..N$, $F=(F_1,...F_N)$. Therefore they are uniquely determined assuming that the sets of values of each F_i is bounded from below. In nontrivial multicriteria decision-making problems the ideal point does not belong to the set of admissible values of criteria $F(U)$, because in such a case the set of nondominated points would reduce to x^*. Instead, elements of the criteria space with the coordinates equal or better than those of x^* - so called totally dominating points - serve often as reference points in various multicriteria decision-making algorithms. As a result of a numerical procedure calculating the minima of nonconvex criterion functions one can derive a set of points of the criteria space associated to the local minima (or infima) of F_i, $i=1,..N$. These points, called here the local ideal points, may play an important role in improving the accuracy of the nondominated set approximation using distance scalarization techniques.

In this paper we will investigate the properties of ideal points, totally dominating points, and local ideal points for the vector optimization problem

$$(F : U \to E) \to \min (\theta), \tag{1}$$

where U is the set of admissible decisions, E is the space of criteria - a Banach space partially ordered by a closed, convex and pointed cone θ, and F is a vector objective to be minimized with respect to the partial order introduced by θ. In the sequel we will use the notation $X := F(U)$.

1.1 Basic definitions

Let us recall that a subset θ of a linear space E is called a <u>convex cone</u> iff each positive linear combination of elements of θ belongs to θ, i.e. iff

$$\forall\, x,y \in \theta \quad \forall\, s,t \in \mathbb{R}_+ \quad sx + ty \in \theta\ .$$

Each convex cone θ defines the following partial order, $\le_\theta$, in E

$$x \le_\theta y \Leftrightarrow y - x \in \theta.$$

If two points, x and y fulfill the above relation then we say that x _dominates_ y, or, equivalently that y _is dominated by_ x.

A cone $\theta \subset E$ is called _non-degenerated_ iff it contains a base of E. We say that θ is _pointed_ iff

$$\theta \cap (-\theta) = \{0\}.$$

An element y of X fulfilling the condition

$$(y - \theta) \cap X = \{y\} \tag{2}$$

is called _θ-minimal_ or _nondominated_ in X. Thus, $(-\theta)$-minimal points are called θ-maximal. The set of all θ-minimal points in X is denoted by $P(X,\theta)$.

Following [6], now we present several definitions related to the ideal points.

Definition 1.1. A point $x \in E$ such that

$$X \subset x + \theta$$

is called a totally dominating point for X. The set of totally dominating points is denoted by $TD(X,\theta)$. ■

As we have already mentioned, the coordinates of ideal points in multicriteria optimization problems with $\theta=\mathbb{R}^N_+$, express the best values of criteria on the same set of admissible decisions calculated separately. Here we present a more abstract definition based on the notion of totally dominating points.

Definition 1.2. Any $(-\theta)$-optimal element of $TD(X,\theta)$ is called an ideal point for X. The set of ideal points is denoted by $x^*(X,\theta)$. ■

This notion is attractive because each solution of the distance scalarization problem

$$\inf\{\|x^* - F(q)\| : q \in U\},$$

is nondominated (even properly nondominated - see e.g. [4]) for the vector optimization problem (1), provided that the norm $\|.\|$ in E is strongly monotonically increasing with respect to the partial order introduced by θ, i.e. if $0 \leq_\theta x \leq_\theta y$, $x \neq y$ implies $\|x\| < \|y\|$.

The notion of <u>strictly dominating points</u>, $SD(X,\theta)$, defined in [6] as

$$SD(X,\theta) := \{x \in E : P((x+\theta) \cap X,\theta) = P(X,\theta) \cap (x+\theta)\}$$

plays an important role in distance scalarization as well as in the theory of local ideal points presented in Sec.4. The reader is referred to [4], [7], or [8] for more details concerning distance scalarization.

Throughout this paper we will assume that X is θ-closed and θ-complete, i.e. the set $X + \theta$ is assumed closed and

$$\forall\ x \in X \quad \exists\ y \in P(X,\theta) : y \leq_{\theta} x. \tag{3}$$

To guarantee the existence of totally dominating points we will make the assumption that the set X is <u>θ-bounded</u>, by definition it means that there exists a point $y \in E$ such that $X \subset y + \theta$. Hence, it follows immediately that X is θ-bounded iff $TD(X,\theta)$ is non-empty.

2. PROPERTIES OF TOTALLY DOMINATING POINTS

Before studying the properties of ideal points we prove the following lemma.

Lemma 2.1. The multifunction

$$T : P(X,\theta) \ni a \to a - \theta \subset E$$

is continuous in the topology generated by the Hausdorff distance in the family of closed subsets of E.

Proof. Let us take an arbitrary $\delta > 0$. If a and b are such that $\|a - b\| < \delta$ then the Hausdorff distance, d_H, of T(a) and T(b) can be estimated as follows:

$$\begin{aligned} d_H(T(a),T(b)) &= d_H(a-\theta,\ b-\theta) = \\ &= d_H(a-\theta,\ (b-a) + (a-\theta)) \leq \|a-b\| \leq \delta \end{aligned}$$

Hence T is Hausdorff - continuous. ∎

Now we can show the following

Lemma 2.2. If θ is closed then $TD(X,\theta)$ is closed.

Proof. Let us take a sequence of totally dominating points $\{x_i\}_{i\in\mathbb{N}}$ convergent to an $x \in E$. By Lemma 2.1 the multifunction

$$S(y) := (y + \theta) \cap (X + \theta)$$

is upper semicontinuous as an intersection of two continuous, closed-valued multifunctions and

$$S(x_i) = (x_i+\theta) \cap (X+\theta) = X+\theta$$

by the definition of $TD(X,\theta)$. Hence, it follows that

$$S(x) = (x+\theta) \cap (X+\theta)$$

equals to $X+\theta$ which is equivalent that $X \subset x+\theta$, or $x \in TD(X,\theta)$. ∎

From Lemma 2.1 we derive a useful characterization of the set of totally dominating points.

Theorem 2.1. If θ is closed and pointed, and $TD(X,\theta)$ is non-empty, then the latter can be expressed in the form

$$TD(X,\theta) = x^*(X,\theta) - \theta. \tag{4}$$

Proof. If $X \subset x + X$ then for each $z \in x - \theta$

$$x + \theta \subset z + \theta,$$

consequently, if $x \in TD(X,\theta)$ then $x - \theta \subset TD(X,\theta)$ which proves the inclusion "$\supset$". To prove that for each $z \in TD(X,\theta)$ there exists $v \in P(TD(X,\theta),(-\theta))$ such that $z \in v - \theta$, which is equivalent to stating that $TD(X,\theta)$ is $(-\theta)$-complete, it is sufficient to observe that $TD(X,\theta)$ is closed and $(-\theta)$-bounded. The latter fact - as justified by Sawaragi, Nakayama and Tanino ([5], Theorem 3.2.10)) - implies θ-completness of $TD(X,\theta)$. The first property is a consequence of the fact that both sets θ and $X + \theta$ were assumed closed, while any element of X can be taken as a $(-\theta)$-totally dominating point for $TD(X,\theta)$. Therefore the inclusion "$\subset$" holds as well which ends the proof. ∎

The above theorem implies immediately the existence of ideal points for non-empty $TD(X,\theta)$.

Corollary 2.1. Under the assumptions of Thm. 2.1 the set of ideal points is non-empty. ■

3. UNIQUENESS OF IDEAL POINTS

The theorem given below relates the uniqueness of ideal points to the properties of θ.

Theorem 3.1. Suppose that E is a linear space partially ordered by a closed, convex and pointed cone θ. Then the following conditions are equivalent :

a) If the subset X of E is θ-bounded then the set of ideal points for X, $x^*(X,\theta)$, consists of a single point.
b) For every pair of points $x_1, x_2 \in E$ there exists the unique $z \in E$ such that

$$(x_1+ \theta) \cap (x_2+ \theta) = z + \theta \tag{5}$$

c) (E,θ) is a vector lattice, i.e.

$$\forall\ a,b \in E\ \exists\ a \wedge b := x^*(\{a,b\},\theta).$$

Proof. Assume first that the condition (b) is satisfied and suppose that for a set $X \subset E$ there exists a pair of distinct totally dominating points, z_1 and z_2, which both are ideal, i.e. θ-maximal in $TD(X,\theta)$. However, if

$$X \subset z_1 + \theta \quad \text{and} \quad X \subset z_2 + \theta$$

then, of course,

$$X \subset (z_1+ \theta) \cap (z_2+ \theta) = (z + \theta),$$

therefore $z \in TD(X,\theta)$. Thus we have found an element of $TD(X,\theta)$ which $(-\theta)$-dominates z_1 and z_2 - a contradiction with the assumption that both points are ideal.

To prove the remaining implications observe that (c) is immediately implied by (a), so it suffices to prove that (c) $\Rightarrow$ (b). Since for each pair of points $x_1, x_2 \in E$ there exists the unique ideal point x for the set $\{x_1, x_2\}$, then by Prop. 2.2

$$TD(\{x_1, x_2\}, \theta) = x - \theta. \tag{6}$$

On the other hand, $TD(\{x_i\}, \theta) = x_i - \theta$ for i=1,2, hence

$$TD(\{x_1, x_2\}, \theta) = (x_1 - \theta) \cap (x_2 - \theta). \tag{7}$$

Replacing θ by $(-\theta)$ we conclude from (6) and (7) that (5) is fulfilled for $(-\theta)$. Applying the already proved implication (b) $\Rightarrow$ (a), we observe that there exists the unique ideal point y for $\{x_1, x_2\}$ with respect to $(-\theta)$. Similarly as in (6) and (7) we observe that

$$TD(\{x_1, x_2\}, (-\theta)) = x - (-\theta) = x + \theta$$

and

$$TD(\{x_1, x_2\}, (-\theta)) = (x_1 + \theta) \cap (x_2 + \theta).$$

Therefore

$$(x_1 + \theta) \cap (x_2 + \theta) = (x + \theta)$$

and x is unique with this property. ∎

From Theorem 3.1 it follows that to have the unique ideal point in a vector optimization problem, the intersection of any two translations of the ordering cone must exist and be congruent to this cone. It is easy to see that in $\mathbb{R}^N_+$ only the polyhedral cones generated by exactly N linearly independent vectors, so called simplicial cones (cf. e.g. [1] or [2]), possess this property. This class of convex cones can be equivalently characterized as the isomorphic transformations of the natural positive cone $\mathbb{R}^N_+$.

Hence we conclude that in all situations where the ideal point is unique, its determination can be reduced to finding N optimal (minimal or maximal) values of N scalar functions G_i, i=1,..N, obtained as the transformation of the vector criterion F according to the formula

$$G = I^{-1} \circ F, \tag{8}$$

where $G=(G_1,\ldots G_N)$ and I is an automorphism of $\mathbb{R}^N$ such that $\Theta = I\,(\mathbb{R}^N_+)$.

The above proved Theorem 3.1 implies that the class of convex cones which do not satisfy (5) is large in linear spaces of dimension greater than 2, e.g. (5) is not satisfied by the "ice-cream cone"

$$\Theta := \{(x,y,z) \in \mathbb{R}^3 : x^2 + y^2 \le z\}.$$

Specifically, one can see that the set

$$Z := \Theta \cap ((1,0,0) + \Theta)$$

does not have a unique ideal point with respect to Θ since the points (0,0,0) and (1,0,0) are evidently totally dominating, non-comparable, and the cone translated to a point belonging to $\Theta \cup ((1,0,0) + \Theta)$ and different from (0,0,0) and (1,0,0) may not contain Z.

4. PROPERTIES OF LOCALLY IDEAL POINTS

From Theorem 3.1 it follows that to find the unique ideal point for the problem (1) satisfying the assumptions of Theorem 3.1 it is necessary to calculate the global optimal values of the criterion functions (perhaps transformed according to (8)). However, while performing the numerical optimization procedures for nonconvex functions we often deal with the local minima rather than with the global ones. It is natural to expect that the elements of the criteria space with all coordinates being local minima of the criterion functions will constitute local surrogates of ideal points to a similar extent as are the local minima in the scalar optimization with respect to the global optimum. Unfortunately, it comes out that such points can even be attainable. Consequently, some elements of the set

$$L(X) := \{y=(y_1,\ldots,y_N)\in\mathbb{R}^N : y_i\in V(F_i,U) \text{ for } i=1,..N\} = \\ = \prod_{i=1}^{N} V(F_i,U), \qquad (9)$$

where $X = F(U)$ and

$$V(F_i,U):=\{r\in\mathbb{R} : r \text{ is a local minimal value of } F_i \text{ on } U\},$$

cannot be admitted as reference points in distance minimization procedures without loosing the guarantee that the least-distance solutions are nondominated. Therefore we are interested in finding a subset $L^*(X)$ of $L(X)$ consisting exclusively of strictly dominating points, because the minimization of a strongly monotonically increasing norm with respect to a strictly dominating reference point will yield a nondominated solution (cf. [6]). We call such points <u>locally ideal points</u>.

From a decision-maker point of view the most interested locally ideal points may be those located as close as possible to the set $F(U)$.

Definition 4.1. The elements of the set

$$PL(X) := P(L^*(X), (-\theta))$$

are called the proper local ideal points. ■

Before giving a filtering criterion for locally ideal points let us define the set $R(X)$ of <u>local edge points</u> in the criteria space for the vector optimization problem (1) with $E = \mathbb{R}^N$:

$$R(X) := \{x\in\mathbb{R}^N : x_j\in V(F_j,U) \text{ for exactly one } j\in\{1,..N\} \\ \text{and } x_i = \inf\{F_i(u) : u \in F_j^{-1}(x_j)\cap U\} \text{ for } i\neq j\}. \qquad (10)$$

Let us observe that for each i to every element v of $V(F_i,U)$ there corresponds exactly one local edge point p, and this correspondence is one-to-one.

Now we give the following without proof :

Theorem 4.1. Let $X \subset \mathbb{R}^N$ and let $S(X)$ be the nondominated subset of $R(X)$,

$$S(X) := P(R(X), \theta),$$

and $T_i(X)$ - the set of those locally optimal values of F_i which are represented in $S(X)$, i.e.

$$T_i(X) := \{v \in V(F_i, U) : \exists\ j \in \{1,..N\}\ \exists\ x = (x_1,..x_N) \in S(X) \text{ so that } x_j = v\}.$$

Then the local ideal points $L^*(X)$ can be obtained by the following construction :

$$L^*(X) = \{x \in \mathbb{R}^N : \forall\ i \in \{1,..N\}\ x_i \in T_i(X) \text{ and } x_j \le r_j(x_i)\}, \quad (11)$$

where $r(x_i)$ is the local edge point corresponding to the local optimal value x_i with the coordinates

$$r_i(x_i) = x_i,$$
$$r_j(x_i) = \inf\{F_j(u) : u \in F_i^{-1}(x_i) \cap U\} \quad \text{for } i \ne j \qquad ■$$

From the above Theorem 4.1 one can be directly derive an algorithm for finding the local ideal points for the criteria functions with a finite number of local minima. This algorithm employs N independent global minimization procedures for each coordinate of F and a selection of a nondominated subset from a discrete set.

Example 4.1. Let us consider the following bicriteria linear programming problem with disjunctive constraints :

$$((F_1, F_2) : \mathbb{R}^2 \supset U \to \mathbb{R}^2) \to \min(\mathbb{R}^2_+),$$

where $F_1(x_1, x_2) = x_1$, $F_2(x_1, x_2) = x_2$, and the decision set U is determined by the following inequalities

$$x_1 \ge 0, \quad x_2 \ge 0, \quad x_1 + x_2 \ge 5$$

and

$$[x_1 - 1 \le 0, \quad \text{or} \quad x_2 - 1 \le 0,$$
$$\text{or} \quad (x_2 - x_1 - 1 \le 0 \quad \text{and} \quad x_1 - x_2 - 1 \le 0),$$

or $(-x_1-x_2+10\leq 0$ and $x_2-x_1-4\leq 0$ and $x_1-x_2+2\leq 0)]$.

It is easy to see that U = F(U) and U consists of four disjoint regions U_1, U_2, U_3, U_4

$$U_1=\{x\in\mathbb{R}^2 : x_1\geq 0,\ \ x_1+x_2\geq 5,\ \ x_1-1\leq 0\},$$
$$U_2=\{x\in\mathbb{R}^2 : x_1+x_2\geq 5,\ \ x_2-x_1-1\leq 0,\ \ x_1-x_2-1\leq 0\},$$
$$U_3=\{x\in\mathbb{R}^2 : x_2\geq 0,\ \ x_1+x_2\geq 5,\ \ x_2-1\leq 0,$$
$$U_4=\{x\in\mathbb{R}^2 : -x_1-x_2+10\leq 0,\ \ x_2-x_1-4\leq 0,\ \ x_1-x_2+2\leq 0\}.$$

The procedure to find the sets of local ideal and proper ideal points can be presented as follows. We start from calculating all local minima of F_1, and F_2,

$$V(F_1,U)=\{4,3,2,0\} \quad\text{and}\quad V(F_2,U)=\{6,4,2,0\}.$$

Thus we get the (global) ideal point $x^*(U,\mathbb{R}^2_+)=(0,0)$ and the set

$$L(U) = V(F_1,U)\times V(F_2,U) =$$
$$= \{(4,6),..,(4,0),(3,6),...,(2,0),(0,6),..,(0,0)\}$$

consisting of 16 points.

At each local minimum of F_i we calculate the minimal admissible values of the remaining criteria F_j, $j\neq i$, (in our case $i,j=1,2$) in order to determine the set of local edge points

$$R(U)=\{(4,1),(3,7),(2,3),(0,5),(4,6),(1,4),(3,2),(5,0)\}.$$

The nondominated part of R(U) consists of six points, namely

$$S(U)=\{(4,1),(2,3),(0,5),(1,4),(3,2),(5,0)\}.$$

Observe that the points (3,7) and (4,6) have thus been eliminated from R(U).

According to Theorem 4.1 now we have to find the sets $T_i(U)$ for $i=1,2$, containing those local minima of F_i which are represented in S(U), i.e.

$$T_1(U)=\{4,2,0\} \quad\text{and}\quad T_2(U)=\{4,2,0\}.$$

Following the formula (11) we find the set of local ideal points of F on U :

$$L^*(U) = \{(4,0),(2,0),(0,0),(2,2),(0,2),(0,4)\}.$$

Three elements of $L^*(U)$ are moreover proper local ideal points (cf. Definition 4.1), namely

$$PL(U) := \{(4,0),(2,2),(0,4)\}.$$ ■

A straightforward field of applications of local ideal points are distance scalarization methods. An approximation of the set of nondominated points by finding the least-distance solutions from a fixed totally dominating point with respect to a parametrized family of norms has some unpleasant features, namely :

- assuming a fixed step of the norm parameter discretization the accuracy of approximation depends on the distance between the reference point and the nondominated set : the more distant parts of it are surveyed, the lower is the accuracy;
- for nonconvex problems the least-distance solutions for some values of the norm parameters may not be unique, in particular they may belong to different connected components of the set $P(X,\theta)$;
- it is difficult or even impossible to restrict a priori the family of norms so that the search for a compromise solution would be confined to the prescribed subset of $P(X,\theta)$ being of a special importance to the decision-maker.

The disadvantages of distance scalarization with respect to a single reference point may be eliminated by decomposing the set F(U) into subregions associated to the proper local ideals and using a distance scalarization technique for the approximation of each subregion. Below we propose an algorithm of such scalarization procedure.

Algorithm 4.1.

Step 0. Transform, if necessary, the coordinate system in the criteria space according to (8) in order to obtain the natural $\mathbb{R}^N_+$- ordering in the transformed space.

Step 1. Calculate the local minimal values of the functions F_i, for i=1,..N

Step 2. Find those local minimal values which correspond to the nondominated edge points applying Theorem 4.1.

Step 3. Find the set $L^*(X)$ according to the construction presented in Theorem 4.1.

Step 4. Find PL(X) - the $(-\Theta)$-optimal part of $L^*(X)$.

Step 5. Perform the distance scalarization procedure for a parametrized family of norms with respect to each element of PL(X).

A variation of the Algorithm 4.1. can also be applied as a two-stage interactive multicriteria decision-making method : first, the decision-maker selects interactively this proper local ideal point whose coordinates fits best his preferences, then the point thus selected is used as the reference point in an interactive distance minimization procedure.

5. CONCLUSIONS - IMPACT ON MULTICRITERIA DECISION AID

In Sec.3 we have shown that the existence of unique ideal points may not be taken for granted for multicriteria decision-making problems with the preference structure other than congruent to a simplicial cone. In particular, applying a distance scalarization procedure with respect to different ideal points will usually result in different least-distance decisions. This means that the notion of compromise decision based on ideal points is in this case inconsistent. However, this inconsistency does not touch upon the bicriteria problems because in $\mathbb{R}^2$ all nontrivial convex cones which are not degenerated to a half-line are simplicial.

Furthermore, we concluded that in all situations where the ideal point is unique, its determination can be reduced to finding global optima of N scalar functions using the transformation (8). The notion of local ideal points introduced in this paper can replace the ideal point in distance minimization procedures for multicriteria problems with nonconvex criteria iff they are previously filtered as proposed in Theorem

4.1. Local ideal points may be used in the general framework of multiple reference points method as well as they may increase the accuracy of the approximation of the nondominated set applying the distance scalarization techniques.

Finally, let us notice that to find all local ideal points for a non-convex vector optimization problem one has to determine all local minima of the objectives considered separately. Thus to achieve this goal on a parallel machine one can assign the execution of each scalar minimization procedure to N different processors. Similarly, the decomposition of the nondominated set into subregions proposed in Algorithm 4.1 allows for a parallel approximation of each subregion which should essentially increase the numerical efficiency of MCDM methods applying ideal points.

REFERENCES

[1] G.P. Barker (1981). Theory of Cones. Linear Algebra Appl., 39, 263-291.

[2] B. Fuchsteiner, W. Lusky (1981). Convex Cones. North Holland Mathematics Studies, 56, North-Holland, Amsterdam, p.429.

[3] H. Gorecki, A.M.J. Skulimowski (1986). A Joint Consideration of Multiple Reference Points in Vector Optimization Problems. Found. Contr. Engrg., 11, No.2, 81-94.

[4] J. Jahn (1984). Scalarization in Vector Optimizazation, Math. Programming, 29, 203-218.

[5] Y. Sawaragi, H. Nakayama, T. Tanino (1985). Theory of Multi Objective Optimization. Academic Press, New York.

[6] A.M.J. Skulimowski (1989). Classification and Properties of Dominating Points in Vector Optimization. Methods of Oper. Res., 58, 99-112.

[7] A.P. Wierzbicki (1986). On the Completness and Constructiveness of Parametric Characterizations to Vector Optimization Problems. Oper. Res. Spectrum, 8, No. 2, 73-88.

[8] P.L. Yu, G. Leitmann (1974). Compromise Solutions, Domination Structures and Salukvadze's Solution. J. Optimiz. Theory and Appl., 13, 362-378.

A GLOBAL SEARCH FOR MULTICRITERIAL PROBLEMS

Ilya M. Sobol'

Keldysh Institute of Applied Mathematics
USSR Academy of Sciences
4, Miusskaya Square
Moscow 125047, USSR

Abstract

Various methods for solving multicriterial problems include multidimensional search procedures. As a rule, maximal and minimal values of several objective functions must be estimated and the accuracy requirements at the first stage of investigation are moderate. Often more information about the resources of these functions is required. Therefore a crude search can be applied so that all objective functions are computed at the same trial points that cover uniformly the domain of definition.

Curiously enough, in practice trial points selected according to the theory of optimal algorithms ("optimal points") perform poorly. Much better results are obtained using points of uniformly distributed LP_τ-sequences. New error estimates that lead to an explanation of that situation are presented.

1. Problem statement

Various methods for solving multicriterial problems include traditional (single-criterial) optimization. Theoretically, one may use arbitrary optimization procedures. But there are several features specific for multicriteria problems that suggest a more careful choice of optimization procedures.

First, several objective functions must be optimized.

Second, a global optimization rather than a local one is necessary.

Third, the accuracy requirements (at least at an early stage of investigation) are moderate since the best choices are as a rule not extremes but compromises.

And fourth, often more information is required, not only the best value (sometimes the decision maker wants to know the worst value or to have an estimate of the steepness of the function).

In that situation a crude search looks attractive: we select N trial points, calculate the values of all objective functions at these points and declare the best ones for each function to be approximately optimal. Still there is the question how to select the trial points?

2. Convergence estimates

Assume that the objective functions are defined in the n-dimensional unit cube

$$I^n = \{ x \mid 0 \leq x_1 \leq 1, \ldots, 0 \leq x_n \leq 1 \},$$

$x = (x_1, \ldots, x_n)$, and have piece-wise continuous derivatives. Let $C_1 = C_1(L_1, \ldots, L_n)$ be the set of all such functions $f(x)$ that satisfy the conditions

$$|\partial f / \partial x_i| \leq L_i, \quad 1 \leq i \leq n.$$

As an approximation to

$$f^* = \sup_{x \in I^n} f(x)$$

we consider the computed value

$$f_N^* = \max_{1 \leq k \leq N} f(x^{(k)}),$$

where $x^{(1)}, \ldots, x^{(N)}$ are the trial points. The approximation error for the set C_1 is defined as

$$d_N = \sup_{f \in C_1} (f^* - f_N^*)$$

and depends both on $L_1, \ldots, L_n$ and on the trial points.

According to the theory of optimal algorithms (Traub and Vozniakowski [10]) the optimal set of trial points for C_1 is the one satisfying

$$d_N \to \min. \tag{1}$$

Theorem 1. For arbitrary trial points in I^n

$$d_N \geqslant c_N,$$

where the lower bound

$$c_N = \frac{1}{2} \max \left(s!\, L_{i_1} \cdots L_{i_s} / N \right)^{1/s} \tag{2}$$

and the maximum is extended over all groups $1 \leqslant i_1 < \ldots < i_s \leqslant n$; $s = 1, 2, \ldots n$.

The expression (2) was introduced for the first time in Sobol' [8]. Earlier in all papers concerning optimization only the case $L_1 = \ldots = L_n = L$ was considered. Indeed, for estimating the error d_N we can always use as L the largest amont all the constants L_i. But when we intend to select good trial points the situation is changed dramatically.

For $L_i = L$ the order of convergences of the lower bound (as $N \to \infty$) is defined by

$$c_N = \frac{1}{2}(n!)^{1/n} L N^{-1/n} \sim N^{-1/n}. \qquad (3)$$

This is a well known result. And often from (3) a conclusion is drawn that the crude searching algorithm is inefficient when n is large.

But if among the L_i there are only t positive constants, $t < n$, then $c_N \sim N^{-1/t}$ which may be much better than (3). One may easily expect that in the case when the L_i are of different orders of magnitude, (2) may be much lower than (3).

3. Uselessness of optimal sets

So, optimal sets of trial points obtained from (1) with the assumption that all the L_i are equal, quarantee an order of convergence $N^{-1/n}$ only, and in the particular case of a function with very different L_i it is often a poor choice.

An attempt could be made to obtain optimal trial points from (1) for nonequal values $L_1, \ldots, L_n$. Unfortunately, in most practical problems we do not know the exact (i.e. least possible) values of these constants, and to calculate precise estimates for all L_i is much more difficult than to estimate f^*.

Thus we arrive at a paradoxical conclusion that optimal algorithms for classes of functions may in practice turn out useless... (They are surely good algorithms in the case $n = 1$ only).

I am sure that a typical situation is the following: most of the objective functions depend on all variables $x_1, \ldots, x_n$ so that the total number n of variables cannot be reduced; but each of these functions depends strongly on a small number of its own "leading variables"; so the best approximation errors are much better than (3).

Mathematicians have more than once encountered the situation described above, e.g. Piliavski et al [4]. But they had no explanation to the fact that non-optimal algorithms perform better than optimal ones though the functions under investigation were clearly from the considered class.

4. Uniform estimates for P_τ-nets

We recall the definitions of dyadic intervals, dyadic boxes and P_τ-nets.

Subintervals $[(j-1)2^{-m}, j2^{-m})$ of the unit interval I are called dyadic intervals; here j and m are integers, $1 \leq j \leq 2^m$, $m \geq 0$; at $j = 2^m$ the dyadic interval is by definition closed. So, a fixed integer m defines a partition of I into a sum of 2^m equal dyadic intervals.

A dyadic box (parallelepiped) Π is the Cartesian product of dyadic intervals. A set of integers $(m_1,\ldots,m_n) \neq (0,\ldots,0)$ defines a partition of I^n into a sum of 2^m equal dyadic boxes Π whose volumes $V(\Pi) = 2^{-m}$, $m = m_1+\ldots m_n$.

Let $0 \leq \tau < \nu$ be integers. A point set of $N = 2^\nu$ points in I^n is called a P_τ-net if every dyadic box Π with $V(\Pi) = 2^\tau/N$ contains exactly 2^τ points of the set.

The smaller is τ the better is the uniformity of P_τ-nets. It was shown by Sobol' [5, 6] that P_0-nets exist only in I, I^2 and I^3. As $n \to \infty$ the least possible $\tau \sim n\log_2 n$.

Theorem 2. For an arbitrary P_τ-net in I^n

$$d_N \leqslant A(n, \tau)\, c_N,$$

where $A(n,\tau)$ depends neither on N nor on $L_1,\ldots,L_n$.

Two important conclusions can be drawn from theorems 1 and 2. First, that c_N defines the best possible order of convergence (as $N \to \infty$) of the error d_N. Second, that there exist nets with uniformly optimal orders of convergence: they realize the best order of convergence whatever the constants $L_1, \ldots, L_n$ may be. If we use such sets as sets of trial points we need not bother which are the essential variables for this or that function $f(x)$: the best order of convergence will be realized simultaneously for all functions though these orders may be rather different.

In Niederreiter [3] n-dimensional P_τ-nets containing $N = 2^\nu$ points are called (τ, ν, n)-nets.

5. LP_τ-sequences for practical use

Consider an infinite sequence $x^{(0)}, x^{(1)}, \ldots \in I^n$. An initial segment of the sequence is the set of N points $x^{(0)}, \ldots, x^{(N-1)}$. A binary segment is the set of points $x^{(i)}$ with numbers satisfying $(k-1)2^p \leq i < k2^p$ for some integers $k \geq 1$ and $p \geq 1$.

A sequence $x^{(0)}, x^{(1)}, \ldots$ is called an LP_τ-sequence if every binary segment of the sequence with $p > \tau$ is a P_τ-net (Sobol' [5, 6]).

Subroutines for generating such sequences can be found in Sobol' [7], Sobol' and Statnikov [9], Bratley and Fox [1]. For dimensions $n \leq 16$ these LP_τ-sequences possess additional uniformity properties that are especially important for small initial segments (when $N < 2^\tau$).

In Niederreiter [3] n-dimensional LP_τ-sequences are called (τ, n)-sequences.

For practical computations the recommendation is to use as trial points initial segments of LP_τ-sequences containing $N = 2^p$ points ($p = 4, 5, \ldots$). Comparing results obtained at $p = 4, 5, \ldots$ we can make suggestions about the practical rate of convergence that may be different for different objective functions.

The success of Parameter Space Investigation (Sobol' and Statnikov [9]) in dealing with multicriterial optimum design problems can be mainly attributed to the use of trial points obtained from LP_τ-sequences. In Lieberman [2] the Parameter Space Investigation is called LP_τ-method.

6. A simple example

Consider a two-criterial problem

$$f_1(x) \to \min, \qquad f_2(x) \to \min,$$

where $x \in I$ and the functions are shown in Fig. 1. Assume that the first criterion is much more important than the second but its values are required only with accuracy δ.

From Fig. 1 one may see that $m_1=f_1(x_1)$ and $m_2=f_2(x_2)$ are the absolute minima of $f_1(x)$ and $f_2(x)$ but the best choice is obviously $x \approx x_0$.

A traditional approach to the problem is to minimize first $f_1(x)$. After having found m_1 the second criterion $f_2(x)$ is minimized in the region

$$\{ x \mid 0 \leq x \leq 1, \quad f_1(x) \leq m_1 + \delta \} = [x', 1].$$

Thus we obtain a point x' near to x_0. But the computation of m_1 may proceed rather slowly (especially if x is multidimensional) though the point x_1 is in fact unnecessary.

With the crude search algorithm we select a moderate number N of trial points $x^{(1)}, \ldots, x^{(N)}$ and calculate all the values $f_1(x^{(k)})$ and $f_2(x^{(k)})$, $1 \leq k \leq N$. Let

$$f^*_{1N} = \min_{1 \leq k \leq N} f_1(x^{(k)})$$

be the crude approximation to m_1. Then among the trial points $x^{(k)}$ that satisfy an additional restriction

$$f_1(x^{(k)}) \leq f^*_{1N} + \delta$$

the one that minimizes $f_2(x^{(k)})$ is chosen. So we easily obtain a trial point in the vicinity of x_0 (and the same procedure applies for multidimensional x).

In the case of Fig. 1 almost N/2 trial points satisfy the additonal restriction.

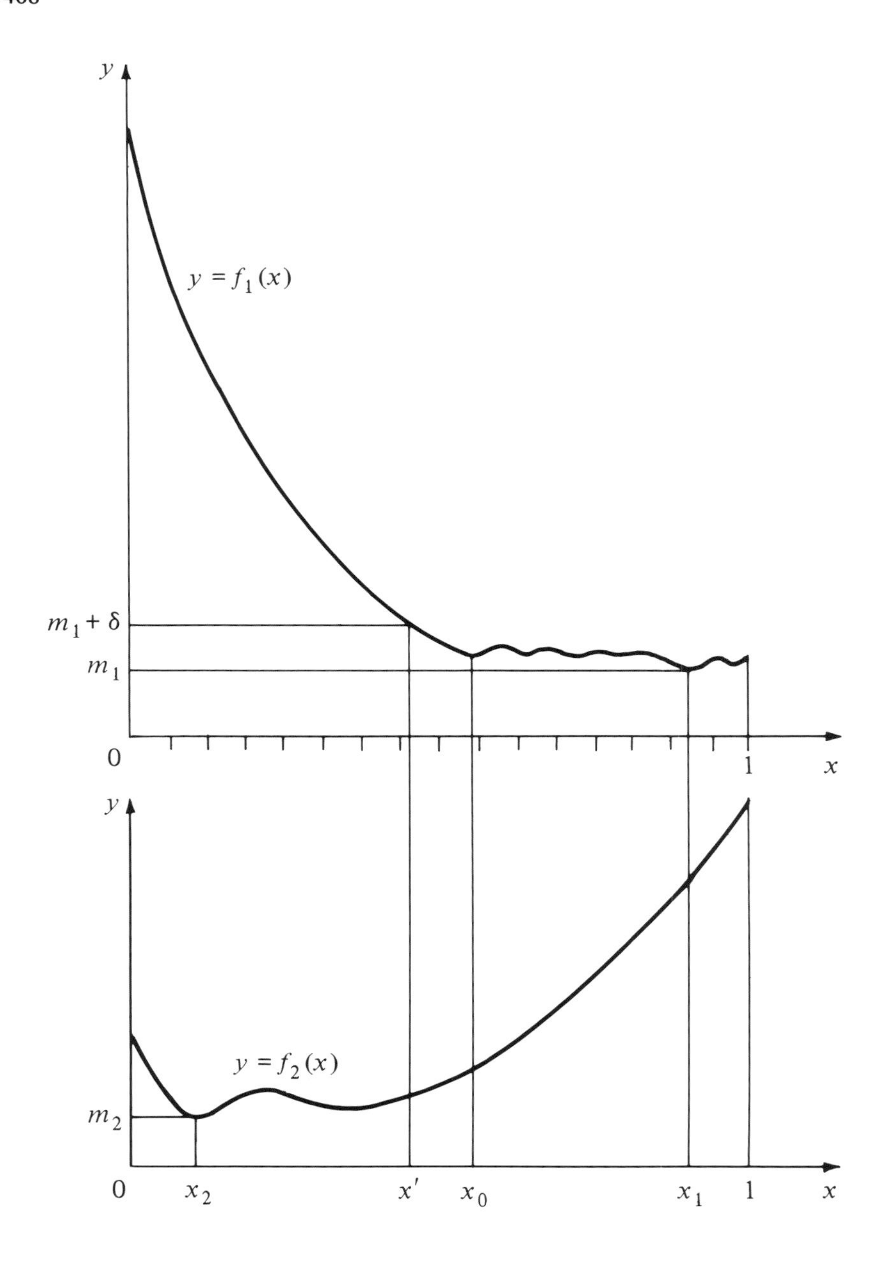

Figure 1. A two-criterial example

7. Numerical example

In the 5-dimensional cube I^5 five sets of constants $L_1, \ldots, L_5$ were considered and a formula from [8] has been used:

$$d_N = \sup_{x \in I^n} \min_{1 \leq k \leq N} \sum_{i=1}^{n} L_i \left| x_i^{(k)} - x_i \right| .$$

With traditional optimization techniques values of d_N have been computed for initial segments of the 5-dimensional LP_τ-sequence at N=32 and N=1024.

At these N rectangular lattices can be set up consisting of $N = M^5$ points with coordinates $((i_1+1/2)/M, \ldots, (i_5 + 1/2)/M)$, where $i_1, \ldots, i_5$ range independently over the values 0, 1,..., M-1. For such lattices values of d_N can be computed analytically [8].

Quotients $q = d_N/c_N$ for trial points of both types are presented in Table 1.

Table 1. Values of L_i, c_N and q.

№	L_1	L_2	L_3	L_4	L_5	N = 32			N = 1024		
						c_N	q_{LP}	q_{RL}	c_N	q_{LP}	q_{RL}
1	1	1	1	1	1	0.651	2.2	1.9	0.326	2.5	1.9
2	1	2^{-1}	2^{-2}	2^{-3}	2^{-4}	0.165	2.4	2.9	0.081	2.5	3.0
3	1	1	5^{-1}	10^{-2}	10^{-3}	0.167	2.6	3.3	0.053	2.3	5.2
4	1	10^{-2}	10^{-2}	10^{-3}	10^{-3}	0.016	3.0	16.0	0.0042	2.6	30.5
5	1	0	0	0	0	0.016	2.0	16.0	0.0005	2.0	256.0

From Table 1 it is easy to see that for the first set (equal constants L_i) the values of d_N for RL are somewhat better than for LP. However, when the L_i differ the situation changes radically and RL may be severalfold inferior to LP. As $N \to \infty$ all the q_{LP} remain bounded while the maximum q_{RL} increase without limit.

Interestingly enough, the question whether the fourth set of constants may be replaced by the fifth does not have a definite answer: it depends on N. E.g., when N = 32 both sets yield the same value $c_N=0.016$, but when N = 1024 the two values of c_N differ by a factor 8 and as N increases so does the difference.

8. Appendix. Proofs of theorems

For classes $\mathrm{Lip}(L_1,\ldots,L_n)$ of functions $f(x)$ that satisfy a general Lipschitz condition

$$|f(x') - f(x)| \leq \sum_{i=1}^{n} L_i \, |x_i' - x_i|$$

theorems 1 and 2 can be found in Sobol' [8]. Theorem 2 is an immediate corollary of the corresponding theorem 2 since $C_1(L_1,\ldots,L_n) \subset \mathrm{Lip}(L_1,\ldots,L_n)$.

To prove theorem 1 we may approximate an arbitrary function $f(x) \in \mathrm{Lip}$ by a function $f_\varepsilon(x) \in C_1$:

$$f_\varepsilon(x) = \frac{1}{(2h_1)\cdots(2h_n)} \int_{x_1-h_1}^{x_1+h_1} \cdots \int_{x_n-h_n}^{x_n+h_n} f(u)\,du_1 \cdots du_n ,$$

where

$$h_i = \begin{cases} x_i & \text{if } x_i \leq \varepsilon , \\ \varepsilon & \text{if } \varepsilon \leq x_i \leq 1-\varepsilon , \\ 1-x_i & \text{if } 1-\varepsilon \leq x_i . \end{cases}$$

One can easily verify that $|f(x) - f_\varepsilon(x)| \leq \frac{1}{2}\varepsilon \sum_{i=1}^{n} L_i \equiv \eta$ and $\eta \to 0$ as $\varepsilon \to 0$.

Let $g(x)$ be the worst function in Lip so that

$$g^* - g_N^* = \sup_{f \in \mathrm{Lip}} (f^* - f_N^*)$$

and according to theorem 1 in Sobol' [8]

$$g^* - g_N^* \geqslant c_N.$$

Then consider the function $g_\varepsilon(x)$:

$$g_\varepsilon^* = \sup_x g_\varepsilon(x) \geqslant \sup_x [g(x) - \eta] = g^* - \eta .$$

At the same time

$$g_{\varepsilon,N}^* = \max_{1 \leqslant k \leqslant N} g_\varepsilon(x^{(k)}) \leqslant \max_{1 \leqslant k \leqslant N} [g(x^{(k)}) + \eta] = g_N^* + \eta .$$

Hence,

$$g_\varepsilon^* - g_{\varepsilon,N}^* \geqslant g^* - g_N^* - 2\eta \geqslant c_N - 2\eta$$

and a fortiori

$$\sup_{f \in C_1} (f^* - f_N^*) \geqslant g_\varepsilon^* - g_{\varepsilon,N}^* \geqslant c_N - 2\eta .$$

As $\varepsilon \to 0$ we conclude that $\sup_{f \in C_1} (f^* - f_N^*) \geqslant c_N$.

REFERENCES

1. Bratley P. and Fox B.L., "Implementing Sobol's quasi-Random Sequence Generator", ACM Trans.Math. Software, Vol. 14, No.1, pp.88-100, 1988.

2. Lieberman E.R., "Soviet Research in Multi-Objective Programming - Overview", Dept. Geogr. and Environ. Engng, Johns Hopkins Univ., Baltimore, 1988.

3. Niederreiter H., "Low-Discrepancy and Low-Dispersion Sequences", J.Number Theory, Vol. 30, No.1, pp.51-70, 1988.

4. Piliavski V., Volfovski I. and Jushkiene E., "On Comparison of the Efficiency of Global Optimization Methods in the Inverse Problems of Laquer-Colour Optics", Theory of Optimal Solutions, No. 11, Vilnius, pp.29-39, 1985 (Russian).

5. Sobol' I.M., "On the Distribution of Points in a Cube and the Approximate Evaluation of Integrals", Zh.Vychisl.Mat. i Mat.Fiz., Vol. 7, No. 4, pp.784-802 = USSR Comput.Math.and Math.Phys., Vol. 7, pp.86-112, 1967.

6. Sobol' I.M., "Multidimensional Quadrature Formulas and Haar Functions", Nauka, Moscow, 1969 (Russian).

7. Sobol' I.M., "On the Systematic Search in a Hypercube", SIAM J.Numer. Analysis, Vol. 16, No.5, pp.790-793 (corr. in No.6), 1979.

8. Sobol' I.M., "On the Search for Extremal Values of Functions of Several Variables Satisfying a General Lipschitz Condition", Zh.Vychisl. Mat. i Mat.Fiz., Vol. 28, No.4, pp.483-491 = USSR Comput. Math. and Math.Phys., Vol. 28, No. 2, pp. 112-118, 1988.

9. Sobol' I.M. and Statnikov R.B., "Selection of Optimal Parameters in Problems with Several Criteria", Nauka, Moscow, 1981 (Russian).

10. Traub J.F. and Woźniakowski H., "A General Theory of Optimal Algorithms", Acad. Press, N.-Y., 1980.

MULTIOBJECTIVE DECISION MAKING

UNDER UNCERTAINTY

John Stansbury
Research Engineer, Civil Engineering Department
University of Nebraska-Lincoln
Lincoln, NE 68588

Istvan Bogardi
Professor, Civil Engineering Department
University of Nebraska-Lincoln

Yong W. Lee
Research Engineer, Civil Engineering Department
University of Nebraska-Lincoln

Wayne Woldt
Research Engineer, Civil Engineering Department
University of Nebraska-Lincoln

Abstract

Contaminants in disposed dredge material may result in various environmental risks (human and ecological) in the ecosystem. Composite Programming (CP), a Multicriteria Decision Making (MCDM) tool, is used to optimize dredge material disposal management decisions involving uncertain criteria (e.g., component risks and costs). CP aggregates criteria into fewer, more general groups or levels and finally computes the trade-off distance for the highest-level criteria (e.g., cost versus risk). Uncertainty in criteria is encoded by means of fuzzy set membership functions. Finally the trade-off distance for each alternative is computed as a fuzzy number so that the alternatives (with their associated uncertainties) can be compared.

Introduction and Purpose

Multicriterion decision making tools such as Composite Programming, ELECTRE, and Multi-Attribute Utility Theory have been used successfully to assist decision makers to deal with problems involving multiple and conflicting objectives (Goichoechea et al., 1982; Keeney and Wood, 1977). However, uncertainties inherent in these problems are often treated separately from the decision process itself. That is, the decision analysis is often carried out using point estimates (no uncertainty) of decision variables, and uncertainty is then characterized by "plausible intervals" (Cothern et al., 1987). The decision maker is then faced with either point estimates, which provide no characterization of uncertainty, or intervals showing bands of uncertainty which are often large and difficult to use in a decision framework.

Figure 1 illustrates the importance and some of the difficulties of considering uncertainty in decision-making. For example, if either management option A or B (in Figure 1) must be selected on the basis of risk, then using point estimates of the risks (a method commonly used today), management option B is superior because it clearly involves less risk. However, when uncertainties associated with risk assessment (which often span several orders of magnitude (Dourson et al., 1983)) are considered, it becomes difficult to determine which option is better. Further, if option B involves other negative aspects (e.g., more costly, necessitates loss of jobs, etc.) society may accept an economically inferior alternative (Option B) that may not provide greater protection from risk. Therefore, a method is developed using fuzzy set theory to rank uncertain numbers i.e. risk and cost estimates in a formal way so that decisions can be made while considering associated uncertainties.

Overview of Fuzzy Set Theory

Fuzzy set theory (Zadeh, 1965; Zimmerman, 1985), is a mathematical method used to characterize and propagate uncertainty and imprecision in data and functional relationships. Fuzzy sets are especially useful when insufficient data exist to characterize uncertainty using standard statistical meassures (e.g., mean, standard deviation and distribution type). For example, fuzzy sets can be used to characterize the uncertainty in the failure rate of a landfill liner (Bogardi et al., 1987) where there is insufficient historical data to use classical statistics. Fuzzy sets have been used to characterize the relationship between hydraulic conductivity and electrical resistivity in aquifers (Bardossy et al., 1987) where too few samples exist for classical statistical analysis.

Fuzzy sets or fuzzy numbers are described by membership functions that characterize their maximum range of uncertainty or fuzziness (0.0-level membership) and their most likely range of values (1.0-level membership). Figure 2 shows a fuzzy number that represents an imprecise or fuzzy value for probability of cancer risk from a disposal site. The 0.0-level membership is the interval between 2.5E-8 and 5.7E-3, while the 1.0-level membership (most likely interval) ranges from 3.0E-5 to 5.0E-5.

A major element of using fuzzy set theory to characterize uncertainty is the assessment of membership functions for the uncertain parameters. That is, many different membership functions (intervals) could be suggested to define the uncertainty in the fuzzy cancer example. Several procedures have been developed to assess membership functions of fuzzy numbers (Chameau and Santamarina, 1987). For the dredge material disposal problem, where fuzzy numbers are used to characterize uncertainty in physical parameters (e.g., population density, water consumption, etc.), "expert judgment" based on experience and observed measurement variability is used to assess membership functions (greatest and most likely intervals) for the uncertain parameters.

Dredge Material Disposal Example

The nation's major rivers and harbors need periodic dredging to facilitate waterborne commerce (U.S. Army Corps of Engineers (COE), 1983). However, sediments from these harbors are often polluted with chemical contaminants from industry, agriculture, sewage, and urban runoff (NOAA, 1988). Disposal of these contaminated sediments can cause risks to ecological species (toxicological and physical) and risks to human health from consumption of contaminated seafood or water (Saucier, 1978; Tetra Tech Inc. 1986).

A case study site consisting of a bay and harbor in a heavily developed metropolitan area is selected (Figure 3). For the study, a total of 66,000 cubic yards (cy) of sediments are to be dredged from the three dredging sites and disposed in any of five disposal sites (Stansbury et al., 1989). The sediments to be dredged are polluted with numerous contaminants (both carcinogenic and non-carcinogenic). The objective is to determine the best disposal site for the project on the basis of a risk-cost trade-off analysis, and to use uncertainties in the risk and cost estimates as an integral part of the analysis.

The disposal alternatives are: unconfined aquatic disposal (UAD) where the sediments are simply discharged to the bottom of the bay, capped aquatic disposal (CAD) which is similar to UAD except that the sediments are capped with several feet of clean sediments to contain any contaminants, confined disposal facility (CDF) where sediments are pumped into a diked area near the shore while effluent water is allowed to escape, upland disposal which is the same as CDF except that it is totally on land, and upland secure disposal which is similar to upland disposal except that liners and other contaminant confinement measures are employed.

Each of these disposal technologies is associated with a specific cost which generally increases with greater degrees of contaminant confinement (U.S. COE, 1988).

Risk Assessment

Risk assessment is conducted for human cancer, human non- cancer, ecological toxicity, and ecological burial risks. U.S. Environmental Protection Agency (U.S. EPA, 1984) guidelines are used for human risk assessment, and methods compatible with procedures developed by the U.S. Army Corps of Engineers Waterways Experiment Station (Peddicord et al., 1986) are employed for ecological risk assessment. The following four steps comprise the risk assessment:

1) Hazard Identification
2) Dose Response Assessment
3) Exposure Assessment
4) Risk Characterization

Cost Analysis

Costs of dredging and dredge material disposal are calculated in 1986 dollars by the Corps of Engineers (U.S. COE, 1988). Costs include transportation (barge or truck), disposal site acquisition, design, disposal site construction, sediment placement, and contaminant monitoring.

Sources of Uncertainty

Uncertainties are associated with all phases of risk-cost assessment. However, for this paper only uncertainties associated with exposure assessment are used. A partial list of these sources of uncertainty might include:

1) inaccurate modeling of fate and transport of contaminants in disposed material,
2) errors in estimating ecological population densities,
3) errors in estimating migratory and feeding habits of ecological species,
4) errors in estimating concentrations of contaminants in seafood and drinking water, and
5) errors in estimating consumption rates of seafood and drinking water.

Assessment of Membership Values for Fuzzy Parameters

High, central, and low values of the parameters used in the risk (from exposure assessment) and cost assessments are estimated. For example, the uncertainty in contaminant concentrations is characterized by the highest, lowest, and average recorded concentrations in the dredge material. Other parameters such as population densities, migratory habits, consumption rates, etc. are also characterized in this manner. As a result, the uncertainty associated with the exposure to contaminants from disposed dredge material is characterized by intervals at the 0.0 and 1.0-level memberships.

Using the interval values representing uncertain exposure to contaminants, high, central and low estimates are calculated for cancer, non-cancer, and ecological risks (toxicity and burial) at each disposal site (management alternative). These risk intervals then are described as fuzzy numbers having trapezoidal membership functions. That is, the high and low calculated risk values are used for the 0.0-level membership interval and a plausible range is assumed around the central calculated risk estimate as the 1.0-level membership interval. Figure 2 shows the fuzzy number representing cancer risk from the CAD site.

Fuzzy numbers representing uncertainties in the costs are constructed in a similar manner. Table 1 shows the 0.0-level and the 1.0-level membership intervals for the risks and costs for each disposal site.

Risk-Cost Trade-Off Using Fuzzy Composite Programming

To determine the "best" disposal alternative, that is the one that best satisfies both environmental and economic concerns, a risk-cost trade-off is conducted. Composite Programming (CP), a distance-based MCDM method (Bogardi and Bardossy, 1983a; 1983b) is used for the risk-cost analysis. To incorporate uncertainties associated with the risks and costs into the analysis, fuzzy CP has been developed (Bardossy, 1988) and modified by Lee et al. (1990) to use fuzzy numbers (fuzzy risks and fuzzy costs) as inputs.

Fuzzy CP organizes a problem into the following format: 1) define management alternatives, 2) define basic indicators, 3) group basic indicators into progressively fewer, more general, groups, 4) define weights, balancing factors, and worst and best values for the indicators, and 5) evaluate and rank the alternatives.

For this paper, the management alternatives are defined as the disposal of all the dredge material into each disposal site (UAD, CAD, CDF, Upland, and Upland Secure) alternatively. Each alternative is then associated with a different risk and cost, both being uncertain or fuzzy.

The basic indicators used for this analysis (for each disposal site alternative) are: 1) cancer risk, 2) non-cancer risk, 3) toxicity risk to fish, 4) burial risk to fish, 5) toxicity risk to shellfish, 6) burial risk to shellfish, 7) toxicity risk to terrestrial species, 8) burial risk to terrestrial species, and 9) costs. These basic indicators are grouped into second-level and third-level etc. indicators according to Figure 4.

Weights are assigned to the indicators in each level to represent the relative importance of the indicators as viewed by the decision maker or various other groups (e.g., developers, ecologists, etc.). For example, human health might be considered more important than risks to shellfish, and therefore receive a higher weight. Weights are assigned using a method developed by Saaty (1988). The weights used for this analysis are shown in Figure 4.

Next balancing factors are assigned to each group of indicators. Balancing factors reflect the relative importance that is assigned to the maximal deviations of the indicators and limit the ability of one indicator to substitute for another. In other words, with an appropriately high balancing factor, an indicator that must not be compromised, such as risk to an endangered species, will not be substituted for by one such as risk to an abundant, non-endangered species.

The "best" and "worst" possible values for the basic indicators for the particular study are then assessed. For instance, the "best" value for cost of operation would be the cost associated with the most inexpensive alternative, while the "worst" cost would be that associated with the most expensive alternative.

Fuzzy Composite Programming Algorithm

The numerical values of the basic indicators for the dredging process are given as fuzzy numbers to characterize their uncertainty. Evaluation of the various management alternatives under uncertainty proceeds as follows, let $Z_i(x)$ be a fuzzy number for the ith basic indicator and its membership function $\mu(Z_i(x))$ be a trapezoid (Figure 2), where x is one element of the discrete set of management alternatives. Note that if the trapezoid is reduced to a vertical line, it represents a so-called crisp number. A level cut concept (Dong and Shaw, 1987) is used to decide the interval of each indicator at various levels of "confidence". As shown in

Figure 2, $Z_{i,\alpha}(x)$ is the interval value of the ith basic indicator at level-cut α.

Since units of basic indicators such as cost and risk are different and thus difficult to compare directly, the actual value of each basic indicator should be transformed into an index. Using the best value ($BESZ_i$) of Z_i and the worst value ($WORZ_i$) of Z_i for the ith basic indicator, the index function $S_{i,\alpha}(x)$, used to produce the index value of $Z_{i,\alpha}(x)$, can be represented by the piecewise linear function:

1) If $BESZ_i > WORZ_i$, index function $S_{i,\alpha}(x)$ is defined by:

$$S_{i,\alpha}(x) = \begin{cases} 1 & , \quad Z_{i,\alpha}(x) \geq BESZ_i \\ \dfrac{Z_{i,\alpha}(x) - WORZ_i}{BESZ_i - WORZ_i} & , \quad WORZ_i < Z_{i,\alpha}(x) < BESZ_i. \\ 0 & , \quad Z_{i,\alpha}(x) \leq WORZ_i \end{cases} \tag{1}$$

2) If $BESZ_i < WORZ_i$, index function $S_{i,\alpha}(x)$ is defined by:

$$S_{i,\alpha}(x) = \begin{cases} 1 & , \quad Z_{i,\alpha}(x) \leq BESZ_i \\ \dfrac{Z_{i,\alpha}(x) - WORZ_i}{BESZ_i - WORZ_i} & , \quad BESZ_i < Z_{i,\alpha}(x) < WORZ_i. \\ 0 & , \quad Z_{i,\alpha}(x) \geq WORZ_i \end{cases} \tag{2}$$

This index function takes on values between 0 and 1. As shown in Figure 5, since variable $Z_{i,\alpha}(x)$ is an interval with lower bound a and upper bound b (i.e., $Z_{i,\alpha}(x) \in [a,b]$), the index function, $S_{i,\alpha}(x)$, resulting from $Z_{i,\alpha}(x)$ is also an interval (i.e., $S_{i,\alpha}(x) \in [c,d]$). Symbolically, the interval, $S_{i,\alpha}(x)$, of the index function for the ith basic indicator can be simply expressed by:

$$S_{i,\alpha}(x) = \left[\min S_{i,\alpha}(x),\ \max S_{i,\alpha}(x)\right] \tag{3}$$

where if $c \leq S_{i,\alpha}(x) \leq d$, $\min S_{i,\alpha}(x) = c$ and $\max S_{i,\alpha}(x) = d$.

Next, index functions, $L_{j,\alpha}(x)$, for second-level composite indicators are defined as:

$$L_{j,\alpha}(x) = \left[\sum_{i=1}^{n_j} w_{ij} * (S_{i,\alpha,j}(x))^{p_j}\right]^{(1/p_j)} \tag{4}$$

where:

n_j = the number of elements in the second-level group j,

$S_{i,\alpha,j}(x)$ = the fuzzy interval of the index function for the ith basic indicator in the second-level group j of basic indicators,

w_{ij} = the weight reflecting the importance of each of basic indicators in group j; $\Sigma w_{ij} = 1$, and

p_j = the balancing factor among indicators for group j.

Then, the fuzzy interval, $L_{j,\alpha}(x)$, of the index function for the second-level group j is given as:

$$L_{j,\alpha}(x) = [\min L_{j,\alpha}(x), \max L_{j,\alpha}(x)] \tag{5}$$

Trade-off distances for the other levels are calculated in similar fashion until a final trade-off (composite risk versus cost) distance is determined for each alternative.

Ranking the Alternatives

If there are n management alternatives, $L_\alpha(x)$, x = 1,...,n, represent n fuzzy numbers corresponding to the risk-cost trade-off distances. To determine the ranking of these n fuzzy numbers, the ranking method developed by Chen (1985) is applied. The method decides the ranking of n fuzzy numbers by using a maximizing set and a minimizing set.

The maximizing set M is a fuzzy subset with membership function μM given as:

$$\mu M(L) = \begin{cases} \dfrac{L - L_{min}}{L_{max} - L_{min}}, & L_{min} \le L \le L_{max} \\ 0, & otherwise \end{cases} \tag{6}$$

where for x = 1,...,n, $L_{min} = \min [\min L_{\alpha=0}(x)]$ and $L_{max} = \max [\max L_{\alpha=0}(x)]$.

Then, the right utility value, $U_R(x)$, for alternative x is defined as:

$$U_R(x) = \max\{ \min [\mu M(L), \mu(L(x))]\} \tag{7}$$

The minimizing set G is a fuzzy subset with membership function μG given as:

$$\mu G(L) = \begin{cases} \dfrac{L - L_{max}}{L_{min} - L_{max}}, & L_{min} \le L \le L_{max} \\ 0, & otherwise \end{cases} \tag{8}$$

The left utility value, $U_L(x)$, for alternative x is defined as:

$$U_L(x) = \max\{ \min [\mu G(L), \mu(L(x))]\} \tag{9}$$

Then the total utility or ordering value for alternative x is:

$$U(x) = (U_R(x) + 1 - U_L(x))/2 \tag{10}$$

As a result, the final trade-off distances of the disposal alternatives, even though they are uncertain (fuzzy), can be compared and ranked. Thus the alternative that has the highest ordering value in the discrete set of management alternatives is then selected as the superior alternative.

Results

Trade-offs are made for various levels in the CP structure (Figures 6, 7, and 8). Uncertainties in the trade-offs are represented by the widths of the boxes (all shown at the 0.0-level membership) representing the disposal alternatives. For example, Figure 6 shows much more uncertainty in toxicity risks for shellfish than for burial risks at the UAD site, while more uncertainty is associated with burial than toxicity at the CDF site. An explanation for this might lie in the different contaminant confinement measures taken at the two sites.

Lower-level trade-offs are useful when analyzing effects on components of the system. For example, the effects of burial versus toxicity for shellfish can be seen in Figure 6. This Figure illustrates that the UAD presents relatively high toxicity risks but low burial risks to shellfish.

Figure 7 shows the trade-off between human risks and ecological risks. Uncertainty in ecological risks is decreased for the CAD, Upland, and Upland Secure sites. This is probably due to the low population densities at these sites.

Figure 8 shows the final trade-off between composite risk and cost. Note that there is much greater uncertainty in the risk estimates than in the cost estimates. The UAD site provides the lowest-cost disposal alternative, but it does so at a relatively high risk. The Upland Secure site provides the lowest risk, but it requires a very large cost. The CAD site provides the best trade-off between risk and cost. Table 2 shows the final ranking (representing the fuzzy trade-off distance (L)) of the five disposal alternatives.

Discussion and Conclusions

Fuzzy Composite Programming is used to conduct a risk-cost trade-off analysis under uncertainty for a simplified dredge material disposal problem. The uncertainties inherent in the components of the risk-cost analysis are characterized as fuzzy numbers and incorporated directly into the trade-off analysis. Using this method, not only can the multicriterion dredge material disposal problem be solved, but alternatives can be compared while considering the inherent uncertainties. Consideration of uncertainties while choosing alternatives is especially important for multicriterion problems where objectives are in conflict (e.g., risk and cost).

Several conclusions can be drawn from this study:

1. Uncertainties in criteria can be characterized using the fuzzy set approach.
2. The methodology of modified fuzzy composite programming allows the incorporation of complex social, ecological, and economic information into a fuzzy decision-making process.
3. The modified fuzzy composite programming methodology can be a useful tool where there are conflicting objectives; numerical values of the criteria are uncertain; and the objectives are of varying degrees of importance.

Acknowledgments

The authors gratefully acknowledge the U.S. Army Corps of Engineers and the University of Nebraska Center for Infrastructure Research for supporting this project (contract LWF/66-287-04601), through the Water Center of the University of Nebraska-Lincoln. The cooperation and assistance of Eugene Z. Stakhiv, Corps of Engineers Institute for Water Resources is also gratefully acknowledged.

References

Bardossy, A., I. Bogardi, W.E. Kelly (1987). "Fuzzy Regression Between HydraulicConductivity and Electrical Resistivity: NAFIPS", Purdue University West Lafayette, Indiana.

Bardossy, A. (1988). "Fuzzy Composite Programming", Proceedings, International Conference on the Foundations of Utility and Risk, Budapest, Hungary.

Bogardi, I., and A. Bardossy, (1983a). "Application of MCDM to Geological Exploration", in Essays and Surveys on Multiple Criterion Decision Making, P. Hansen, ed., Springer-Verlag.

Bogardi, I., and A. Bardossy, (1983b). "Regional Management of an Aquifer for Mining under Fuzzy Environmental Objectives", Water Resources Research, 19(6), 1394-1402.

Bogardi, I., W.E. Kelly, (1987). "Principles of Clay Liner Reliability Estimation: working paper: Department of Civil Engineering, University of Nebraska-lincoln, Lincoln, NE.

Chameau, J. and J.C. Santamarina, (1987). "Membership Functions: Comparing Methods of Measurement", International Journal of Approximate Reasoning, 1:287-301.

Chen, S.H. (1985). "Ranking Fuzzy Numbers with Maximizing Set and Minimizing Set", Fuzzy Sets and Systems, 17, 113-129.

Cothern, C.R., W.A. Coniglio , W.L. Marcus, (1987). "Uncertainty in Population Risk Estimates for Environmental Contaminants": Proceedings of the Society for Risk Analysis International Workshop on Uncertainty in Risk Assessment, Risk Management, and Decision Making. September-October. Knoxville, TN.,

Dong, W. and H.C. Shaw, (1987). "Vertex Method for Computing Functions of Fuzzy Variables", Fuzzy Sets and Systems, 24, 65-78.

Dourson, M.L. and F.F. Stara, (1983). "Regulatory History and Experimental Support of Uncertainty (Safety) Factors", Regulatory Toxicology and Pharmacology, 3:224-238.

Goichoechea, A., D. Hansen, L. Duckstein, (1982). *Introduction to Multiobjective Analysis with Engineering and Business Applications*, John Wiley & Sons, New York.

Keeney, R.L., E.F. Wood, (1977). "An Illustrative Example of the Use of Multiattribute Utility Theory for Water Resources Planning," *Water Resources Research,* V.13, No.4.

Lee, Y.W., I. Bogardi, J. Stansbury, (1990). "Fuzzy Decision-Making in Dredged Material Management", draft report.

NOAA. (1988). "The National Coastal Pollutant Discharge Inventory: Estimates for Puget Sound. National Oceanic and Atmospheric Administration, Rockville, Maryland.

Peddicord, R.K., C.R. Lee, M.R. Palermo, N.R. Fancingues, (1986). "General Decisionmaking Framework for Management of Dredged Material: Example Application to Commencement Bay, WA." U.S. Army Corps of Engineers Waterways Experiment Station, Vicksburg, MS.

Saaty, T.L. (1988). "Multicriteria Decision Making: The Analytic Hierarchy Process", University of Pittsburgh.

Saucier, R.T., (1978). "Executive Overview and Detailed Summary, Dredged Material Research Program", Technical Report DS-78-22, Environmental Lab, Vicksburg MS.

Stansbury, J., I. Bogardi, W.E. Kelly, (1989). "Risk-Cost Analysis for Management of Dredging Material", Engineering Foundation Conference, Santa Barbara,

Tetra Tech Inc. (1986). "A Framework for Comparative Risk Analysis of Dredged Material Disposal Options", Task 5b. Draft Report, Prepared for Resources Planning Associates for U.S. Army Corps of Engineers, Seattle District by Tetra Tech, Inc., Bellevue, WA.

U.S. Army Corps of Engineers, (1983). "Dredging and Dredged Material Disposal, EM 1110-2-5025, U.S.COE,Office of the Chief of Engineers.

U.S. Army Corps of Engineers, (1988). "Disposal Site Selection: Technical Appendix -Phase I", Puget Sound Dredged Disposal Analysis Report.

U.S. EPA, (1984). "Proposed Guidelines for Carcinogenic Risk Assessment: Request for Comments, U.S. EPA, Washington, DC, Federal Register, Vol. 49, No. 227, Part VIII, pp 46304-46312.

Zadeh, L.A. (1965). "Fuzzy Sets", Information and Control, 8, 338-353.

Zimmerman, H.J. (1985). "Fuzzy Set Theory and its Applications", Kluwer-Nijhoff Publishing, Hingham, MA.

Table 1: Values of fuzzy indicators for five disposal alternatives.

Disposal Site	Indicator*	Most Likely Interval Low	Most Likely Interval High	Maximum Interval Low	Maximum Interval High
UAD	1	8.5E-4	3.5E-3	3.7E-7	2.4E-1
	2	0.75	1.2	2.7E-4	173.8
	3	40.0	65.0	4.2	230.0
	4	10.0	30.0	0.55	96.4
	5	500	750	150	1930
	6	2000	5000	1160	8000
	7	0.0	0.0	0.0	0.0
	8	0.0	0.0	0.0	0.0
	9	133280	137300	119100	152800
CAD	1	3.0E-5	5.0E-5	2.5E-8	5.7E-3
	2	0.01	0.04	1.8E-5	4.17
	3	5.0	7.0	0.183	38.0
	4	5.0	30.0	0.37	66.4
	5	50	90	7	314
	6	1000	3500	777	5480
	7	0.0	0.0	0.0	0.0
	8	0.0	0.0	0.0	0.0
	9	148870	153410	132970	170840
CDF	1	5.8E-5	7.0E-5	5.8E-9	2.5E-2
	2	0.04	0.06	4.2E-6	18.0
	3	1.0	2.5	0.03	14.0
	4	35.0	55.0	1.3	230.0
	5	10	40	1	114
	6	7500	9000	2750	19000
	7	0.0	0.0	0.0	0.0
	8	0.0	0.0	0.0	0.0
	9	1081110	1114050	965840	1240230
Upland	1	1.0E-6	2.0E-6	1.1E-9	3.4E-4
	2	2.5E-6	5.5E-6	2.7E-9	1.0E-2
	3	0.0	0.0	0.0	0.0
	4	0.0	0.0	0.0	0.0
	5	0.0	0.0	0.0	0.0
	6	0.0	0.0	0.0	0.0
	7	385	385	8	1190
	8	8540	8540	2750	19100
	9	1527740	1574270	1364880	1752630
Upland Secure	1	8.0E-9	4.0E-8	0.0	3.4E-5
	2	2.0E-8	6.0E-8	0.0	1.1E-4
	3	0.0	0.0	0.0	0.0
	4	0.0	0.0	0.0	0.0
	5	0.0	0.0	0.0	0.0
	6	0.0	0.0	0.0	0.0
	7	74	74	1	334
	8	2000	2000	60	5000
	9	16918860	17434140	15115380	19409280

* 1) cancer risk (probability), 2) non-cancer risk (risk index), 3) fish toxicity risk (mortality index), 4) fish burial risk (mortality index), 5) shellfish toxicity risk, 6) shellfish burial risk, 7) terrestrial species toxicity risk, 8) terrestrial species burial risk, 9) cost.

Table 2: Final Ranking of Disposal Alternatives

Ranking	Disposal Site	Ranking Value
1	CAD	0.866
2	UAD	0.816
3	Upland	0.449
4	Upland Secure	0.408
5	CDF	0.315

Ranking Value indicates relative acceptability of alternative where a value of 1.0 indicates an ideal alternative.

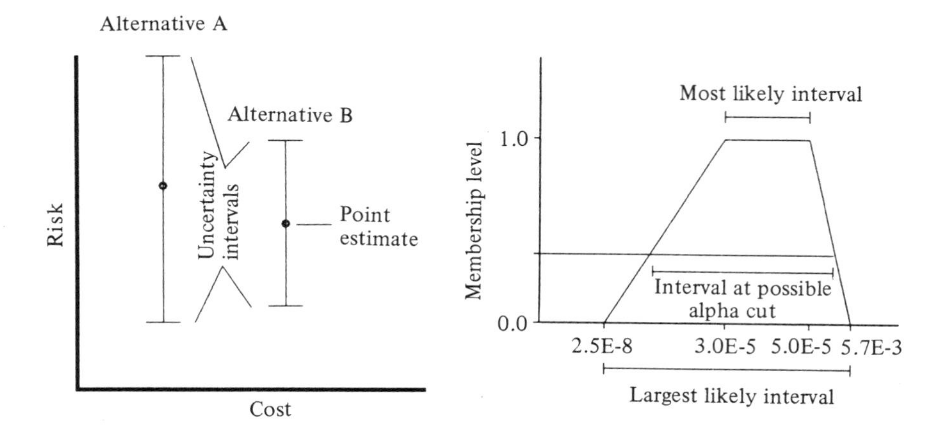

Fig. 1: **Point and interval estimates of risk.**

Fig. 2: **Fuzzy estimate of risk.**

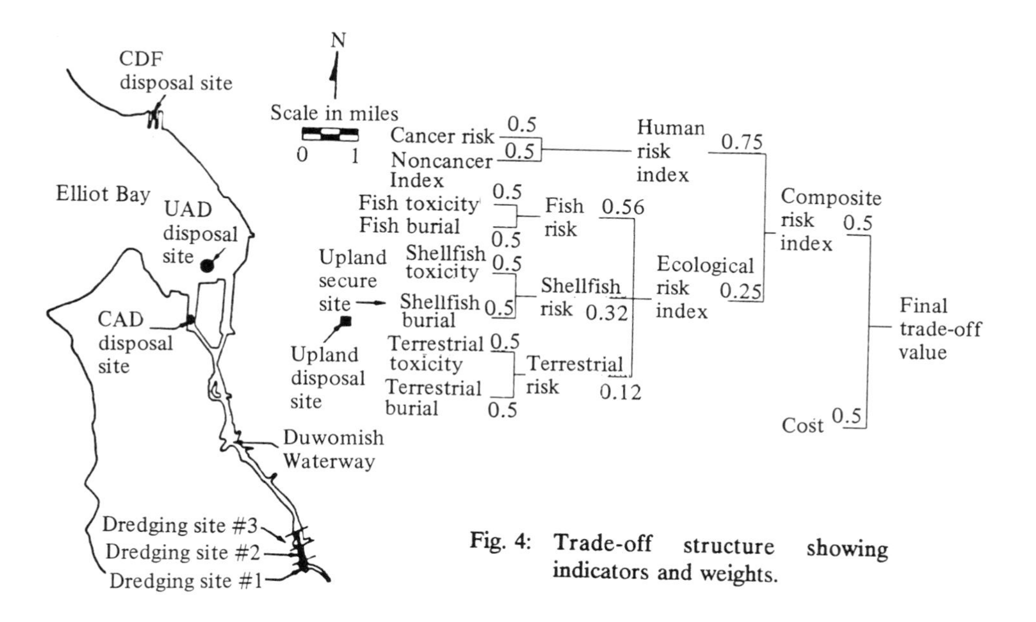

Fig. 4: **Trade-off structure showing indicators and weights.**

Fig. 3: **Map of study area.**

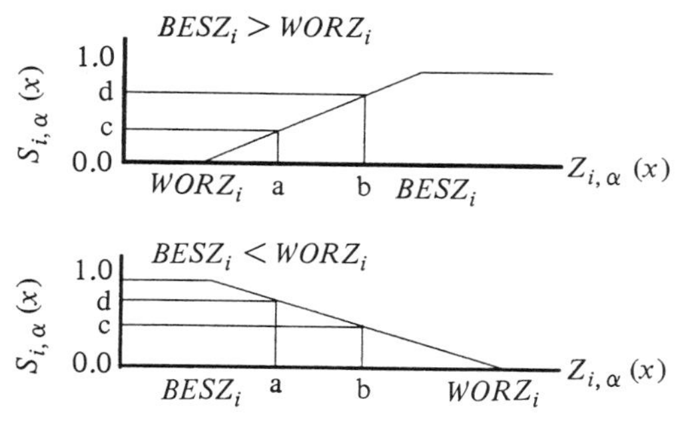

Fig. 5: Index function to produce index of $Z_{i,\alpha}(x)$.

Fig. 6: Trade-off of toxicity versus burial for shellfish.

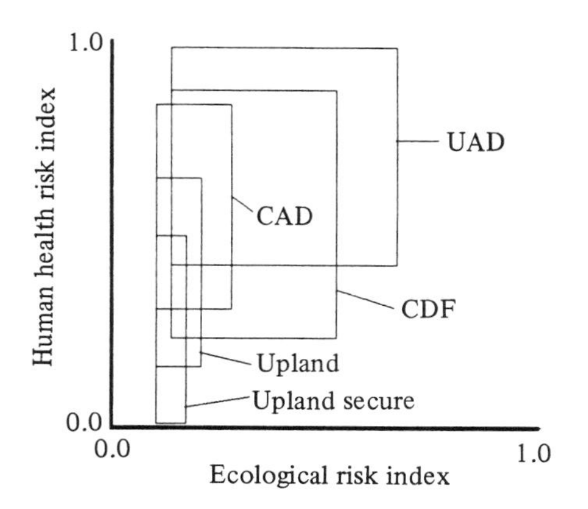

Fig. 7: Trade-off of human health risk versus ecological risk.

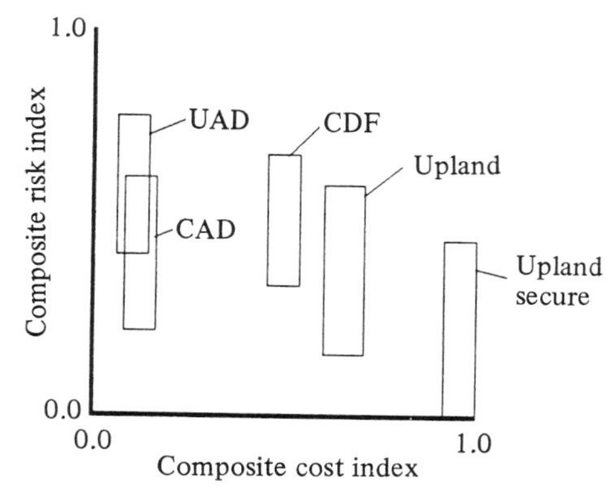

Fig. 8: Final trade-off of composite risk versus cost.

MULTIOBJECTIVE INVESTMENT PLANNING FOR IMPROVING QUALITY OF LIFE : A CASE STUDY FOR TAIPEI CITY

Gwo-Hshiung Tzeng and Junn-Yuan Teng
Energy Research Group, and
Institute of Traffic and Transportation,
National Chiao Tung University
114-4F, Sec. 1, Chung-Hsiao W. Rd., Taipei, Taiwan, R.O.C.
Ching-Luen Ju
Institute of Urban Planning,
National Chung Hsiung University
Taipei, Taiwan, R.O.C.

Abstract *This paper describes a multiobjective decision making model, which aids the Taipei city administration in planning its budget allocation among eighteen items concerning the living environment. The allocation problem is complicated by multiple competing objectives and by uncertainty in the outcome corresponding to a given allocation policy. These characteristics were taken into account by estimating a multiattribute utility function from expressed civic preferences for environmental quality. This model enables the city administration to select an optimal public investment policy for improving the quality of life. The model includes the Analytic Hierarchy Process (AHP) model for estimating the weights of the eighteen items, the additive multiattribute utility function (AMUF) for estimating preferences, and the compromise programming submodel for optimizing an investment plan under governmental budgets and civic satisfaction constraints. Three kinds of indices, namely, a civic minimum index, an ideal aspiration index, and the present comprehensive environmental indices are estimated from civic attitude survey data.*

Keywords : Public opinion, utility function, AHP, AMUF, compromise solution, noninferior solution, multiobjective planning, environmental quality, quality of life.

Introduction

Quality of life is to describe the degree of satisfaction of inhabitants with the environment in different geographic areas. Since the quality of transportation security, pollution, parks/open spaces, medical care, social-welfare varies from place to place and the situation in pollution and social security differ in deferent locations, the positive or negative impact of these factors on quality of life also varies. Similarly, the social needs and values change over time with changes in the economy. Therefore, the urban living environment is no longer a single problem, but a multidimensional one, which includes various complex factors.

When city administrations develop plans and policies, they must give full expression to the needs of city inhabitants (Maeda and Murakami, 1985). Only those policies that reflect the popular opinion will meet with minimum resistance and achieve the best result. However, in policy making the assessment of the quality of life is the weakest part (Milbrath, 1979). The purpose of assessing the quality of life is to build usable indicators to help with the municipal development planning and to develop related policies and possible remedies (Rein and Schon, 1977). In order to formulate effective policies reflecting popular opinion, the indicator should be reliable and comprehensive. Therefore, the indicator and its construction for urban quality of life is an important component in urban planning.

The indicators for urban quality of life are social indicators which cover all aspects of daily life for urban inhabitants and also depict the overall picture of the city life. Since large populations live in urban areas, and each person has his own thinking and values, his evaluation of attitudes on urban functions varies considerably. Even some persons with similar thinking and values, may differ in their evaluation if they live in different districts . Therefore, when the indicators of urban quality of life are built and evaluated, it is necessary to draw a distinction clearly and the items included in quality of life indicators must be centered around living standard (Tzeng and Shiau, 1987; Tzeng, et al, 1991). If it is conducted district by district, the degree of complexity will be reduced and operation will be much easier.

The indicators of the quality of life for inhabitants will have much to do with the statistical survey techniques and utility theorems which are the basis for construction of actual civic preference structures and social welfare functions for forecasting changes in the utility of inhabitants under uncertainty. Budget allocation is a matter of high complexity. Multiobjectives and some uncertain outcomes arising from the given allocation policy must be all taken into account (Joiner and Drake, 1983). This paper tries to help plan the budget allocation among all the indicators by basing itself on a utility function for improving quality of life at different stages.

The purpose of this paper is to construct a multiobjective investment decision model which includes the building of the hierarchy evaluation items for environmental quality, with a questionnaire survey, using the Analytic Hierarchy Process (AHP) to find the weight of the evaluation items, constructing the Additive Multiattribute Utility Function (AMUF), developing the civic utility forecast form, and using compromise programming for investment planning. This paper not only describes evaluation framework, but also applies these approaches to empirical analysis for Taipei city investment planning. In section 2, the concepts and measurements of quality of life will be discussed. In section 3, the evaluation of urban environmental quality will be described. In section 4, multiobjective investment decision for improving the quality of life will be presented. In section 5, the empirical analysis for Taipei city will be described. In the last section, some conclusions are suggested.

Concepts and Measurements of Quality of Life

Urban quality of life represents the living standard and welfare of urban inhabitants. Human desires can never be fully satisfied. The higher the level of economic development is, the greater the needs will be, and desired. Hence, quality of life depends on different values of different people, at different times and places. This section will discuss the development of concepts and measurements for urban quality of life.

Concept development in quality of life

To construct indicators for quality of life, concepts of quality of life in urban society must first be understood. Urban environmental quality is concerned with industry, population, wage, living cost, urban growth, public facilities and services, work and leisure, and urban planning (Myers, 1988).

The interaction between environmental quality and economic development is complex , because higher quality of life can lead to higher immigration and wages, which help to form the potential local labor pool (Porell, 1982). At the same time, the higher quality of life tends to curb wage increase, because amenities compensate for monetary income (Hoch, 1977). Economic development will lead to urban growth, and because the employment opportunity increases, it causes population inflow. In addition, it may also lead to land development expansion and traffic intensity. If a high growth economy or business cycle suddenly accelerates, the urban service facilities will be overloaded. In addition, demand on land increases and land supply is limited. This would cause the land prices to go up, which leads to the increase of rents and housing prices . As a result, living costs will increase, that will force wage to increase. After economic growth has caused wage to increase, economic inducements to incoming businesses will be much weakened, unless they can be made from a long-term point of view of the industry as an integrated whole.

Urban quality of life will be improved as urban amenities (e.g., art, restaurant, and leisure/recreation) and employment opportunities increase. Therefore, by improving the urban amenities the negative impacts can be offset, and urban quality of life can be raised (McNulty et al., 1986). Good planning can help raise the quality, and economic growth can help increase employment opportunity, leading to urban growth.

In regard to infrastructure and service availability, major planning items include transportation, housing, land development, parks, environment, and schools, and urban amenities also include urban design, culture and art, leisure/recreation, and shopping.

Measurements of environmental quality

The quality of life was generally defined as the following:

1. Accurately defined the components of quality of life, e.g., happiness, satisfaction, and living style. Scholars have not reached a consensus definition, but all agree that quality of life should include mental and physical life, which can be divided into: (a) personal consciousness, psychology and subjective judgment, i.e., the satisfaction that a person has (Dalkey & Rourke,1973; Mitchell, et al., 1973; Katona, 1975; Henshaw, 1975). (b) the social environment a person directly makes contact with (Joun, 1973), and (c) personal satisfaction and the objective social environment (Lowchen, 1973).
2. Cost items or factors that have a bearing on quality of life. Major factors include geographical features, transportation, situation, weather, land use, ecology, and regional economics demography .

3. The quality of life is defined by social indicators, namely, gross national product (GNP), health and welfare index, education index, family income, population growth, number of health and medical facility and dwelling space (Michalos, 1974; Liu, 1974; Flanagan,1978). This definition emphasizes on objective quantifiable data, but fails to indicate the subjective evaluation. That is a defect.

When a social science approach has been applied to assess the quality of life, it has been a static approach. A social science approach can also perform trend analysis, but no literature has been found. In regard to the assessment method for quality of life, Myers divides it into five different types (Myers, 1988) : origins of professional approach, measurement focus, statistical approach, conclusions from past studies, political/economic implications of past studies. Measurement for quality of life is divided into four areas, namely, comparison in living standard, wage difference, personal welfare benefit, and social development tendency.

Evaluation of Urban Quality of Life

System analysis approach has been widely applied to urban/regional planning. In early years, the focus was on establishing models for forecasting the environmental changes in urban areas. Then, as decision making theory was developed, some scholars began to proceed from the municipal administrative point of view and incorporated the decision making theory into evaluation process in district planning. The district planning needs to be evaluated from the inhabitant view-point for the purpose of achieving even development and fair welfare benefit (Maeda and Murakami, 1985). Therefore, the indicators of urban quality of life were developed by questionnaire survey to find the inhabitant view-points for quality of life, and then the preference structure of inhabitants in different districts was constructed by using above results for satisfying the civic needs inferred from public opinions.

Theoretical framework

When urban quality of life is evaluated, objective hierarchy can be applied and at the lowest level there are n items (attributes), evaluated (Maeda and Murakami, 1985). Therefore, in this paper the inhabitants preference structure of attributes is assumed to be additive multiattribute utility function as follows :

$$U(\mathbf{X}) = \Sigma\, \alpha_i u_i(x_i),\ \ \Sigma\, \alpha_i = 1,\ \ 0 \leq \alpha_i \leq 1,\ \ 0 \leq u_i(x_i) \leq 100 \qquad (1)$$

where, $\mathbf{X} = (x_1, \dots, x_n)$ is a vector of n attributes, α_i is a weight of attribute i, $u_i(x_i)$ is one dimension of attribute i, and the range of utility function is defined between 0 and 100 points. Thus the utility value level presents the satisfactory degree for attribute of inhabitants.

Based on the practical approach by Edward, questionnaire survey is feasible for establishing a utility function for quality of life (Edward, 1977). Therefore, the difference of preference structure in each district inhabitant can be combined into one dimension utility by using attribute weights. If an urban area can be divided into m districts, the average responses of all inhabitants in each district can be used to present the general situation, i.e.,

$$U_j(\mathbf{X}) = \Sigma \beta_i u_{ij}(x_i), \ \Sigma \beta_i = 1, \ 0 \leq \beta_i \leq 1, \ 0 \leq u_{ij}(x_i) \leq 100 \qquad (2)$$

where, $U_j(\mathbf{X})$ is the multiattribute utility function of average reflection in district j, β_i is the average weight of attribute i in district j, $u_{ij}(x_i)$ is the utility value of attribute i in district j.

Finally, according to the preference structure of urban inhabitants, their total utility function can be constructed, i.e.,

$$U(\mathbf{X}) = \Sigma r_j U_j(\mathbf{X}), \ \Sigma r_j = 1, \ 0 \leq r_j \leq 1, \ 0 \leq U_j(\mathbf{X}) \leq 100 \qquad (3)$$

where, $U(\mathbf{X})$ is the total utility function of urban-inhabitant reflection, r_j is the total urban population ratio of inhabitant population in district j. For each attribute, the utility value reflects the inhabitant's degree of satisfaction. Therefore, the utility value can be used to evaluate the index of each attribute, and can compare their relative importance for every attribute, and to indicate important items on living environment. This result can be provided as reference for urban planning . Therefore, the utility function of all inhabitants of an urban area can be found through the multiattribute function.

Constructing process of environmental quality value

The process of constructing the urban environmental quality includes five steps as follows :

Step 1 Constructing goal hierarchical structure **:** Hierarchical analysis, which is a basic tool in human thinking, will be used for constructing the goal hierarchical structure (Roubens and Vincke, 1985). According to MacCrimmon, in making goal hierarchical structure, the approach of related literature review, system analysis and empirical study can be used (MacCrimmon, 1969). Based on this concept, for making the goal hierarchical structure, a related literature review was conducted by considering : (a) Components in each level must be mutually exclusive to avoid measuring the same objects. (b) Components need to be complete in order to contain urban environmental quality items. For example, evaluating the quality life of Taipei city, the goal hierarchical structure includes three levels, and the evaluation items in the third level have 18 attribute-items as shown Figure 1.

Step 2 Questionnaire survey process **:** Questionnaires should be designed according to the components of the hierarchical evaluation structure for quality of urban life. The contents of questionnaire include three parts : (a) utility values of 18 attributes and civic minimum and ideal aspirations, (b) relative importance among hierarchical components, and (c) basic attribute questionnaire data of interviewee. The purpose of the survey is to obtain one dimension utility function of Eq. (1) and find the weights by using pairwise comparison. In the instance of Taipei city, the survey area includes 16 administration districts. The sample size of each administration district computed by the ratio of that district shared in the total Taipei city population. Then the samples were taken randomly in each district. Surveys are conducted through home interviews with samples of 2000 households.

Step 3 Finding the weights **:** According to pairwise comparison data among evaluation items in questionnaires, the weights (i.e., α_i of Eq. (1)) of each evaluation items in evaluation-hierarchical structure are obtained by the

Analytic Hierarchy Process (AHP) (Saaty, 1980). If a consistency test can not be satisfied, the questionnaire would be used to double-check. For reducing checking-time in consistency test this research adopts a flexible attitude, i.e., if the consistency index (C.I.) is less than 0.2, the consistency of the interviewee to the comparison of evaluation items is considered satisfactory.

Step 4 Finding the utility value : This research will give the utility value of 18 attributes directly to interviewee who answers the questionnaires. First, the interviewee answers the ideal aspiration and the civic minimum; secondly, the interviewee fills in his present degree of satisfaction on a scale between 0 and 100 points and each 10 points on the scale form one interval. These intervals can be summed up into five degrees of satisfaction such as:

0 - 20 : unsatisfactory 20 - 40: rather unsatisfactory
40 - 60: neutral 60 - 80: rather satisfactory
80 - 100: satisfactory

Step 5 Finding the total utility function : According to utility functions in each district and ratios in each district-population, total utility functions for urban inhabitants can be constructed as Eq. (3).

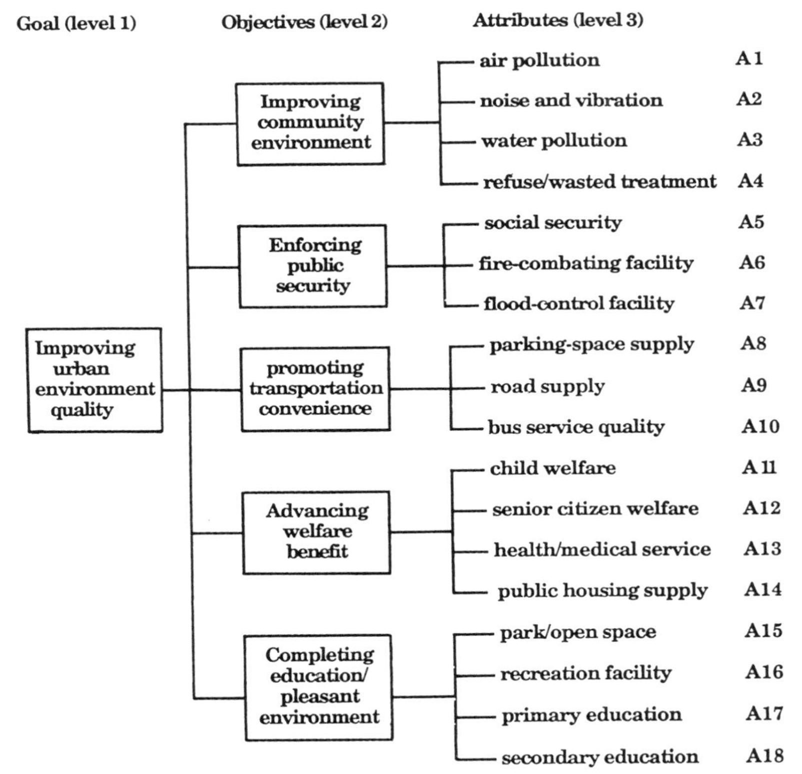

Figure 1 ***Evaluation hierarchical structure of urban environmental quality***

Forecast of utility values

The impact of economic growth on the quality of life is very complex. The trend in various countries has been that personal income rises with economic growth. Moreover, empirical study done by this team found that the utility function of all items in the quality of life is of concern to urban inhabitants. Based on these concepts this paper assumes the utility function of individual attributes (evaluation items of quality of life) is a function of income to simplify problems and apply the utility functions to forecasting the changes of inhabitant-utilities.

In this research the relationship between household (family) income and 18 attributes was firstly tested. After analysis, the results are shown as Table 1. If the significant level is under α =0.05 , all attributes, except bus-service quality and parking-space supply, are related to household income. If under α =0.15, all attributes, including bus-service and parking-space, all related to household income.

Table 1 Analysis of the relationship between household income and18 attributes.

attributes	χ^2	p-value	attributes	χ^2	p-value
1. air pollution	156.3	0.00	10. bus-service quality	94.2	0.13
2. noise pollution	222.2	0.00	11. child welfare	106.4	0.03
3. water pollution	222.2	0.00	12. Senior citizen welfare	116.5	0.01
4. refuse/wasted treatment	206.7	0.00	13. health/medical service facility	146.9	0.00
5. social security	206.1	0.00	14. public housing supply	101.6	0.04
6. fire-combating facility	123.4	0.00	15. park area	112.6	0.01
7. flood control	125.3	0.00	16. recreation facility	113.4	0.01
8. parking-space supply	100.0	0.07	17. primary education	119.8	0.00
9.road capacity	148.9	0.00	18.secondary education	124.7	0.00

Based on the relationship between household income and 18 attribute utility functions, the utility forecasting equations were constructed as follows,

$$U_{ti} = f(Y_t) \qquad (4)$$

where,U_{ti} is an average utility value forecast . Y is household income. There are numerous kinds of potential utility functions. However, they can be divided into three basic types : risk aversion, risk neutral and risk prone (Keeney and Raiffa, 1976). Seven potential types of equations are fitted to forecast utility functions, and a highest R^2 is used to select the equation in each attribute, i.e, $U_{ti} = \alpha Y_t + \kappa$, $U_{ti} = \alpha Y_t + \beta Y_t^2 + \kappa$, $U_{ti} = \alpha Y_t + \beta Y_t^2 + \gamma Y_t^3 + \kappa$, $U_{ti} = \beta \alpha^{Y_t}$, $U_{ti} = \beta \exp(\alpha Y_t)$, $U_{ti} = \kappa \exp(\alpha Y_t + \beta Y_t^2)$, and $U_{ti} = \kappa \exp(\alpha Y_t + \beta Y_t^2 + \gamma Y_t^3)$. Based on the fitted results, the equations in all attributes can be made into two kinds as : $U_{ti} = \alpha Y_t + \beta Y_t^2 + \gamma Y_t^3 + \kappa$ and $U_{ti} = \kappa \exp(\alpha Y_t + \beta Y_t^2 + \gamma)$.

Multiobjective Investment Planning for Improving Quality of Life

Multiobjective decision problems

Investment planning for improving the quality of life is a typical application of multiobjective decision problems. Multiobjective decision problems have more than one objective, yet the objectives can not directly be combined by using addition. In mathematics form, multiobjective problems can be expressed as a vector optimization problem(Cohon, 1978; Dinkel & Erickson, 1978) :

$$\text{Max} \quad \mathbf{Z}(\mathbf{X}) = [\, Z_1(\mathbf{X}), Z_2(\mathbf{X}), \ldots, Z_p(\mathbf{X}) \,] \qquad (5)$$

$$\text{s.t.} \quad g_i(\mathbf{X}) \leq 0, \quad i = 1,2,\ldots,m; \quad X_j \geq 0, \quad j = 1,2,\ldots,n$$

where, $\mathbf{Z}(\mathbf{X})$ is the vector of objective functions $(Z_1(\mathbf{X}), Z_2(\mathbf{X}), \ldots, Z_p(\mathbf{X}))$, $g_i(\mathbf{X})$ is a constraint, and $\mathbf{X}$ is a vector of decision variables $(X_1, X_2, \ldots, X_n)$.

There does not exist a single optimal solution to multiobjective decision problems, because objectives conflict with one another and their compensatory tradeoffs can not be established. The objectives conflict, i.e., an optimal value of $Z_R(\mathbf{X})$ usually implies that other p-1 objectives are of non-optimal value. If objectives do not conflict , the weight of objectives can be established and the vector optimization problems can be reduced to a single objective optimize problem. If objectives are complementary, then Max $Z(\mathbf{X})$ = Max $Z_1(\mathbf{X})$ = Max $Z_2(\mathbf{X})$ etc., and $Z_R(\mathbf{X})$ can be selected to optimization. If objectives can be combined, the objective functions can be expressed by using $(Z_1(\mathbf{X})+Z_2(\mathbf{X})+ \ldots +Z_p(\mathbf{X}))$ and multiobjective functions also can become a single objective function (Giuliano, 1985). The multiobjective decision problems are characterized by lack of a single optimal solution.

Multiobjective decision methods can find noninferior solutions and evaluate multiobjective problems. Their purpose is to identify the most possible solution set. The most possible solution set can achieve all objectives. Multiobjective problems may have a feasible solution set, and in that feasible solution region all alternatives satisfy the constraints (i.e., $g_i(\mathbf{X})$). So multiobjective methods are used to find out the most possible solution set, where no further improvement is possible in any objective without worsening others. It can let one objective value improve, but this solution set doesn't make at least another objective value reduce. These most possible solutions are called noninferior solutions, or Pareto optimal solutions. When $Z(\mathbf{X})$ becomes a noninferior solution, the infeasible solution $Z(\mathbf{X'})$ will surely exist, such as $Z(\mathbf{X'}) \geq Z(\mathbf{X})$ and $Z_i(\mathbf{X}) \neq Z_i(\mathbf{X})$ for at least one k.

Compromise programming

Except for finding noninferior solutions, multiobjective decision methods are expected to reduce a compromise solution set from a noninferior solution set, so that decision makers can make the final choice from compromise solution. Finding a compromise solution is based on the concepts of ideal (aspiration) solution. In investment planning for improving quality of life, compromise programming is applied to the analysis (Zeleny, 1982).

The purpose of compromise programming is to find compromise solutions that account for all objectives. Therefore, compromise solutions are the closest to ideal solutions, and they are noninferior solutions themselves.

Compromise programming is to find a solution according to the geometrical definition, i.e., to find from a feasible solution set the nearest solution to the ideal solution. This is called a compromise solution. The distance is defined as follows :

$$d_\alpha = \{ \Sigma \mid Z_i^*(\mathbf{X}) - Z_i(\mathbf{X}) \mid^\alpha \}^{1/\alpha}, \quad i = 1,2, \dots, p \qquad (6)$$

In Eq.(6), $Z_i(\mathbf{X})$ represents the objective value of ith attribute, $Z_i^*(\mathbf{X})$ represents the optimal value in the ith attribute, and α is a scale parameter between 1 and ∞. Therefore, the compromise programming model can be shown as bellows :

$$\text{Min } d_\alpha = \{ \Sigma \mid Z_i^*(\mathbf{X}) - Z_i(\mathbf{X}) \mid^\alpha \}^{1/\alpha}, \qquad 1 \leq \alpha \leq \infty \qquad (7)$$

Since $d_\infty \leq d_\alpha \leq d_1$, we need merely to find out d_1 and d_∞, and the convex hulls $\mathcal{H}(\{d_1, d_\infty\})$ that are made up by these two compromise solutions. All of the sets constituting compromise solutions are included in the convex hull.

Based on this concept, the steps needed in applying compromise programming to solve multiobjective investment for improving quality of life can be summarized as follows :

Step 1 : Find the ideal solution

$$Z^*(\mathbf{X}) = [Z_1^*(\mathbf{X}), Z_2^*(\mathbf{X}), \dots, Z_p^*(\mathbf{X})]$$

where, $Zi^*(\mathbf{X})$ is an ideal (aspiration) index (e.g., park area (m^2/person)) of environmental quality in attribute i.

Step 2 : Find the compromise solution 1, i.e.,

$$\text{Min } d_1 = \{ \Sigma \mid Z_i^*(\mathbf{X}) - Z_i(\mathbf{X}) \mid^1 \}^1$$

$$= \Sigma\, [Z_i^*(\mathbf{X}) - Z_i(\mathbf{X})], \qquad \text{if } Z_i^*(\mathbf{X}) \geq Z_i(\mathbf{X}) \qquad (8)$$

s.t. $\mathbf{X} \in \mathbf{Fd}$

where, $Z_i(\mathbf{X})$ is a plan index (unknown) of environmental quality in attribute i and $Z_i(\mathbf{X})$ can not be lower than the civic minimum index $\underline{Z}_i(\mathbf{X})$ (i.e., minimum limit of environmental quality), and **Fd** is a decision space in feasible solution. Because $Z^*(\mathbf{X})$ is given, $Z^*(\mathbf{X})$ can be regarded as a constant set and found from the utility function.

Step 3 : Find the compromise solution 2, i.e.,

$$\text{Min } d_\infty = \{ \Sigma \mid Z_i^*(\mathbf{X}) - Z_i(\mathbf{X}) \mid^\infty \}^{1/\infty} \qquad (9)$$

When $\alpha = \infty$, only the largest deviation objective (the largest distance from ideal solution) has the absolute affection. Hence, the above equation is :

$$\text{Min } d = \max\, [Z_i^*(\mathbf{X}) - Z_i(\mathbf{X})], \qquad i = 1, 2, \dots, p \qquad (10)$$

s.t. $\mathbf{X} \in \mathbf{Fd}$

Step 4 : Decide the compromise solution set, i.e.,
$\mathcal{H}(\{d_1, d_\infty\}$

Planning for improving quality of Life

A questionnaire survey can find the utility and relative importance (weight) to urban inhabitants of objective hierarchical evaluation attributes and utility forecasting equations can estimate the inhabitant utility in plans for the future. When the question of improving the quality of life is studied, the following two major objectives are taken into account:

The first objective is maximum degree of satisfaction of inhabitants with the quality of life in a plan year. Utility forecasting equations can be used to estimate the future utility value under different household incomes for each evaluation attribute. This research assumed that inhabitants with lower living standards have a relatively high sensitivity in the same environment index situation. In other words, higher physical indices are required in higher living standards (regarded as a future plan year index $Z_{ti}(\mathbf{X})$ (higher income)) than in lower living standard (i.e., present actual index $Z_{oi}(\mathbf{X})$). Generally, inhabitants in lower income levels can easily satisfy the physical environment index level, and inhabitants in higher income levels require higher levels of quality of life. Based on this concept, the utility value and the quality index can be obtained. The utility value of civic minimum index ($Z_{ti}(\mathbf{X})$) was set at 40 points (between slightly unsatisfactory and ordinary) and in ideal (aspiration) index ($Z_{ti}^*(\mathbf{X})$ was set at 80 points (between slightly satisfactory and satisfactory).
Based on above concepts, we want to minimize the distance between the aspiration index ($Z_{ti}^*(\mathbf{X})$) and plan index ($Z_{ti}(\mathbf{X})$) in all environmental quality attributes, and to have the plan index greater than the civic minimum index (i.e., $Z_{ti}(\mathbf{X}) \geq \underline{Z}_{ti}(\mathbf{X})$), i.e.,

$$\text{Min } \mathbf{Z}_1 = \{\Sigma\, \delta_i \,|Z_{ti}^*(\mathbf{X}) - Z_{ti}(\mathbf{X})|^{\alpha}\}^{1/\alpha}, \quad \delta_i = \Sigma\, \beta_{j\,i} r_j \qquad (14)$$
$$\text{s.t. } Z_{ti}(\mathbf{X}) \geq \underline{Z}_{ti}(\mathbf{X})$$

where δ_i is a comprehensive responding weight for the urban inhabitant for attribute i, $\beta_{j\,i}$ is the average weight in district j for responding to attribute i, r_j is a ratio of the population in district j compared to total city population, and $Z_{ti}^*(\mathbf{X})$, $Z_{ti}(\mathbf{X})$ and $\underline{Z}_{ti}(\mathbf{X})$ is respectively an aspiration index, a plan index, and a civic minimum index of attribute i in plan year t (It should be noted that an index is measured per person).

The second objective is to minimize the investment cost, and the total investment cost can not exceed government budget. Investment items include services and facilities that are related to all evaluation attributes. When investment allocations are made, the relative importances (weights) which urban inhabitants attach to environment quality attributes need to be taken into consideration. This can be expressed with the following mathematical formulation :

$$\text{Min } \mathbf{Z_2} = \Sigma\, \delta_i c_i P_t (Z_{ti}(\mathbf{X}) - Z_{oi}(\mathbf{X})) \tag{15}$$
$$\text{s.t. } \Sigma\, c_i\, P_t (Z_{ti}(\mathbf{X}) - Z_{oi}(\mathbf{X})) \leq B_t$$

where, c_i is a unit improvement cost in attribute i, P_t is a total population (given for the study area) in year t and $Z_{oi}(\mathbf{X})$ is an actual physical index in attribute i in the base year.

Combining the above two objectives and constraints, multiobjective planning problems for improving urban quality of life can be explained as follows.

$$\text{Min } \mathbf{Z} = (\mathbf{Z_1}, \mathbf{Z_2}) \tag{16}$$
$$\text{s.t. } Z_{ti}(\mathbf{X}) \geq \underline{Z}_{ti}(\mathbf{X})$$
$$\Sigma\, c_i\, P_t (Z_{ti}(\mathbf{X}) - Z_{oi}(\mathbf{X})) \leq B_t$$

The process of solving this problem can be divided as follows : (a) let $\alpha = 1$ and $\alpha = \infty$ in Eq.(14) respectively to calculate two objective problems and obtain two compromise solutions, (b) find the compromise solution set, i.e., the noninferior solution, and (c) then provide the decision maker with the preferred solution.

Empirical Analysis : A Case of Taipei City

During the past decade and more, because of continuous rapid economic growth and increases in income and living standard of inhabitants, great changes are taking place in people's value concepts. Alongside this, the social value and the level of needs for quality of life are being raised, especially in urban areas where the gaps between people's ideal aspiration and status quo are more significant. How to make investment allocations to reduce that gap is an important question facing the city administration. Hence, it is necessary to evaluate the current urban quality of life to know the extent of acceptance by inhabitants for the present quality of life, and also know their civic minimum and ideal aspiration.

Taipei city is located in northern Taiwan, and it is the political, economic and cultural center of Taiwan. Over the past year, urban growth accelerated and population increased each year. It reached 2.5 million in 1986. But the area is only 205 square kilometers, divided into 16 administrative districts (merged into 12 districts in 1990). The working population keeps expanding, the number of vehicles grows every year, housing prices and rents continue to rise, and security becomes increasingly serious. The Taipei city government will undertake a five-year program to improve the quality of life for inhabitants. It is against this background that the investment planning for improving quality of life is conducted in the hope of making the allocation appropriate and reasonable under budgets constraints.

Questionnaire Survey and Estimation of Inhabitant Utility

The questionnaire design and survey were conducted in Taipei city, according to the evaluation framework, and based on these data, the utility values and weights of the 18 attributes were obtained. Again, because of the close ties between economic development and household income are related to the 18 attributes, the forecasting equations of inhabitant response to 18 attributes utility were built. Then, according to a household income forecasts for 1994 in Taipei city, the inhabitant utility value corresponding to the civic minimum index, aspiration index, and actual physical index in environmental quality were estimated.

According to the average inhabitant response on 18 attributes, it is revealed that the satisfaction degree is lower than 30 points in air pollution(A1), noise pollution(A2), and parking space supply(A8). This proves that traffic problem to Taipei city inhabitants is a pressing issue to be addressed. For their response to weight, the most important items are social security (A5) and parking space supply (A8).

Investment Planning for Improving Quality of Life

The inhabitant response tells that inhabitants care for the overall improvement of the quality of life. Therefore, to improve each item in quality of life is to be improved, two objectives must be taken into account. One is to minimize the difference between the aspiration index and the plan index, and to minimize improvement cost of future quality of life. Therefore, it is a multiobjective planning problem. If heavy investments can be made for improvements, the gap in quality of life will be narrowed. However, given government budget constraints, large-scale improvements are not possible. If only limited funds can be invested for improving quality of life . In a nutshell, the two objectives in question contradict and are incompatible with each other.

The investment budget for the second mid-term plan in Taipei city is NT $141,108 million. This research classifies all items into 18 attributes and unit improvement costs in each attribute are combined and estimated. This research applies compromise programming to investment allocation. The parameters of objective 1 are from degree of importance responded by the inhabitants (i.e., weights in each attribute) and the parameters of objective 2 are the unit improvement cost multiplied by the weight in each attribute. The upper bounds in each attribute are from the aspiration index value reduced by the actual index value. The lower bounds in each attribute are from the civic minimum index value minus the actual index value. If the actual index value is more than the civic minimum value, the gap is zero.

According to the computational steps of compromise programming presented in the previous section, we find the ideal solution of two objectives, i.e., $\mathbf{Z}^* = (\mathbf{Z}_1^*, \mathbf{Z}_2^*) = (13.73, 7667715)$ and calculate the minimum investment capital requirement. Then, the compromise solutions in step 2, step 3 and step 4 and the capital investments in each attribute were also obtained.

From obtaining the compromise solution, it is revealed that the attributes of air quality (A1), noise pollution (A2), water pollution (A3), social security (A5), parking space supply (A8), bus service (A10), child welfare (A11), senior citizen welfare (A12) and recreation facilities (A16) only reach the civic minimum level; and for the attributes of road capacity (A9), housing supply (A14), park/open space (A15), and flood control (A7) the utility function improvements reach over 30 points. In the budget allocation, the investment in flood control (A7) and road capacity (A9) are the largest, totaling NT$ 30 billion, and housing supply (A15) and park/open space are the second largest, totaling NT$ 10 billion.

Conclusions

The attitude toward the social needs and values over time change with the continuous rapid growth in economic development during the past decades in Taiwan. The urban environment problems are complex. The budget allocation problems are also complicated by multiple competing objectives and by uncertainty in the outcome corresponding to a given allocation policy. These characteristics are taken into account by estimating the utility function concerning physical environment indices from public opinion by using a questionnaire survey. The aspiration index and civic minimum index of the environmental quality in each attribute were obtained from the utility functions. Then the compromise programming and multiobjective investment planning model were established. These results not only satisfy the civic needs by responding to the public opinion, but also can provide the optimum budget allocation and improvement level for raising quality of life.

According to the results of case of Taipei city, we find that the expansion of the public facility and services can not keep pace with those developments. Such a situation makes the improvement of urban quality of life an ever-pressing problem. The city administration must take appropriate measures to address this problem and plan budget allocation for the purpose of improving the living environment and quality of life for urban inhabitants. When plans and policies are to be made, they should give expression to inhabitants' needs and allocate the budget in a most appropriate way.

Acknowledgement : The authors would like to thank the National Science Council of the Republic of China for supporting the research fund (NSC-78-0421-H009-04Z) and thank Mrs. Guilan Dong (Associate Research Professor, National Research Center for Science & Technology for Development, Beijing, China) for helping to check from Chinese translation to English and Dr. Paul M. Schonfeld (Associate Professor, the University of Maryland) for correcting manuscript.

References

Cohon, J. L. 1978. Multiobjective programming and planning. Mathematics in science and engineering : 140. Academic press/ New York.

Dalkey, N. C. and D. Rodurke. 1973. The Delphi procedure and rating quality / The quality of life concept. U.S. Environmental Protection Agency/ Washington.

Dinkel, J. J. and J. E. Erickson. 1978. Multiple objectives in environmental protection programs.*Policy sciences* 9 (1) : 87-96.

Edwards, W. 1977. How to use multiattribute utility measurement for social decision making. *IEEE transactions on systems, man and cybernetics* 7 (5) : 326-340.

Flanagan, J. C. 1978. A research approach to improving our quality of life. *America psychologist* 33 (2) : 138-147.

Giuliano, G. 1985. A multicriteria method for for transportation investment investment planning. *Transportation research* 19A (1) : 29-41.

Henshaw, P. S. 1975. The issue, factors, and question. Ecology and quality of life, edited by S. J. Kaplan. Illinois.

Hoch, I. 1977. Variations in the quality of urban life among cities and regions. Public economic and the quality of life, edited by L. Wingo and A. Evans. Johns Hopkins University press, Baltimore.

Joiner, C. and A. E. Drake. 1983. Governmental planning and budgeting with multiple objective models. *Omga* 11 (1) : 57-66.

Joun, Y. P. 1973. Data requirements for a quality growth policy. The quality of life concept. U.S. Environmental Protection Agency/Washington.

Keerfer, D. L. 1978. Allocation planning for R & D with uncertainty and multiple objectives. *IEEE transactions on engineering management* 25 (1) : 8-14.

Keeney, R. L. and H. Raiffa. 1976. Decision with multiple objectives : Preference and value tradeoffs. John Wiley & Sons /New York.

Liu, B. C. 1974. Quality of life indicators in U.S. metropolitan areas. U.S. Environmental Protection Agency / Washington.

Lowchen, W. 1973. The quality of life toward a microeconomic definition. *Urbanstudies* 10 (1) : 3-18.

MacCrimmon, K. 1969. Improving the system design and evaluation process by the use of trade-off information : an application to northeast corridor transportation planning. RM-5877-DOT / The Rand Corporation / Santa California.

Maeda, H. and S. Murakami. 1985. Population's urban environment evaluation model and its application. *Journal of regional science* 25 (2) : 273-290.

McNulty, R. H., et al.. 1986. The return of the livable city : Learning from America's best. Acropolis book/Washington.

Milbrath, L.W. 1979. Policy relevant quality of life research. *Annals of the American academy of political and social science 444* : 32-44.

Mitchell, A., et al. 1973. An approach to measuring the quality of life / The quality of life concept. U.S. Environmental Protection Agency / Washington.

Myers, D. 1988. Building knowledge about quality of life for urban planning. *journal of American planningassociation* 54(3) : 347-357.

Porell, F. W. 1982. Inter metropolitan migration and quality of life. *Journal of regional science* 22 (1) : 137-158.

Rein, M. and D. Schon. 1977. Problem setting in policy research . Weiss, C. (eds) / Lexington Books.

Roubens, M. and Ph. Vincke. 1985. Preference modeling. Springer-Verlag / New York.

Saaty, T. L. 1980. The analytic hierarchy process. McGraw-Hill /New York.

Tzeng, G. H., et al.. 1991. Urban environmental evaluation and improvement: Application of multiattribute utility and compromise programming. *Behaviormetrika*. 29 :71-87.

Tzeng, G. H. and T.A. Shiau. 1987. Urban environmental evaluation : Application of multiattribute utility function. *Traffic and transportation* (National Chaio Tung University) 9 : 25-36.

Zeleny, M. 1982. Multiple criteria decision making. McGraw-Hill /New York.

THREE BASIC CONCEPTIONS UNDERLYING MULTIPLE CRITERIA INTERACTIVE PROCEDURES

Daniel Vanderpooten
Lamsade, Université de Paris-Dauphine
Place du Maréchal de Lattre de Tassigny
75775 Paris Cedex 16 - France

Abstract

From a formal viewpoint, any multiple criteria interactive procedure may be seen as an iterative algorithm which alternates calculation phases for constructing proposals with dialogue phases actively involving the decision maker. In spite of this common formal structure, we can distinguish several conceptions justifying the use of interactivity. In this paper, we introduce three basic conceptions and compare them with respect to several important aspects. We believe that this distinction provides an interesting framework for classifying existing multiple criteria interactive procedures and gives some useful guidelines when designing a new procedure.

Keywords : Multiple Criteria Decision Aid, Interactive procedures, Decision making process.

1. Introduction

Multiple Criteria Interactive Procedures (MCIPs) form the largest class of multiple criteria methods (see *e.g.* the surveys by Hwang and Masud (1979, chap. 3.3), Goicoechea *et al.* (1982, chap. 6), White (1983), Teghem and Kunsch (1986), Steuer (1986, chap. 13), Vanderpooten and Vincke (1989),...). From a formal viewpoint, they can be considered as iterative algorithms which alternate two kinds of phases:

- *calculation phases* which aim at constructing a proposal (which usually consists of one or several alternatives),

- *dialogue phases* where the current proposal is submitted to the decision maker (DM) who may react and provide preference information which is incorporated into a new calculation phase.

The set of alternatives is iteratively explored in this way until a proposal is considered as satisfactory (either by the DM or by the procedure).

The diversity of MCIPs clearly results from the technical details of each method and from the type of preference information required from the DM. A more fundamental distinction issues from the role devoted to interactivity in the decision aid process. We distinguish in this paper three basic conceptions which may underly the use of interactivity in MCIPs. In order to make this distinction clear, we first present a simple model of the decision making process (Section 2). The three conceptions are then introduced (Section 3) and compared with respect to some important aspects (Section 4).

2. A simple model of the decision making process

The type of exploration implemented in MCIPs, based on an iterative investigation of alternatives, is rather close to the exploration spontaneously used by any person faced with a decision problem. A MCIP can thus be viewed as a - more or less normative - tool for amplifying the decision making process (or, more precisely, the part of this process which consists in exploring the set of alternatives). It is then interesting to examine the nature of this exploration which is evoked or even described in some models of the decision making process (see *e.g.* the models proposed by Simon (1977), Janis and Mann (1977), Yu (1980), Zeleny (1982, chap. 3)).

We present a very simple model, highly inspired from the above-mentioned models, which identifies two types of exploration which seem to coexist within the decision making process. This model also emphasizes the concept of preference structure which is central in a multiple criteria perspective.

The DM, when approaching his decision problem, is usually unable to express clear-cut preferences. However, he often has prior references resulting from previous experiences with similar problems and from general ideas about the current problem. This concept of *reference structure* is close to the concept of *schemata* in cognitive psychology (see *e.g.* Horton and Mills (1984)) and to the *habitual domains* of Yu (1980).

A *preference structure* is progressively elaborated through a natural interaction between the reference structure and an improved perception of the decision problem. This is achieved by the DM through a *free exploration* of the set of potential alternatives. Possible dissonances occuring during this free exploration lead the DM to revise his perceptions and beliefs. The reference structure progressively evolves towards a more established preference structure. This process corresponds to a *learning stage*, the DM both learning about his problem and his preferences.

When his preference structure becomes well-established, the DM may engage in a *directed exploration* of the set of alternatives aiming at detecting a prescription, *i.e.* a best compromise solution. This is what we refer to as the *search stage*.

The previous description shows the evolving nature of the decision making process. However, the search stage may also induce modifications of the preference structure rather than an immediate detection of a best compromise solution. This retroactive aspect indicates that the decision process is not sequential, but should be seen as a succession of learning stages and search stages.

Our model of the decision making process is then summarized as follows (see also Figure 1):

The decision making process is a retroactive process aiming at:

- *strengthening, or even creating, the DM's preference structure through a free exploration of the set of alternatives (learning stage),*
- *making use of this preference structure so as to detect a satisfactory compromise solution through a directed exploration of the set of alternatives (search stage).*

Figure 1: **A model of the decision making process**

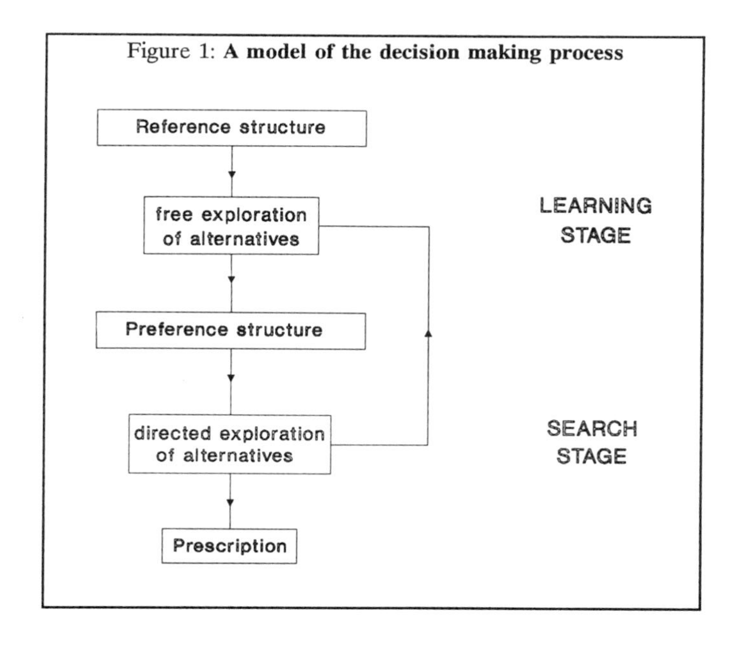

3. Three basic conceptions

Considering the model introduced in the previous section on the one hand, the MCIPs proposed in the literature on the other hand, it can be noticed that most of these procedures favour, in a rather exclusive way, either the *search stage* (implementing a *directed* exploration of the set of alternatives), or the *learning stage* (implementing a *free* exploration of the set of alternatives).

Indeed, a first class of methods aims at implementing converging processes which guide the *search* for prescriptions. Let us note that MCIPs designed according to this conception are among the oldest ones. For some procedures, it is clearly assumed that the DM's preference structure is pre-existing and represented by an implicit value (or utility) function (Geoffrion *et al.* (1972), Zionts and Wallenius (1976)). For the others, no specific assumption is made with respect to the type of preference structure (Benayoun *et al.* (1971), Steuer and Choo (1983)). In both cases however, this structure is supposed to remain stable. The main interest of this **search-oriented conception** is that it really supports the DM in determining a prescription. Search-oriented procedures are thus highly normative. The main drawback of this conception is that it is based on questionable assumptions which reflect a very partial perception of the decision making process.

A second class of methods has been developing rejecting the assumptions of pre-existence and stability of the DM's preference structure. Primarily designed so as to support the *learning* of preferences, such procedures help the DM to explore, in a trial-and-error fashion, the set of alternatives. The methods proposed by Roy (1976), Vincke (1976), Wierzbicki (1980),... were designed according to this **learning-oriented conception**. The interest of this conception is that no assumption is made about the DM's preference structure and his rationality and consistency. Rather than being normative, learning-oriented procedures are conceived in a constructivist perspective, as flexible investigation tools. The main drawback of this conception is that the DM is not guided when he wishes to determine a prescription.

It is clearly interesting (observing Figure 1, for instance) to consider procedures which both support learning stages and search stages, *i.e.* which try to reconcile the previous conceptions. The resulting **learning and search-oriented conception** (or **mixed conception**) aims at guiding the search for compromise solutions while allowing the DM to control this exploration. Such a conception follows a constructivist perspective, but it also recognizes the necessity of supporting the DM in the determination of prescriptions. A way of implementing this conception consists in designing a 'guiding by default' mechanism which may be freely interrupted and reoriented by the DM.

4. A comparison between the different conceptions

In the previous section, three basic conceptions underlying the design of a MCIP were distinguished:

(a) a *search-oriented conception*,
(b) a *learning-oriented conception*,
(c) a *mixed conception*.

In order to make this distinction clear, we propose to compare these conceptions with regard to some important aspects. The comments that follows are made considering pure versions of each conception.

Assumptions about the DM's preference structure:

(a) The DM is supposed to behave according to a pre-existing preference structure (often specified as a value or utility function). Moreover, this structure is assumed to remain stable during the interactive decision aid process.

(b) No assumption is made about the DM's preference structure which may be more or less clearly established at the beginning of the interactive decision aid process. Above all, this (p)reference structure may widely evolve during this process.

(c) Same as (b).

Assumptions about the DM's behaviour:

(a) The DM is supposed to be consistent with his underlying stable preference structure. He cannot express information in contradiction with this structure and thus with pieces of information he previously supplied (in particular, he cannot change his mind).

(b) The DM is supposed to behave in accordance with a desire to progress towards the solution of his problem: his reactions and the information he provides are not arbitrary. However, he is free to express any judgment. In particular, he may change his mind without restriction during the interactive decision aid process.

(c) Same as (b).

Type of exploration implemented:

(a) The exploration is directed by the procedure. This exploration is designed so as to provide, at each iteration, a new proposal which is at least as good as the previous one (considering the stable underlying preference structure). Thus, the procedure aims at generating a converging process.

(b) The exploration is directed by the DM. Exploration strategies are primarily based on trial and error.

(c) By default, the exploration is directed by the procedure which tries to make use of what seems to be stable in the DM's reactions. However, the DM may freely take the control of the exploration.

Type of dialogue:

(a) A rigid dialogue controlled by the procedure which consists of a questioning mechanism allowing to obtain the information required by the procedure.

(b) A free dialogue under the control of the DM.

(c) A flexible dialogue:
- directed by the DM, when he explicitly wishes to express some information or to control the exploration,
- directed by the procedure, when the DM does not know how to conduct his exploration.

Exploitation of the preference information:

(a) The preference information supplied by the DM at any iteration is considered as consistent and valid throughout the whole interactive decision aid process. Thus, the procedure may exploit any piece of information obtained since the beginning of this process.

(b) Since the DM's preference structure may evolve, preference information may not be considered as valid throughout the whole interactive decision aid process. There is no objective way of identifying the pieces of information previously obtained which remain valid at a specific iteration. The most obvious strategy consists then in only retaining the information obtained at the current iteration.

(c) Here again, it is impossible to know which pieces of information remain valid at a specific iteration. However, more elaborated exploitation strategies may be devised (see Vanderpooten (1989)). Roughly speaking, such strategies consist in retaining the part of preference information which does not contradict the information currently obtained.

Stopping rule and the final proposal:

(a) The stopping rule is controlled by the procedure; it is similar to the stopping rules used in classical iterative algorithms (*e.g.* identity between two consecutive proposals). The final proposal appears as a local optimal solution with respect to the DM's preference structure (which may be proved globally optimal under additional convenient assumptions). Let us note that the interest of this optimality concept is limited by the fact that it is often directly based on the strong assumptions concerning the DM's preference structure and his behaviour.

(b) The DM stops the procedure when he is convinced from his exploration that the current proposal is satisfactory. In this case, no formal guarantee is provided as to the interest of this proposal.

(c) The exploration is also stopped by the DM. However, the procedure must try to reinforce the DM's conviction regarding the interest of the final proposal. For instance, it may induce him to look for an improved proposal, which results either in a new exploration or in an improved confidence in the current final proposal.

References

Benayoun, R., de Montgolfier, J., Tergny, J., Larichev, O., (1971), "Linear Programming with Multiple Objective Functions: STEP Method (STEM)", *Mathematical Programming* **1**, pp. 366-375.

Geoffrion, A.M., Dyer, J.S., Feinberg, A., (1972), "An Interactive Approach for Multi-Criterion Optimization, with an Application to the Operation of an Academic Department", *Management Science* **19**, 4, pp. 357-368.

Goicoechea, A., Hansen, D.R., Duckstein, L., (1982), *Multiobjective Decision Analysis with Engineering and Business Applications*, Wiley, New York.

Horton, D.L., Mills, C.B., (1984), "Human Learning and Memory", *Annual Review of Psychology* **35**, pp. 361-394.

Hwang, C.L., Masud, A.S.M., (1979), *Multiple Objective Decision Making - Methods and Applications*, LNEMS 164, Springer-Verlag, Berlin.

Janis, I.L., Mann, L., (1977), *Decision Making - A Psychological Analysis of Conflict, Choice and Commitment*, The Free Press, New York.

Roy, B., (1976), "From Optimization to Multicriteria Decision Aid: Three Main Operational Attitudes", in *MCDM*, Proceedings Jouy-en-Josas France 1975, H. Thiriez and S. Zionts (Eds.), LNEMS 130, Springer-Verlag, Berlin, pp. 1-32.

Simon, H.A., (1977), *The New Science of Management Decision*, Prentice-Hall, New-Jersey.

Steuer, R.E., (1986), *Multiple Criteria Optimization: Theory, Computation and Application*, Wiley, New York.

Steuer, R.E., Choo, E.U., (1983), "An Interactive Weighted Tchebycheff Procedure for Multiple Objective Programming", *Mathematical Programming* **26**, pp. 326-344.

Teghem Jr, J., and Kunsch, P.L., (1986), "Interactive Methods for Multi-Objective Integer Linear Programming", in *Large-Scale Modelling and Interactive Decision Analysis*, Proceedings Eisenach GDR 1985, G. Fandel, M. Grauer, A. Kurzhanski and A.P. Wierzbicki (Eds.), LNEMS 273, Springer-Verlag, Berlin, pp. 75-87.

Vanderpooten, D., (1989), "The Use of Preference Information in Multiple Criteria Interactive Procedures", in *Improving Decision Making in Organisations*, Proceedings Manchester, UK, 1988, A.G. Lockett and G. Islei (Eds.), LNEMS 335, Springer-Verlag, Berlin, pp. 390-399.

Vanderpooten, D., Vincke, Ph., (1989), "Description and Analysis of Some Representative Interactive Multicriteria Procedures", *Mathematical and Computer Modelling*, **12**, 10/11, pp. 1221-1238.

Vincke, Ph., (1976), "Une méthode interactive en programmation linéaire à plusieurs fonctions économiques", *RAIRO* **10**, 6, pp. 5-20.

White, D.J., (1983), "A selection of Multi-Objective Interactive Programming Methods", in *Multi-Objective Decision Making*, S. French, R. Hartley, L.C. Thomas and D.J. White (Eds.), Academic Press, London, pp. 99-126.

Wierzbicki, A.P., (1980), "The use of Reference Objectives in Multiobjective Optimization", in *MCDM Theory and Application*, Proceedings Hagen/ Königswinter 1979, G. Fandel and T. Gal (Eds.), LNEMS 177, Springer-Verlag, Berlin, pp. 468-486.

Yu, P.L., (1980), "Behavior Bases and Habitual Domains of Human Decision/ Behavior - Concepts and Applications", in *MCDM Theory and Application*, Proceedings Hagen/Königswinter 1979, G. Fandel and T. Gal (Eds.), LNEMS 177, Springer-Verlag, Berlin, pp. 511-539.

Zeleny, M., (1982), *Multiple Criteria Decision Making*, Mc Graw-Hill, New York.

Zionts, S., Wallenius, J., (1976), "An Interactive Programming Method for Solving the Multiple Criteria Problem", *Management Science* **22**, 6, pp. 652-663.

WIDELY-IGNORED SUBTLETIES THAT ARE CRITICAL TO DECISION-MAKING

John N. Warfield
University Professor and Director of the
Institute for Advanced Study in the Integrative
Sciences (IASIS)
George Mason University
Fairfax, Virginia 22030-4444

ABSTRACT

Decision-making theories and practice typically ignore several factors that are critical to decision-making about complex matters. This conclusion follows from analysis of extensive data on group work involving numerous organizations and a wide variety of issues.

Credibility of decisions may rest on combining these factors into an integrated process system for group support:

- The critical importance of learning during the processes
- The extent of framebreaking and remodeling that may be required
- The predominance of logic cycles in analyses and designs which seldom is recognized in the absence of the kind of support needed to produce them
- The extent to which remodeling can occur in a modest time period, if suitable processes are used
- The importance of integrative processes that combine organically the anthropological, the technological, and the formal logical

INTRODUCTION

It is clear that there is widespread interest in enhancing the process of decisionmaking. Not only is the literature growing rapidly, but also the spread of special-purpose facilities for supporting group decisionmaking is quite visible.

The author began his own work in group decision support in 1970. This work started as part of an in-house project at the Battelle Memorial Institute in Columbus, Ohio, as part of a program called "Science and Human Affairs". It was recognized that the scale and complexity of issues was growing rapidly, and that little attention had been given to how to approach such issues scientifically. It was clear that the knowledge of the issues was distributed, so that it would be necessary to involve **groups of actors** in efforts to resolve complex issues, rather than isolated investigators, if progress was to be made on a reasonable time scale.

The line of activity started in 1970 has continued until today, and is ongoing. This work has produced many publications, including two books [2,4] and two bibliographies [5,6].

Virtually all of the source documents, along with many of the applications studies, have been placed in the IASIS Reserve File of the Fenwick Library of George Mason University, which accepts orders for copies and makes the indexed file available for inspection by visitors to the Fenwick Library.

Dozens of sponsored projects using our research results have involved facilitated groups striving to come to grips with complex issues. Such issues typically are poorly organized, involve multiple objectives, involve multiple decision-making jurisdictions, and frequently reflect long-standing issues which grow more severe with time.

Having interacted with many such groups in the specially-designed DEMOSOPHIA situation room, following clearly demarcated and rigorously-applied methodologies, it has been possible to collect significant amounts of data on the outcomes of this work that relate especially to the philosophy and conduct of it. These data allow comparisons to be made with other approaches that appear to be applying different philosophy and methodology.

Because many of the other approaches are not well-documented, and frequently are proprietary, and because they furnish little or no data on their results, such comparisons are necessarily anecdotal and subjective. Hopefully some day it may become possible to get better documentation and data, but until that time comes subjective, anecdotal comparison is all that is possible.

What is the value in such a subjective comparison? At least it may serve to focus upon some issues that seem to be critical to good decision-making, yet seem to be ignored in many of the presently chronicled systems for providing group decision support.

In order to try to place this work in perspective, a four-cell matrix is used to break up the domain of consideration into four subdomains. The two headings for this matrix are: The Situation (taken as Coherent or Incoherent) and the Posture (taken as Descriptive or Prescriptive).

COHERENT SITUATIONS

Coherent situations are those for which the prevailing viewpoint is that the situation is well understood. This implies a good organization of the logic and the description of the situation. Distinctions made for coherent situations have to do with whether the work to be done by the group involves both descriptive and prescriptive components. The latter is inherent if a decision is to be made, but the former may be optional.

Cell A. Descriptive Work Involving a Coherent Situation. If the situation with which the group is working is coherent, then the group may or may not feel any real need to provide a formal description of that situation. Instead, as is often the case, the situation will be replaced with a surrogate called a "problem" or a "decision", the assumption being that the situation is so well understood that no formality is needed with regard to its comprehensive definition. Instead, a formal statement of the problem or of the decision to be made will often be taken as adequate.

Cell B. Prescriptive Work Involving a Coherent Situation. It seems that most of today's decision analysis and decision support is tailored to the domain of Cell B. The situation is generally taken as coherent, and the prescription consists of arriving at the particular decision to be adopted, through a process involving the use of data, numerical algorithms realized in computers, and discussion of the feasibility and approach to implementing the decision.

Work in this cell may often involve concepts from expert systems, wherein part of the presumption that the situation is coherent is that there exist experts whose knowledge (even if not yet articulated by them) can be extracted through expertise in probing, and reduced to formal information, whereupon it can be applied to decisionmaking.

INCOHERENT SITUATIONS

Incoherent situations are those for which the presumption is that the situation is not well understood by anybody. It is generally true that there is a sense of dissatisfaction with the situation, often accompanied by a clearly expressed sense of need for change, but not necessarily accompanied by a clear image of possible alternatives from which choices could be made.

Cell C. Descriptive Work Involving an Incoherent Situation It is often true that, even when a situation is incoherent, the desire of the group is to try to reach a decision or resolution through a "short-cut" route that bypasses the descriptive work whereby that situation might be adequately and comprehensively articulated. Perhaps the greatest hazard demonstrated empirically in many cases of failing to do the descriptive work is that this precludes the opportunity for critical appraisal of the relevant conceptualization of the situation, thereby preventing possible corrections of errors to be made at early stages. In other words, if people can't be informed about the nature of the situation, they are likely to make significant errors in their presumptions about its nature.

Cell D. Prescriptive Work Involving an Incoherent Situation Prescriptive work involving an incoherent situation is inherent in the decisionmaking process. The goal is to reach a decision about what to do to correct the dysfunctions perceived to be present in that situation. Unfortunately, because the situation is incoherent, the concept of decision support itself may be too narrow. Experience and habits gained in working with coherent situations tend to be automatically carried over, without question, leading to attempts to formulate the incoherent situation in terms of a particular decision to be made. It may well be advised to replace the concept of group decision support with the broader concept of group design support. The latter has three advantages: (a) it provides a broader framework within which to advance ideas, (b) it focuses the work on the production of a broad concept that may be implementable, rather than just one decision, and (c) if the situation really can be dealt with by just making a decision, the "design" in question can be reduced conceptually to the special case of a decision or a response to a decision question.

DEFECTIVE STRATEGIES

Strategies for carrying out group decision support that do not reflect consideration of the four cells described above are likely to be responsible for very extensive difficulties in arriving at good decisions or designs.

Perhaps the most evident source of difficulty is to aggregate all decision or design questions into one category and to develop the archival literature as though what is being offered is applicable to all situations.

Perhaps the second most evident source of difficulty is to mistakenly assume that a situation is coherent, when it is truly incoherent. The effect of this miscalculation is likely to be the same as the effect of lumping all situations into one aggregate category.

Even if the distinction is made between coherent and incoherent situations, and strategy is adjusted accordingly, it may be that the third most evident source of difficulty is to bypass the descriptive work needed to congeal adequately the understanding of the situation.

In any case, it is the incoherent situations where the most grievous mistakes get made, and where many important subtleties that relate to effective group work are ignored. Whether the situations are recognized as incoherent by the groups is largely irrelevant to this particular point, because the effects will be the same whether the situation is correctly categorized or not unless attention is given to these subtleties.

FRAMEBREAKING

Case studies suggest that the vast majority of decisions being made with respect to systems that are large in scale (and which are almost always incoherent) are bad. The reasons the decisions are mostly bad have been discussed under two headings: "underconceptualization" [7] and "presuppositions" [8]. Data from numerous workshops provide considerable insight into the origins of the bad decisions [7]. Usually the origins are not what people seem intuitively to think they are.

The initial point of attack is to break the frames of reference that furnish the information leading to bad decisions, for such frames are invariably too narrow and invariably contain bad information. Often they are based on generic misconceptions.

The consequence of overlooking the specific requirement of framebreaking is to erect a system of decisions founded in bad information.

METHODOLOGY FOR REMODELING

If the multiple frames that animate multiple decisionmakers are successfully broken, then remodeling is required to develop a new and higher-quality frame. If this frame is developed in a group process, the entire group may share a single frame. This has the great advantage that it will not be necessary to go through a new framebreaking exercise that would be required if the remodeling produced a new set of numerous different frames.

EFFICIENT REMODELING

There is a great need for efficiency in remodeling. This need goes well beyond the normal idea that it is good if things are efficient. Instead, efficiency is needed because groups of people typically are not willing, not able, or not interested in working together long enough to do the necessary remodeling, once a frame has been broken. To overcome this destructive posture, it is necessary to apply processes that are extremely efficient in carrying out the remodeling. This has been made possible by the development, test, and application of the "consensus methodologies" [4] that are a set of methodologies associated with the practice of "Interactive Management."[3]

By using inefficient processes, remodeling cannot succeed, and people are left possibly with a broken framework and nothing substantive to replace it.

PREDOMINANCE OF LOGIC CYCLES

The data [4,7] from numerous workshops on complex issues show very clearly that the logic of complex issues is literally awash in logic cycles. Because of the predominance of these cycles, one would think that in every situation involving complex decisions about complex issues, the identification, analysis, and interpretation of cycles would be a key feature, if not the primary feature of such studies.

On the contrary, most decisionmaking ignores the possibility that cycles might exist, displays no strategy for discovery of the cycles, offers no way to analyze them and interpret the larger logic in the light of the cycles and, in fact, proceeds merrily toward poorly-conceived outcomes in a responsibility-free posture of "what we don't know won't bother us".

CORRELATION OF BEFORE AND AFTER (MEASURING LEARNING)

When decisionmaking is largely keyed to predetermined quantitative algorithms, in effect most of the underlying framework has been taken as a given (in spite of the fact that it is usually wrong).

The possibility or significance of measuring logic frameworks and beliefs at the outset of a study, and doing the same thing at the end of a study (i.e., exploring the before and after views and structure of issues, the "extremes"), and then correlating these two extreme patterns of ideas is seldom considered and almost never done in working with complex issues.

Yet in numerous instances, there is essentially no correlation between these extreme views, illustrating that substantial learning has occurred enroute to the development of systems of decisions about issues [1].

Learning does not even seem to be construed as a major necessity for any process of decisionmaking that involves complex issues. Rather the common posture is that there is an expert lurking somewhere, and all that has to be done is to find that expert and articulate what the expert has, so far, left unarticulated.

TRIPLY-INTEGRATIVE PROCESSES

What accounts for the fact that most processes don't reflect any attention to the matters discussed above? A short answer is that the people who promote, advocate, or conduct such processes have a point of view which causes their thinking and their processes to reflect unidimensional reasoning. A longer answer is that "professional groupthink" permeates much of the technical literature that deals with decisionmaking, policy-setting, and related topics.

A more direct answer is that their processes lack the feature of being triply-integrative. Triply integrative processes integrate three things: (a) the anthropological, (b) the technological, and (c) the formal logical.

Most processes are based only in the anthropological, or only in the technological. Those that somehow merge these two can often be superior to the single-basis processes. However until the processes also integrate the formal logical components, as the evidence clearly shows, the outcomes cannot be expected to reflect adequate use of human knowledge. Even the integration must be done subject to a level of quality control that recognizes the depth of quality needed to get a suitable organic integration.

SUMMARY AND CONCLUSIONS

Decisionmaking can be disaggregated. It can be described in overview in terms of situations and postures. Situations may be coherent or incoherent. Postures may be descriptive or prescriptive.

An approach to group decision support that disregards this disaggregation and lumps all such work into a single category will not provide adequate focus or definition to the requirements for high-quality group work.

The tendency is to work with all situations as though they were coherent and only need to be dealt with prescriptively by experts who are provided with group decision support. Following this tendency will often lead to ignoring critically-important aspects of group work, and lead to low-quality, ineffective outcomes.

Today, the large-scale system or large-scale issue typically should be approached through group activity as though it were incoherent, requiring careful descriptive work followed by prescriptive activity. A better way to describe the support system for such work may be in terms of group design support rather than group decision support.

Because of the nature of the situations being dealt with at this point in history, one must not overlook the importance of a number of critical factors in group work. These include framebreaking, remodeling (efficiently), discovery and interpretation of logic cycles relevant to issues, correlation of group perceptions before and after passage through discovery processes, and the careful choice of triply-integrative processes for carrying out facilitated group activity.

The newly-developed science of generic design provides a sound and tested basis for dealing with these and other often-overlooked factors in group work.

REFERENCES

1. I. B. Kapelouzos, "The Impact of Structural Modeling on the Creation of New Perspectives in Problem Solving Situations", Proceedings of the 1989 European Congress on Systems Science, Lausanne: AFCET, October, 1989, 915-932.
2. J. N. Warfield, Societal Systems: Planning, Policy, and Complexity, New York: Wiley, 1976 (reprinted, Salinas, CA: Intersystems, 1989).
3. J. N. Warfield, "Principles of Interactive Management", Proceedings of the International Conference on Cybernetics and Society, New York: IEEE, January, 1984, 746-750.
4. J. N. Warfield, A Science of Generic Design: Managing Complexity Through Systems Design, Salinas, CA: Intersystems, 1990.
5. J. N. Warfield, Interpretive Structural Modeling and Related Work (Annotated Bibliography), Fairfax, VA: IASIS, George Mason University, 1990.
6. J. N. Warfield, Generic Systems Design and Interactive Management (Annotated Bibliography), Fairfax, VA: IASIS, George Mason University, 1990.
7. J. N. Warfield, "Underconceptualization", Systematica 1-6/8, 1990 (Edited by G. De Zeeuw and R. Glanville).
8. J. N. Warfield, "Presuppositions", in Cybernetics and Systems '90, Robert Trappl, Editor, Singapore: World Publishing, 1990.

MODELING PREFERENCE TRADE-OFFS
IN MULTIPLE OBJECTIVE LINEAR PROGRAMMING

Malgorzata Wiecek
Assistant Professor

Matthew L. TenHuisen
Graduate Student

Department of Mathematical Sciences
Clemson University
Clemson, South Carolina 29634-1907
U.S.A.

Abstract

The paper presents a methodology for solving multiple objective linear programming (MOLP) problems. The approach is based on compromise programming and explores the efficient set with respect to the class structure associated with ordering objectives according to their distance from the utopia point. The class structure and related trade-offs are analyzed in terms of changing the location of the utopia point in the objective space. An interactive procedure for the MOLP decision making process is proposed. The procedure allows the decision maker to concentrate only on those criteria whose current distance from the utopia point (or any reference point) is the farthest.

1. Introduction.

Early work in the field of multiple objective linear programming (MOLP) problems and the fundamental concept of efficient solutions may be found in Geoffrion (1968), Zeleny (1974a), and Yu and Zeleny(1975). A great variety of techniques for generating and exploring the efficient set has been developed along with many methods to support the decision maker (DM) in choosing an optimal solution.

One of the first techniques introduced for multiple criteria analysis utilizes the idea of a compromise solution developed by Zeleny (1973, 1974b, 1975) and studied later by several authors (Freimer and Yu, 1976; Yu and Leitmann, 1974; and others). In particular, an extensive theoretical study of the relationship between the efficient set and the compromise set was developed by Gearhart (1979). Ecker and Shoemaker (1981) published a unique, according to the authors, paper in which they introduced the concept of a trade-off compromise set as a subset of the efficient set. They also presented a method for generating the trade-off compromise set and indicated its interesting properties which gave valuable information about the trade-offs among the objectives.

The results of the theory and methods of compromise programming as well as some extensions have been included in recent extensive studies of Chankong and Haimes (1983), Sawaragi et. al. (1985), Yu (1985), and Steuer (1986).

Independently of theoretical research on the structure of the efficient set, work on interactive methods in Multiple Criteria Decision Making (MCDM) has been conducted. The literature in this area includes dozens of methods, many of them using the compromise programming approach. In fact, the STEP method (STEM), perhaps the first interactive procedure for MOLP, elaborated by Benayoun et. al. (1971), used the concept of an efficient solution which is nearest, in the minimax sense, to the utopia solution. Interactive MOLP methods are collected and classified in a survey paper of Shin and Ravindran (1987). The compromise programming still seems to attract researchers and contributes to new interactive procedures in MOLP (e.g. Jacquet-Lagrèze et. al., 1987).

This paper further develops the ideas of Ecker and Shoemaker mentioned above. The objective is to develop an interactive methodology that would help the decision maker (DM) concentrate on improving current values of some objective functions rather than all. The subset includes the criteria whose distance from the utopia point is the farthest and is referred to as the poor class. Fundamental definitions and concepts are introduced in Section 2. In Section 3 sensitivity analysis leads to some new results in predicting trade-offs among objective functions. Poor class properties are studied in Section 4. The subsequent section presents an interactive method for solving MOLP problems in which special interest is given to improving poor class criteria. The method is based on decision rules resulting from the preceding developments. A sample decision making process in Section 6 shows an application of the method. Finally, in Section 7 conclusions on results are presented.

2. Fundamental Concepts.

Our interest lies in the MOLP problem with $k \geq 2$ objective functions. The coefficients of each objective function are stored as a row vector c_i of a matrix C_{kxn} so that the problem can be expressed as

$$\text{MOLP:} \quad \max \; Cx \quad \text{s.t.} \; x \in X$$

where X is a convex polyhedron in R^m. Since there is usually not a unique solution $x^* \in X$ which is the optimal solution to the maximization of each of the objective functions independently, it is necessary to examine the efficient (nondominated, Pareto) solutions.

Definition 1. A point $x^0 \in X$ is called an efficient solution to MOLP if there is no other point $x \in X$ such that $c_i x \geq c_i x^0$ for $i=1,...,k$ with strict inequality holding for at least one i.

The set of efficient solutions, N, is a subset of the set of weakly efficient solutions, which is also of our interest.

Definition 2. A point $x^0 \in X$ is called a weakly efficient solution if there is no other point $x \in X$ such that $c_i x > c_i x^0$ for $i=1,...,k$.

Another basic concept in the MOLP theory is the utopia point (ideal vector) which was defined by Zeleny (1973).

Definition 3. The utopia point, M^*, is the vector such that $M_i^* = \max_{x \in X} c_i x, \; i=1,...,k.$

Given the utopia point a compromise solution is defined as follows (Zeleny (1973)).

Definition 4. A point $x^0 \in X$ is called a compromise solution with respect to the l_∞-norm if it is optimal for the program

$$Q(X): \quad \min_{x \in X} \max_{1 \leq i \leq k} (M_i^* - c_i x).$$

In general, a compromise solution is not necessarily an efficient solution. The formulation of the following program introduces the concept of an efficient compromise solution.

$$Q(N): \quad \min_{x \in N} \max_{1 \leq i \leq k} (M_i^* - c_i x).$$

Definition 5. The set of efficient compromise solutions, N_C, is the set of alternate optimal solutions to the linear program Q(N).

Programs Q(X) and Q(N) are not equivalent. It has been proven (Dinkelbach and Durr, 1972), however, that if x^* is a unique optimal solution of program Q(X), then x^* is also a unique optimal solution of program Q(N), and $N_C=\{x^*\}$. We are interested in the case when program Q(N) has multiple optima.

Another known result in the MOLP theory is that Q(X) is equivalent to the following linear program:

$$\begin{array}{llll} Q: & \min & w & \\ & \text{s.t.} & M_i^* - c_i x \le w & i=1,\dots,k \\ & & x \in X & \end{array}$$

where $w \in R^1$. If (w^0, x^0) is an optimal solution of program Q then x^0 is a weakly efficient solution of MOLP.

The k constraints of program Q of the form $M_i^* - c_i x \le w$ are of particular interest to us. All future references to objective function constraints pertain to these constraints.

The next concept introduced by Ecker and Shoemaker (1981) defines a subset of N.

Definition 6. Let d(x) denote the deviation vector where $d_i(x) = M_i^* - c_i x$, $i=1,\dots,k$. A point $\bar{x}$ is called a trade-off compromise point if $c_j x > c_j \bar{x}$ for some $x \in X$ implies that there is an index i such that $c_i x < c_i \bar{x}$ and $d_i(\bar{x}) \ge d_j(\bar{x})$.

It is clear from the definition that a trade-off compromise point is necessarily an efficient solution. In fact, Ecker and Shoemaker (1981) showed that if T denotes the set of all trade-off compromise points, then $N_C \supseteq T$.

Henceforth our primary interest is in the trade-off compromise set, T, which is a subset of the efficient set. Algorithm A given below was developed by Ecker and Shoemaker (1981) and generates the trade-off compromise set T after solving no more than k linear programs.

Algorithm A.

Step 0. Let $J_1=\{1,\dots,k\}$, $X^1=X$, $n=1$.

Step 1. Solve the linear program

$$\begin{array}{llll} Q_n: & \min & w & \\ & \text{s.t.} & M_i^* - c_i x \le w & i \in J_n \\ & & x \in X^n & \\ & & w \ge 0 & \end{array}$$

Let w^n be the minimal value.

Step 2. Let $X^{n+1}=\{x \in X^n: c_i x \ge M_i^* - w^n, i \in J_n\}$

Let $J_{n+1}=\{j \in J_n$: there exists $x \in X^{n+1}$ with $c_j x > M_j^* - w^n\}$

If $J_{n+1}=\emptyset$, stop. Otherwise, let $n=n+1$, and go to step 1.

The reader is asked to refer to the quoted paper for additional characteristics of Algorithm A. The analysis in the next section will reveal potential benefits of the algorithm and lead to the final results.

3. Sensitivity Analysis.

Algorithm A presented in the previous section provides a classification of the objective functions based on how close their values at termination of the algorithm are to their corresponding values of the utopia point. At each iteration of Algorithm A at least one of the objective function constraints becomes active. Those objective function constraints which became active at the first iteration of the algorithm are the farthest from their utopia point values and are, therefore, deemed as being poor. It is these poor objective function constraints that we are most interested in.

Consider the program Q_1 which is solved during the first iteration of Algorithm A. Q_1 as a linear program can be formulated as follows:

$$
\begin{aligned}
Q_1': \quad & \min \quad w \\
& \text{s.t.} \quad -c_i x - w \leq -M_i^* \quad i=1,\ldots,k \\
& \qquad\quad x \in X \\
& \qquad\quad w \geq 0 .
\end{aligned}
$$

Assume that the optimal solution of Q_1' is $x=(x_B^1, w^1, 0)$ and B is the optimal basis matrix. The optimal value of the objective function w^1 would then be

$$w^1 = [\,0 \mid 1\,] \begin{bmatrix} x_B \\ \hline w \end{bmatrix}. \tag{1}$$

The optimal dual solution of Q_1 is

$$\lambda^T = [\,0 \mid 1\,]\; B^{-1} . \tag{2}$$

Assuming nondegeneracy, we know from sensitivity analysis of linear programming (Bazaraa et. al., 1990) that small changes ΔM_i^*, $i \in J_1 \setminus J_2$, in the right-hand side (RHS) vector M^* will not cause the optimal basis to change, but consequently, will change the optimal solution to $x=(x_B^1+\Delta x_B^1, w^1+\Delta w^1, 0)$. Utilizing (1) and (2) we can see that the change ΔM_i, $i \in J_1 \setminus J_2$, will induce the increment of the objective function value

$$\Delta w^1 = \sum_{i \in J_1 \setminus J_2} \lambda_i \, \Delta M_i . \tag{3}$$

Assume now that all the utopia point coordinates stay unchanged but one. Then

$$\Delta w^1 = \lambda_i \, \Delta M_i . \tag{4}$$

Let Z_i be the value of the ith objective function; $Z_i = c_i x$. For the constraints which became active at the solution of Q_1', $i \in J_1 \setminus J_2$, we have $w^1 = M_i^* - c_i x^1$ which implies that

$$\Delta w^1 = \Delta M_i^* - \Delta Z_i \ . \tag{5}$$

Combining (4) and (5) we get

$$\Delta Z_i = \Delta M_i^*(1-\lambda_i) \ . \tag{6}$$

On the other hand, the Kuhn-Tucker conditions for optimality (Bazaraa et. al., 1990) applied to Q_1' yield

$$1 - \sum_{i=1}^{k} \lambda_i - \mu = 0, \qquad \lambda_i \geq 0,\ i=1,\ldots,k,\ \mu \geq 0 \tag{7}$$

where λ_i is the Lagrange multiplier associated with the ith objective function constraint, and μ is the Lagrange multiplier associated with the nonnegativity constraint for w. Recalling the fact that program Q(X) requires $w > 0$, we know that at optimality $\mu = 0$ in order to satisfy the Kuhn-Tucker conditions. Thus, the equation in (7) gives us

$$\sum_{i=1}^{k} \lambda_i = 1 \tag{8}$$

Given equations (4), (6), and (8) we also see that

$$\frac{\Delta Z_i}{\Delta w^1} \geq 1 \qquad \text{if } 0 \leq \lambda_i \leq \frac{1}{2} \tag{9a}$$

$$i \in J_1 \setminus J_2$$

$$\frac{\Delta Z_i}{\Delta w^1} \leq 1 \qquad \text{if } \frac{1}{2} \leq \lambda_i \leq 1 \ . \tag{9b}$$

Those objective functions which become active at each successive iteration are closer to their utopia point values than are those which became active previously. Thus, those objective function constraints which do not become active until the final iteration of the algorithm are the closest to their utopia point values and are, therefore, deemed as being rich. Those becoming active during the middle iteration(s) are classified as being (different levels of) middle class. Thus, performing Algorithm A leads to a classification of the objective functions, namely, ordering the objectives according to their distance (in the sense of the l_∞-norm) from the utopia point M*. Objectives with the same distance belong to the same class. The sensitivity analysis given above for program Q_1' can be applied at each of the successive iterations of Algorithm A, that is, to each class of the objectives.

4. Poor Class Properties.

Let Z(M) denote the vector of the objective function values for each point in the trade-off compromise set T(M) found for some $M \geq M^*$. We will refer to M as a reference point in the objective space. Given Z(M), the classification of the objective functions is available. Let PC(M) be the set of indices of the poor class objectives, that is $PC(M) = \{i: i \in J_1 \setminus J_2, i=1,...,P\}$ where $2 \leq P \leq k$.

Theorem 1. Let $[M^L, M^U]$ be the interval of allowable changes for M available through the RHS-sensitivity analysis performed for $Q_1(M)$. If $Q_1(M)$ has a nondegenerate solution, and if $M+\Delta M \in [M^L, M^U]$ then $PC(M)=PC(M+\Delta M)$.

Proof: Assuming nondegeneracy, if $M+\Delta M \in [M^L, M^U]$, then the optimal basis for the linear program Q_1 solved with M is also optimal for this program solved with $M+\Delta M$. Thus, both programs have the same optimal dual solution which yields $PC(M) = PC(M+\Delta M)$.

Theorem 1 shows how to change the location of the reference point in the objective space and protect the poor class structure at the same time.

Let I(M) be the set of poor class indices whose reference point values will be increased, that is, $I(M) = \{i \in PC(M): \Delta M_i > 0, 1 \leq i < P\}$. Theorem 2 below gives the possibility of predicting any changes of Z_i, $i \in PC(M)$, and satisfying a requirement of keeping the current poor class structure.

Theorem 2.

If $PC(M) = PC(M + \sum_{i \in I(M)} \Delta M_i)$ then

$$\Delta Z_i = \Delta M_i - \sum_{i \in I(M)} \lambda_i \Delta M_i \quad \text{for } i \in I(M), \tag{10a}$$

$$\Delta Z_j = - \sum_{i \in I(M)} \lambda_i \Delta M_i < 0 \quad \text{for } j \in PC(M),\ j \notin I(M), \tag{10b}$$

$$\Delta Z_j = -\Delta w^1 \quad \text{for } j \in PC(M),\ j \notin I(M), \tag{10c}$$

where λ_i is the optimal dual solution for $Q_1(M)$. In particular, if all ΔM_i are equal to each other, then $\Delta Z_i > 0$ for $i \in I(M)$.

Proof:

a) For $i \in PC(M)$ we have $M_i - Z_i = w^1$ and for $i \in PC(M+\Delta M)$ there is $(M_i+\Delta M_i) - (Z_i+\Delta Z_i) = w^1+\Delta w^1$. Thus $\Delta Z_i = \Delta M_i - \Delta w^1$, and using equation (3) we obtain $\Delta Z_i = \Delta M_i - \sum_{i \in I(M)} \lambda_i \Delta M_i$ for $i \in I(M)$. If all ΔM_i are equal to each other, then equation (8) yields $\Delta Z_i > 0$.

b) and c) For $j \in PC(M)$ we have $M_j - Z_j = w^1$. For $j \in PC(M)$ and $j \notin I(M)$ there is

$M_j - (Z_j+\Delta Z_j) = w^1+\Delta w^1$. Thus $\Delta Z_j = -\Delta w^1$, and again applying equation (3) we get

$$\Delta Z_j = -\sum_{i\in I(M)} \lambda_i \, \Delta M_i \quad \text{for } j\in PC(M),\ j\notin I(M), \text{ and obviously } \Delta Z_j<0.$$

5. An Interactive MOLP Method.

The theoretical results presented in the previous section allow for developing an interactive method for finding a solution to an MOLP problem that would be considered optimal by a DM. At each stage of the procedure a trade-off compromise set T(M) and the corresponding vector of current criteria values Z(M) are found. If these values are not accepted by the DM, decision rules are applied to predict any possible improvements of the objectives, especially of those in the poor class. According to this information a new reference point is chosen and the process moves on to the next stage.

The interactive method as well as the decision rules are given below. Rules A1-A4 refer to any objective function of any class while rules B1-B2 help get improvement only within the poor class.

INTERACTIVE METHOD

Step 0. Find M*, and let M=M*.
Step 1. Solve for Z(M), and apply RHS-sensitivity analysis.
Step 2. Present Z(M) to DM. If DM accepts Z(M), stop. Otherwise, go to step 3.
Step 3. Present PC(M) to DM. If DM wants to keep PC(M), apply decision rules A1-A4, and go to step 4. Otherwise, apply decision rules B1-B2, and go to step 4.
Step 4. Choose ΔM. Let $M=M+\Delta M$, and go to step 1.

DECISION RULES

A1: Given a trade-off compromise point, if the DM wishes to improve any given objective function, then he/she must be willing to accept a deterioration of at least one of the objective functions which is in the same or lower class(es) as the one for which the improvement is sought. If no such deterioration is tolerable, then no such improvement is possible.

A2: The change of the deviation w^n in iteration n of Algorithm A (n=1,...,N) is $\Delta w^n = \sum_s \lambda_s^n \, \Delta M_s$ if ΔM_s is in the allowable range of RHS change. Index s identifies the objective functions that belong to the criteria class established in iteration n, and λ_s^n is the optimal dual solution associated with the sth objective function constraint.

A3: The change of the current value of the sth objective function in iteration n, ΔZ_s^n, caused by an allowable change ΔM_s is $\Delta Z_s^n = (1-\lambda_s^n)\, \Delta M_s$, where s is the index of one of the objective functions that established the criteria class in iteration n, and λ_s^n is the optimal dual solution associated with the sth objective function constraint.

A4: The ratio $\frac{\Delta Z_s^n}{\Delta w^n}$ is approximated as $\frac{\Delta Z_s^n}{\Delta w^n} \begin{cases} \geq 1 & \text{if } 0 \leq \lambda_s^n \leq \frac{1}{2} \\ < 1 & \text{if } \frac{1}{2} < \lambda_s^n \leq 1 \end{cases}$ where the changes are caused by an allowable change ΔM_s of M_s, s and λ_s^n are defined as in rule A3, and $\Delta w^n = \lambda_s^n \, \Delta M_s$.

B1: Relocation of the reference point M within the allowable RHS-ranges available from the sensitivity analysis performed in the first iteration of algorithm A protects the poor class structure.

B2: Protecting the poor class structure allows for precise prediction of all improvements and/or deteriorations of the objectives in this class (according to(10a-c)).

At each stage of the interactive method a sequence of linear programs Q_n has to be solved. None of those stages, however, is redundant in the sense of learning about the existing efficient solutions since at every iteration of each stage of the method a weakly efficient solution of the MOLP problem is generated.

The method does not annoy the DM with difficult questions and allows for special examination of the poor class criteria. Thus, the DM can concentrate on a few criteria rather than all. This greatly facilitates the decision making process.

6. A Sample Decision Making Process.

In this section we solve a MOLP problem and apply the interactive procedure of the previous section. Thus, we simulate a decision making process that is ended once the DM finds an optimal solution.

Consider the MOLP given as

$$\begin{array}{ll} \max & Cx \\ \text{s.t} & Ax \leq b \\ & x \geq 0 \end{array}$$

where $C=\begin{bmatrix} 3 & -7 & 4 & 1 & 0 & -1 & -1 & 8 \\ 2 & 5 & 1 & -1 & 6 & 8 & 3 & -2 \\ 5 & -2 & 5 & 0 & 6 & 7 & 2 & 6 \\ 0 & 4 & -1 & -1 & -3 & 0 & 0 & 1 \\ 1 & 1 & 1 & 1 & 1 & 1 & 1 & 1 \end{bmatrix}$, $A=\begin{bmatrix} 8 & -12 & -3 & 4 & -1 & 0 & 0 & 0 \\ -3 & -4 & 8 & 2 & 3 & -4 & 5 & -1 \\ 12 & 8 & -1 & 4 & 0 & 1 & 1 & 0 \\ 15 & -6 & 13 & 1 & 0 & 0 & -1 & 1 \end{bmatrix}$, and $b=\begin{bmatrix} 30 \\ 100 \\ 40 \\ 100 \end{bmatrix}$.

The utopia point for this problem is $M^*=[1080, 847.6923, 1600, 150, 260]^T$. Solving Q_1 we find out from its final simplex tableau that this linear program has multiple optima. Letting $M=M^*$ and solving for T(M), we get the solution which results in the schematic below, where the column headed as "allowable increase" gives the maximum allowable increase in the corresponding component of M without forcing a change in the optimal basis. This information was obtained from the sensitivity analysis on the RHS vector.

$$Z = \begin{bmatrix} 223.9169 \\ \hline 1360.6897 \\ \hline 796.4987 \\ 564.1910 \\ -133.5013 \end{bmatrix}$$

objective	deviation	class	allowable increase
5	36.0831	rich	247.418
3	239.3103	middle	44.191
1			77.428
2	283.5013	poor	132.308
4			66.154

Suppose that the DM does not want to accept this solution. Say, for example, that he/she considers the value of Z_4 to be far too small, and therefore, wishes to increase it. In order to do so, a decrease in the value of at least either Z_1 or Z_2 must be acceptable since these two objective functions are not members of a higher class than is Z_4 (rule A1).

Granting that such decreases are acceptable, and in fact, agreeing that even changing the class structure is permissible in order to more greatly increase the value of Z_4, suppose that the DM requests the following alteration. Let $M=M^*+500e_4$ where e_4 is the identity vector with all components being zero except for the 4th component, which is one.

Such a large increase in the value of M_4 will greatly increase the value of Z_4. However, the allowable increase in M_4 without changing the optimal basis is only 66.154, so the new optimal basis will not be the same, and the classification of the objective functions is likely to change. In fact, solving for T(M) with this new M results in the following:

$$Z = \begin{bmatrix} 158.4912 \\ \hline 752.3077 \\ \hline 238.6392 \\ 990.9469 \\ 40.9469 \end{bmatrix}$$

objective	deviation	class	allowable increase
5	101.5088	rich	507.544
1	327.6923	middle	281.361
2			407.906
3	609.0531	poor	7.692
4			173.960

Notice that, indeed, the value of Z_4 is greatly improved. On the other hand, the values of all four of the other objective functions actually decreased. Note also that objectives 1 and 3 exchanged classes. Although objective 4 now appears to have a larger deviation from its current reference value, this new deviation of 609.05306 is with respect to the altered value of M_4+500. The current deviation of Z_4 from M_4^* is now only 109.05306 which is an improvement from the first solution.

Suppose next that, although this solution is more favorable to the DM than the first one, he/she would like to increase the value of Z_4 even more. This time, however, the DM wishes to retain the present class structure. Assume also, that the DM is willing to allow further decrease in the value of Z_3 but would like to maintain a value for Z_2 as high as possible. In order to accommodate these wishes we must examine the Lagrange multipliers obtained in iteration 1 of Algorithm A since it was during this iteration that the poor class was established. This vector of Lagrange multipliers is $\lambda^1=[0, 0.3102, 0.0231, 0.6667, 0]^T$.

From the allowable increases obtained from the sensitivity analysis on the last problem we see that if M_2 and M_4 are both increased by 150, the optimal basis will not change. From decision rule B2 we know that such a relocation of the reference point would result in the following changes:

$$\Delta Z_2 = \Delta M_2 - (\lambda_2 \Delta M_2 + \lambda_4 \Delta M_4) = 3.47$$
$$\Delta Z_4 = \Delta M_4 - (\lambda_2 \Delta M_2 + \lambda_4 \Delta M_4) = 3.47$$
$$\Delta Z_3 = - (\lambda_2 \Delta M_2 + \lambda_4 \Delta M_4) = -146.53$$
$$\Delta w^1 = -\Delta Z_3 = 146.53$$

Assuming that these projected changes are favorable to the DM, the value of M would then be changed to $M=M+150e_2+150e_4=M^*+150e_2+650e_4$. The results obtained from solving this problem using Algorithm A are given below.

Z =	objective	deviation	class	allowable increase
134.0687	5	125.9313	rich	629.656
602.3077	1	477.6923	middle	277.896
242.1046	2			402.882
844.4123	3	755.588	poor	157.692
44.4123	4			101.188

The new values of Z_2, Z_3, and Z_4 indeed match the calculations obtained by applying (10a-c) according to rule B2. The current vector λ^1 of Lagrange multipliers after the first iteration is the same as the one obtained with the previous value of M.

Observe that if the DM wishes to further increase Z_4 at the expense of Z_3 without deteriorating the value of Z_2, setting $\Delta M_2=\Delta M_4=100$ would achieve this goal. The sensitivity analysis performed on the RHS allows us to increase M_4 by 100 more at this stage without changing the basis even though we increased M_4 by almost the maximum allowable amount the last time.

Suppose, instead, that at this point the DM has decided to allow a slight decrease in the value of Z_2 as well as Z_3 in order to more effectively increase the value of Z_4. A change of basis (and the class structure) is again not permitted.

The greatest such improvement can be accomplished by letting ΔM_4 approach its maximum allowable change of 101.188 without changing any of the other components of M. Since $\lambda^1=[0, 0.3102, 0.0231, 0.6667, 0]^T$, letting $M=M+100e_4=M^*+150e_2+750e_4$ should result in the following changes according to decision rules A3, A4, and B2:

$$\Delta Z_4 = (1-\lambda_4)\Delta M_4 = 33.33$$

$$\frac{\Delta Z_4}{\Delta w^1} < 1$$

$$\Delta Z_2 = \Delta Z_3 = -\lambda_4 \Delta M_4 = -66.67$$

$$\Delta w^1 = -\Delta Z_2 = 66.67$$

The complete table for the results of this problem are given below:

Z	objective	deviation	class	allowable increase
122.9576	5	137.0424	rich	685.212
602.3077	1	477.6923	middle	344.562
175.4379	2			499.533
777.7456	3	822.2544	poor	157.692
77.7456	4			1.188

At this point, the DM sees that there is very little improvement which can be made to Z_4 while keeping the current poor class structure, so he/she, let us say, accepts this solution as the optimal solution to the MOLP problem.

Conclusions

In this paper the concept of the trade-off compromise set for MOLP has been further studied. The sensitivity analysis and Kuhn-Tucker conditions applied to the special sequence of linear programs allowed for developing the interactive MOLP method that ranks criteria functions with respect to the distance (in the sense of the l_∞-norm) from any reference point M, ($M \geq M^*$).

The poor class properties presented in Section 4, except for Theorem 1, cannot be easily transferred to any higher class. This is a consequence of the fact that if constraint i becomes active in iteration n of Algorithm A, then an allowable change of M_i in this iteration will not affect the structure of the criteria class established in this iteration. Such a relocation of the reference point, however, will not bring any valuable information about the trade-offs among the objectives. More interesting would be predicting changes caused by RHS perturbations in higher classes after a trade-off compromise set is found. We intend to investigate "sequential" sensitivity properties preserved through iterations and stages of the interactive method in our forthcoming research.

Acknowledgement

This research was supported in part by Grant No 60NANB0D1023 from the Center for Fire Research, U.S. National Institute of Standards and Technology, Gaithersburg, Maryland, U.S.A.

References

Bazaraa, M. S., Jarvis, J. J., and Sherali, H. D., Linear Programming and Network Flows, John Wiley and Sons, New York, 1990.

Benayoun, R. J., de Montgolfier, J., Tergny, J. and Larichev, O., "Linear Programming with Multiple Objective Functions: STEP Method (STEM)", Mathematical Programming, Vol. 1, pp. 366-375, 1971.

Chankong, V. and Haimes, Y. Y., Multiobjective Decision Making - Theory and Methodology, North-Holland, New York, 1983.

Dinkelbach,W. and Durr, W., "Effizienzaussagen bei Ersatzprogrammen zum Vektormaximumproblem", Multiple Criteria Decision Making, Operations Research Verfabren, H. P. Kunzi and H. Shubert, eds., Verlag Anton Hain, Meisenheim, Germany, pp. 117-123, 1972.

Ecker, J. G. and Shoemaker, N. E., "Selecting Subsets from the Set of Nondominated Vectors in Multiple Objective Linear Programming", SIAM Journal of Control and Optimization, Vol. 19, No. 4, pp. 505-515, 1981.

Freimer, M. and Yu, P. L., "Some New Results on Compromise Solutions for Group Decision Problems", Management Science, Vol. 22, pp. 688-693, 1976.

Gearhart, W. B., "Compromise Solutions and Estimation of the Noninferior Set", Journal of Optimization Theory and Application, Vol. 28, pp. 29-47, 1979.

Geoffrion, A. M., "Proper Efficiency and Theory of Vector Maximization", Journal of Mathematical Analysis and Applications, Vol. 22, pp. 618-630, 1968.

Jacquet-Lagrèze, E., Meziani, R. and Slowinski, R., "MOLP with an Interactive Assessment of a Piecewise Linear Utility Function", European Journal of Operational Research, Vol. 31, pp. 350-357, 1987.

Sawaragi, Y., Nakayama, H. and Tanino, T., Theory of Multiobjective Optimization, Academic Press, Orlando, Florida, 1985.

Shin, W. S. and Ravindran, A., "Interactive Multiple Objective Mathematical Programming (MOMP) Methods: A Survey", University of Oklahoma, School of Industrial Engineering, Working Paper No. 87-15, 1987.

Steuer, R., Multiple Criteria Optimization: Theory, Computation, and Application, John Wiley, New York, 1986.

Yu, P. L., Multiple-Criteria Decision Making, Plenum Press, New York, 1985.

Yu, P. L. and Leitmann, G., "Compromise Solutions, Domination Structures and Salukvadze's Solution", Journal of Optimization Theory and Applications, Vol. 13, pp. 362-378, 1974.

Yu, P. L. and Zeleny, M., "The Set of All Nondominated Solutions in Linear Cases and a Multicriteria Simplex Method", Journal of Mathematical Analysis and Applications, Vol. 49, pp. 430-468, 1975.

Zeleny, M., "Compromise Programming", Multiple Criteria Decision Making, J. L. Cochrane and M. Zeleny, eds., University of South Carolina Press, Columbia, SC, pp. 262-301, 1973.

Zeleny, M., Linear Multiobjective Programming, Springer Verlag, Berlin/Heidelberg, 1974a.

Zeleny, M., "A Concept of Compromise Solutions and the Method of the Displaced Ideal", Computers and Operations Research, Vol. 1, pp. 479-496, 1974b.

Zeleny, M., "The Theory of the Displaced Ideal", Multiple Criteria Decision Making: Kyoto 1975, M. Zeleny, ed., Springer Verlag, New York, pp. 151-205, 1975.

COMPETENCE SET ANALYSIS AND EFFECTIVE DECISION SUPPORT SYSTEMS

Po L. Yu

Carl A. Scupin Distinguished Professor,
School of Business, University of Kansas,
Lawrence, Kansas 66045, U.S.A.

Dazhi Zhang

Assistant Professor,
Hagan School of Business, Machine Intelligence Institute,
Iona College, New Rochelle, New York 10801, U.S.A.

Shude Huang

Associate professor
Dept. of Management Science, Jiaozuo Mining Institute,
Jiaozuo, Henan Province, China

Abstract: In this article, we study how to expand one's competence set as to cover the truly needed competence set when it is discrete and finite. Minimal spanning tree concept will be employed to construct optimal expansion processes. Various optimality concepts of expansion processes will be discussed. These include the lexicographical optimality, the next best optimality and the total cost optimality. An algorithm for finding optimal expansion processes is provided and an example is discussed to illustrate the applicability of the proposed framework. Finally, effective decision support systems based on competence set analysis will be addressed.

1. INTRODUCTION

For each problem, denoted by E, there is a *competence set* consisting of ideas, knowledge, information and skills for its satisfactory solution. When the decision maker has actually acquired and mastered the competence set, he/she will be able to make the decision and solve the problem.

For example, buying a house is an important decision problem to most people. We usually take time and effort to recall our own experiences, talk to our friends, visit brokers and look at advertisements in order to expand our competence set, which includes identifying preferences, digesting market information, making financial arrangement, knowing available houses and their quality, etc. We will not be ready to make a purchase until our competence set becomes large enough. Therefore, the expansion of competence sets plays an important role in the process and quality of decision making. From a decision consultant's (including computer expert systems) point of view, how do we help the decision maker effectively expand his/her competence set as to make good decisions will remain a major challenge to our success.

The goal of *competence set analysis* is to identify the true competence set, the decision maker's competence set, and help the decision maker to effectively expand his/her competence set to make good decisions. The main concept is rooted in set covering and set expansion, in contrast to the traditional numerical ordering which is used to express the objective functions or constraints.

We shall proceed as follows. In Section 2, we discuss cost functions which measure the costs needed in the expansion process. Minimal spanning tree method will be applied to construct an algorithm for finding optimal expansion processes of competence sets. Lexicographical optimality and next best optimality are closely related to the expansion processes based on minimal spanning tree. In Section 3, we discuss effective decision support systems for the proposed framework.

2. OPTIMAL EXPANSIONS

2.1. Introduction

In this section, we study optimal expansion processes. We shall proceed as follows. In Section 2.2, we shall discuss two concepts of cost functions. These are the cost functions among elements and the domains-to-elements cost functions. We then define expansion processes and the related costs in terms of graph concepts in Section 2.3. Finally, in Section 2.4, we discuss various optimality concepts of expansion processes based on the minimal spanning tree concept. These include the lexicographical optimality, the next best optimality and the total cost optimality. An algorithm for finding optimal expansion processes will be provided.

Before we discuss further, let us define the following notation and introduce measurable structures for competence sets.

Given decision problem $\boldsymbol{E}$. We shall use $\mathrm{Tr}(\boldsymbol{E})$ to denote the *truly needed competence set*, which consists of ideas and skills which are truly needed for solving problem $\boldsymbol{E}$. We shall use $\mathrm{Sk}(\boldsymbol{E})$ to denote the decision maker's *acquired skill set*, which consists of the ideas and skills which have been actually acquired by the decision maker. When it is not necessary to specify which competence set (the truly needed competence set or the acquired skill set) is under discussion, notation $\mathrm{Comp}(\boldsymbol{E})$ will be used. When there is no confusion, $\mathrm{Tr}(\boldsymbol{E})$ and $\mathrm{Sk}(\boldsymbol{E})$ will be abbreviated as Tr and Sk respectively.

Assume that the competence set Comp($\boldsymbol{E}$) is known to the experts (it may still be unknown to the less proficient clients or users). Let HD be the discussion universe (potential habitual domain) containing Comp($\boldsymbol{E}$). Then given a set of ideas S, its complement relative to HD can be defined; so is the union and intersection of a countable collection of idea sets $\{S_i\}$. This observation suggests that we could define a *σ-algebra* for HD (or an *algebra* if desired. This makes no difference when HD is discrete and finite). Thus, like real analysis in mathematics, HD may be treated as a *measurable space*

$$\text{HD} = (\text{HD}, \sigma(\text{HD}), \mu),$$

where σ(HD) is a σ-algebra on HD and μ is a meaningful *measure.*

2.2. Cost functions

Recall that HD denotes the discussion universe, i.e., a set consists of all ideas, skills, information, etc., relevant to our discussion. Assume that any idea in HD can be reached from any other one with a finite cost. We can give the following:

DEFINITION 2.2.1 Let m be a real-valued function defined on the Cartesian product HD×HD which satisfies

(i) (Non-negativity) $m(a,b) \geq 0$, $m(a,b) = 0$ iff $a=b$;

(ii) (Triangle Inequality) $m(a,c) \leq m(a,b) + m(b,c)$ $\forall a,b,c \in$ HD.

Then m is called a *cost* or *money-consumption function among elements* and m(a,b) is called the *cost needed to reach idea* b *from idea* a.

The cost function defined above measures the cost needed for reaching one idea from another. If we want to stress that the cost needed in reaching an idea is dependent on the current skill set Sk($\boldsymbol{E}$) (the larger the set Sk($\boldsymbol{E}$), the less cost is needed), then we use the following.

DEFINITION 2.2.2 Let HD be a measurable space (HD,σ(HD),μ). If mapping M: $\sigma(\text{HD}) \times \text{HD} \rightarrow R^1$ satisfies

(i) (non-negativity) $M \geq 0$;

(ii) $M(A,x) = 0$ if and only if $x \in A$;

(iii) (monotonicity) $M(A,x) \geq M(B,x)$ $\forall x \in$ HD, if $A \subseteq B$,

then M is called a *domains-to-elements cost function* and M(A,x) is called the *cost needed to reach idea x given current actual skill set A.*

A special class of domains-to-elements cost functions are of interests. This is what we now define.

DEFINITION 2.2.3 A domains-to-elements cost function M is called *regular* if for any A in σ(HD) and x and y in HD,

$$M(A,y) \le M(A,x) + M(\cap\{B | B \in \sigma(HD), x \in B\}, y).$$

Concerning the relationship between the two types of cost functions defined, we have the following result (Yu and Zhang 1990c).

PROPOSITION 2.2.1 Assume that HD is finite and σ(HD) is the power set of HD. Let m be a domains-to-elements cost function. Then M as defined by

$$M(A,x) = \min\{m(a,x) \mid a \in A\}, \ \forall A \in \sigma(HD), \ x \in HD$$

is a regular domains-to-elements cost function. On the other hand, if M is a regular domains-to-elements cost function, then function m defined by

$$m(a,b) = M(\{a\},b) \ \forall a,b \in HD$$

is a cost function among elements. □

2.3. Expansion Processes and Expansion Costs

In this section, we define expansion processes and expansion costs based on the concepts of graphs. Let us first introduce the following terminologies which are typical in the theory of graphs. See, for example, Hillier and Lieberman (1980).

A *graph* consists of a set of junction points which are called *nodes*, with certain pairs of nodes being joined by lines, called *branches*. A *network* is defined to be a graph with a flow of some type in its branches. A *chain* between two nodes is a sequence of branches connecting these two nodes. A chain is called a *path* if the direction of travel along the chain is specified. A *circle* is a chain connecting a node to itself. A graph is *connected* if there is a chain connecting every pair of nodes. A *tree* is a connected graph containing no circles. Given a set of nodes, a *spanning tree* is a tree that connects all the given nodes.

The *minimal spanning tree problem*, a variation of the so-called *shortest route problem*, is well known in network analysis. The problem can be described as given a set of nodes and the distances between pairs of nodes, how to choose the branches for the network so that the network has the shortest total length while providing a route between each pair of nodes.

The algorithm for minimal spanning tree problem is straightforward:

(1) Select any node arbitrarily, then connect it to the nearest distinct node.
(2) Identify the unconnected node that is closest to a connected node, and then connect these two nodes.
(3) Repeat (2) until all nodes have been connected.

Let us now assume that (i) once an idea is reached, it belongs to and remains in the skill set Sk(***E***). (ii) HD is discrete and finite. and (iii) σ(HD) = $\mathcal{P}$(HD), the power set of HD.

Note that Tr(***E***)\Sk(***E***) is the unacquired yet needed competence set.

DEFINITION 2.3.1 By an *expansion process*, we mean a path Ψ which spans Tr(***E***)\Sk(***E***) and contains no circles when regarded as a graph.

REMARK 2.3.1 Assume that Tr(***E***)\Sk(***E***) = $\{x_1, x_2, ..., x_n\}$. Then an expansion process can be expressed as $(x_{k_1}, x_{k_2}, ..., x_{k_n})$, an arrangement of $\{x_1, x_2, ..., x_n\}$.

For the convenience of presentation, for any expansion process $\Psi = (x_{k_1}, x_{k_2}, ..., x_{k_n})$, let us denote $Sk_0(\Psi)$=Sk(***E***), and $Sk_i(\Psi) = Sk_{i-1}(\Psi) \cup \{x_{k_i}\}$ (i = 1,2,...,n). Set $Sk_i(\Psi)$ is called the *i-th step skill set by expansion process* Ψ.

DEFINITION 2.3.2 Assume that Tr(***E***)\Sk(***E***) = $\{x_1, x_2, ..., x_n\}$. Let $\Psi = (x_{k_1}, x_{k_2}, ..., x_{k_n})$ be an expansion process. Assume that a domains-to-elements cost function M is given. Then

$$C_i(\Psi) = M(Sk_{i-1}(\Psi), x_{ki})$$

is called the *cost needed for acquiring* x_{ki} or the *i-th step expansion cost*; and

$$TC(\Psi) = C_1(\Psi) + C_2(\Psi) + ... + C_n(\Psi)$$

is called the *total cost needed in the expansion process* Ψ.

DEFINITION 2.3.3 If expansion process Ψ minimizes the total cost needed for expansion, then we call Ψ a *total cost optimal expansion process.*

2.4. Lexicographical Optimality and the Next Best Method

In this subsection, we shall discuss how to find a total cost optimal expansion process. The method is the next best rule which we now define. Recall that HD is the discussion universe.

DEFINITION 2.4.1 Assume that HD $\supseteq$ Tr(***E***)$\cup$Sk(***E***), Tr(***E***)\Sk(***E***) = $\{x_1, x_2, ..., x_n\}$ and that the cost function m is given on HD. Expansion process $\Psi = (x_{k_1}, x_{k_2}, ..., x_{k_n})$ is called a *next best expansion process* if x_{k_i}, the i-th element to acquire, is the element in Tr(***E***)\[Sk(***E***)$\cup\{x_{k_1}, x_{k_2}, ..., x_{k_{i-1}}\}$] which minimizes the i-th step expansion cost. That is,

$$M(Sk\cup\{x_{k_1}, x_{k_2}, ..., x_{k_{i-1}}\}, x_{k_i})$$
$$= \min\{M(Sk\cup\{x_{k_1},x_{k_2},...,x_{k_{i-1}}\},x) \mid x \in Tr\backslash(Sk\cup\{x_{k_1},x_{k_2},...,x_{k_{i-1}}\})\},$$
$$i = 1, 2, ..., n.$$

where

$$M(A,x) = \min \{m(s,x) \mid s \in A\}, \; \forall A \subseteq HD.$$

If Ψ is a next best expansion process, we shall say that Ψ is *next best optimal.*

The next best optimality of an expansion process is closely related to the lexicographical optimality which we now define. For the convenience of the reader, let us first introduce the lexicographical preference in a Euclidean space $\mathbb{R}^n$.

Let $y = (y_1, y_2, ..., y_n)$ be indexed so that the k-th component is overwhelmingly more important than the (k+1)-th component for k = 1, 2, ..., n-1. A *lexicographic ordering preference* is defined as follows: the outcome $y^1 = (y_1^1, y_2^1, ..., y_n^1)$ is preferred to $y^2 = (y_1^2, y_2^2, ..., y_n^2)$ if and only if $y_1^1 > y_1^2$ or there is some k so that $y_k^1 > y_k^2$ and $y_j^1 = y_j^2$ for j = 1, 2, ..., k-1.

Recall that $C_i(\Psi)$ is the i-th step expansion cost for process Ψ.

DEFINITION 2.4.2 Assume that the cost function m is given. Denote by $\mathbb{W}$ the set of all expansion processes. Define mapping C: $\mathbb{W} \rightarrow \mathbb{R}^n$ by

$$C(\Psi) = (C_1(\Psi), C_2(\Psi), ..., C_n(\Psi)), \; \forall \Psi \in \mathbb{W}.$$

Define binary relation $\succ$ on $\mathbb{W}$ by:

$$\Psi \succ \Psi' \text{ iff } C(\Psi') \text{ is lexicographically preferred to } C(\Psi).$$

If Ψ^* is not dominated by any other Ψ according to "$\succ$" (i.e., there is no Ψ such that $C(\Psi^*)$ is lexicographically preferred to $C(\Psi)$), then we say that Ψ^* is a *lexicographically optimal expansion process.* For simplicity, "lexico." will be used for "lexicographical" and "lexicographically".

The lexico. preference ordering on the set of expansion processes implies that the i-th step expansion cost is overwhelmingly more important than the (i+1)-th step expansion cost. Thus at each time point of the expansion process, we may only look at the current step of the expansion and need not to worry about the future. This is very similar to the idea of next best optimal. In fact, we have the following result (Yu and Zhang 1990c).

PROPOSITION 2.4.1 If $\Psi^* = (x_{k_1}, x_{k_2}, ..., x_{k_n})$ is a lexico. optimal expansion process, then it is also a next best expansion process. □

We are interested in the expansion process which gives the minimal *total* expansion cost. This is the total cost optimal expansion process defined by Definition 2.3.3. How is this optimality concept related to the lexico. optimality and next best optimality? This is answered by the following theorem (Yu and Zhang 1990c).

THEOREM 2.4.1 If expansion process Ψ is lexico. optimal or next best optimal, then Ψ is also total cost optimal. That is, process Ψ minimizes the total expansion costs. □

The results of Proposition 2.4.1 and Theorem 2.4.1 can be summarized by the following figure:

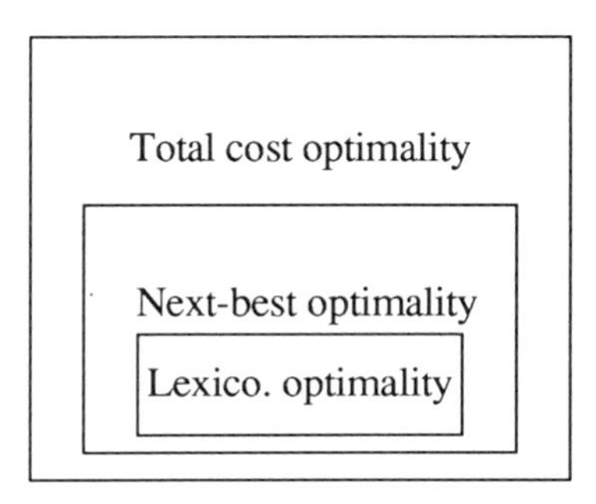

Fig. 1. Relationship among various optimality concepts

To give an algorithm for finding optimal expansion processes, we need the following.

DEFINITION 2.4.3 (1) Given any expansion process $\Psi = (x_{k_1}, x_{k_2}, ..., x_{k_n})$. Define a graph $\Gamma(\Psi)$ as follows: the set of junction points G consists of n+1 elements: $G = \{x_1, x_2, ..., x_n\} \cup \{\Delta\}$ where Δ is an aggregate node representing Sk; the branches of the graph $\Gamma(\Psi)$ are obtained according to the following rules: Let s_{0_i} be the element (if there are several alternatives, choose one arbitrarily) such that

$$m(s_{0_i}, x_{k_i}) = \min\{m(s, x_{k_i}) \mid s \in Sk \cup \{x_{k_1}, x_{k_2}, ..., x_{k_{i-1}}\}\},$$
$$i = 1, 2, ..., n.$$

(i) If $s_{0_i} \in Sk$, then (Δ, x_{k_i}) is a branch of the graph;

(ii) If $s_{0_i} \in \{x_{k_1}, x_{k_2}, ..., x_{k_{i-1}}\}$, then (s_{0_i}, x_{k_i}) is a branch of the graph;

(iii) Graph $\Gamma(\Psi)$ consists of only these branches.

(2) Set $G = (Tr\backslash Sk) \cup \{\Delta\}$. Define mapping $h: G \times G \to \mathbb{R}^1$ by:

$$h(x,y) = \begin{cases} m(x,y), & \text{if } x, y \in Tr\backslash Sk; \\ 0, & \text{if } x=y=\Delta; \\ \min\{m(x,s) \mid s \in Sk\}, & \text{if } x \in Tr\backslash Sk,\ y=\Delta; \\ \min\{m(s,y) \mid s \in Sk\}, & \text{if } x=\Delta,\ y \in Tr\backslash Sk. \end{cases}$$

h(x,y) is called the *length of branch* (x,y).

The following have been derived (Yu and Zhang 1990c):

THEOREM 2.4.2 (i) Graph $\Gamma(\Psi)$ is a spanning tree for G. (Thus we can call $\Gamma(\Psi)$ a *spanning tree generated by the expansion process* Ψ.)

(ii) An expansion process Ψ is total cost optimal (minimal total cost) if and only if *there is* a generated spanning tree $\Gamma(\Psi)$ which is a minimal spanning tree, which holds if and only if *any* generated tree is a minimal spanning tree.

Finally, we have the following general algorithm for finding alternative optimal expansion processes (Yu and Zhang 1990c).

THEOREM 2.4.3 (*Algorithm for finding total cost optimal expansion processes*) Expansion process $(x_{k_1}, x_{k_2}, ..., x_{k_n})$ obtained in the following way is total cost optimal.

1. Specify a next best expansion process Ψ.
2. Construct a generated spanning tree $\Gamma(\Psi)$ of the next best expansion process Ψ (which is a minimal spanning tree on $\{x_1, x_2, ..., x_n\} \cup \{\Delta\}$).
3. Determine $(x_{k_1}, x_{k_2}, ..., x_{k_n})$ as follows:
   (i) The first element x_{k_1} is any node which is directly connected to node Δ according to the generated spanning tree $\Gamma(\Psi)$;
   (ii) If $x_{k_1}, x_{k_2}, ...,$ and x_{k_i} have been specified, then $x_{k_{i+1}}$ is any node which is directly connected to a node in $\{x_{k_1}, x_{k_2}, ..., x_{k_i}, \Delta\}$ according to the generated spanning tree $\Gamma(\Psi)$;
   (iii) Repeat (ii) until all nodes have been chosen. □

EXAMPLE 2.4.1 Consider the decision to buy a house (***E***). Assume that the truly needed set Tr(***E***) = $\{a, b, c, d, x_1, x_2, x_3, x_4, x_5, x_6\}$ where

a = preferences
b = appearance
c = interior design
d = convenience
x_1 = prices
x_2 = tax amount
x_3 = financial arrangement
x_4 = resale value
x_5 = neighborhood
x_6 = insulation quality

Assume that the decision maker has acquired Sk(***E***) = {a, b, c, d}. Thus the unacquired yet needed skill set is Tr(***E***)\Sk(***E***) = $\{x_1, x_2, x_3, x_4, x_5, x_6\}$. Since the decision maker has not yet mastered the competence set, he is not ready and may not be able to make the decision and solve the problem (buy a house). He thus needs to expand his skill set Sk(***E***). He can either learn by himself or hire someone (an agent) to do it for him. Assume that the symmetric cost function m(x,y) has been assessed as follows:

	a	b	c	d	x_1	x_2	x_3	x_4	x_5	x_6
a	0	3	3	3	4	6	9	11	11	13
b	3	0	2	4	6	9	11	14	14	16
c	3	2	0	2	6	8	11	13	14	15
d	3	4	2	0	5	6	9	11	12	13
x_1	4	6	6	5	0	4	6	9	8	10
x_2	6	9	8	6	4	0	3	5	6	7
x_3	9	11	11	9	6	3	0	3	3	5
x_4	11	14	13	11	9	5	3	0	4	2
x_5	11	14	14	12	8	6	3	4	0	5
x_6	13	16	15	13	10	7	5	2	5	0

(1) Let us first specify the next best expansion process. The first element to acquire is the one which minimizes the distance to the skill set. Note that function M(Sk,x) is given as follows:

x	x_1	x_2	x_3	x_4	x_5	x_6
M(Sk,x)	4	6	9	11	11	13

Thus x_1 is the first element to acquire.

Now construct function $M(Sk \cup \{x_1\}, x)$:

x	x_2	x_3	x_4	x_5	x_6
$M(Sk \cup \{x_1\},x)$	4	6	9	8	10

Therefore we can choose x_2 as the second element to acquire. In this way, we get a next best expansion process $\Psi_0 = (x_1,x_2,x_3,x_4,x_6,x_5)$.

(2) Construct a minimal spanning tree generated by this next best expansion process $\Psi_0 = (x_1,x_2,x_3,x_4,x_6,x_5)$: x_1 is connected to Δ, x_2 is connected to x_1, x_3 is connected to x_2, x_4 is connected to x_3, x_5 is connected to x_3, and x_6 is connected to x_4. This minimal spanning tree is shown as follows:

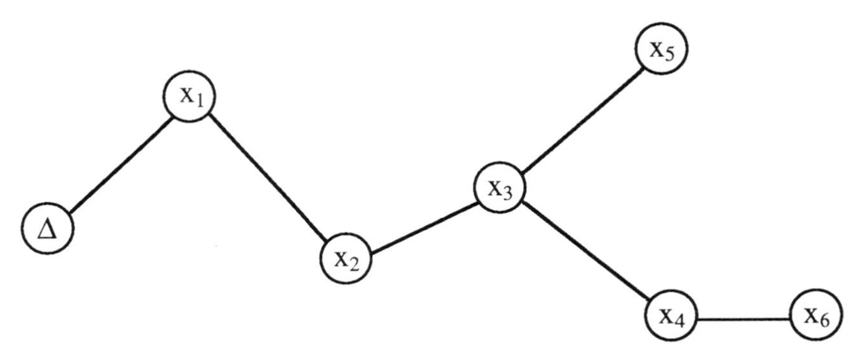

Fig. 2. Minimal spanning tree for Example 2.4.1

(3) The following are total cost optimal expansion processes:

$$\Psi_0 = (x_1,x_2,x_3,x_4,x_6,x_5),$$
$$\Psi_1 = (x_1,x_2,x_3,x_5,x_4,x_6),$$
$$\Psi_2 = (x_1,x_2,x_3,x_4,x_5,x_6).$$

Note that the total cost for these total cost optimal expansion processes is 4+4+3+3+2+3 = 19. It is readily checked that the following are expansion processes with respect to different optimality concepts:

Lexico. Optimal:

$$\Psi_0 = (x_1,x_2,x_3,x_4,x_6,x_5);$$

Next Best Optimal:

$$\Psi_0 = (x_1,x_2,x_3,x_4,x_6,x_5),$$
$$\Psi_1 = (x_1,x_2,x_3,x_5,x_4,x_6);$$

Total Cost Optimal:

$$\Psi_0 = (x_1,x_2,x_3,x_4,x_6,x_5),$$
$$\Psi_1 = (x_1,x_2,x_3,x_5,x_4,x_6),$$
$$\Psi_2 = (x_1,x_2,x_3,x_4,x_5,x_6).$$

Finally, note that the lexico. optimal expansion process $\Psi_0 = (x_1, x_2, x_3, x_4, x_6, x_5)$ means that the decision maker should take the following steps in his acquiring and learning: "prices" $\rightarrow$ "tax amount" $\rightarrow$ "financial arrangement" $\rightarrow$ "resale value" $\rightarrow$ "insulation quality" $\rightarrow$ "neighborhood". Following this process, his total cost will be 19. He cannot do better.

3. EFFECTIVE DECISION SUPPORT SYSTEMS

In order to effectively assist and support the decision maker to reach good decisions, the support system should be able to clearly identify (1) the problem that the decision maker is facing with, (2) the true competence set for solving the problem, (3) the skill set that the decision maker has acquired, and (4) effective plans that can help the decision maker to expand his/her skill set. Let us address these interrelated problems briefly as follows.

(1) Identify the Decision Problem

(i) Decision environments

A. A part of the decision maker's behavior mechanism,
B. Stages of decision process,
C. Players in the decision processes,
D. Unknowns in decision processes.

(ii) Decision elements

A. Decision alternatives,
B. Decision criteria,
C. Decision outcomes,
D. Preferences,
E. Information inputs.

(iii) Decision making methods

A. Centralized and decentralized,
B. Authorization and delegation,
C. Habitual way of ranking and ordering.

(2) Identify the Competence Sets

The following is a comprehensive list for searching competence sets Tr(***E***) and Sk(***E***).

A. Professional skills,
B. People skills,
C. Attitudes,
D. Support systems.

(3) Assessing Tr\Sk

Once Tr(***E***) and Sk(***E***) have been identified, we would have a clear picture of what need to be acquired (i.e., Tr\Sk) in order to make an effective decision for the problem. One of the complex problems involving assessing Sk is the degree of proficiency which is related to the activation probability of a particular skill. This problem may be alleviated by further decomposing the skill into different degree of proficiency or activation propensity. Further discussion on this topic can be found in Yu and Zhang (1990b).

(4) Effective Expansion Plan

How to effectively make suggestions to the decision maker to expand his/her competence sets is an art. It includes implanting, nurturing and habituating the concepts or skills. Like expanding our habitual domains, expanding the competence sets can be accomplished in many ways, including active association, projection and training, etc. The interested reader is referred to Chapters 5-6 of Yu (1990) for a detailed discussion.

Assume that the decision problem is fairly routine (i.e., its competence set is well known and has been acquired and mastered) to the expert, but not to the decision maker, and that once a skill is learned it remains valid and potent all the time during the decision cycle (like a computer program which will remain active all the time once is coded). Then the techniques of lexico. optimality, next best optimality and minimal spanning tree can be used to identify the most effective way to expand the decision maker's skill set. In this case, computer decision support systems can be built.

We summarize the above discussion into the following self-explanatory flow chart.

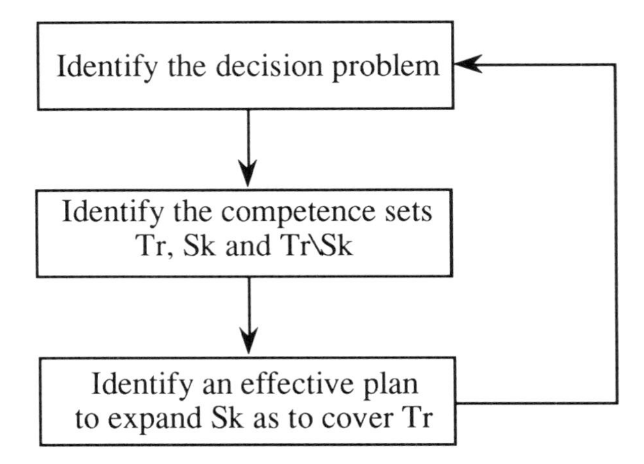

Flow Chart 1. Support system

4. CONCLUSION

We have supplied a framework for helping the decision maker in expanding his/her skill set as to cover the needed competence set. We discussed various optimality concepts of expansion processes including lexicographical optimality, next best optimality and total cost optimality. An algorithm for finding the optimal expansion processes was provided based on the minimal spanning tree concept. We have also sketched the decision support systems for implementing the proposed framework in real life applications.

Some research problems remain open. For instance, (i) How do we utilize competence set analysis to study competitive games or decision making? A detailed description of forming winning strategies in competitive games or decision making can be found in Chapters 11-12 of Yu (1990). However, the mathematical foundation needs to be built. (ii) When the true competence set is unknown, how do we effectively expand our competence set? Some preliminary results on this attempt can be found in Chapter 8 of Yu (1990) and Yu and Zhang (1989, 1990b). (iii) The benefits of expansions have not been considered in this article. To make the proposed framework more realistic in real life applications, we need to study the consequences of expansions and compare the expansion costs with the benefits. A marginal analysis was introduced in Yu and Zhang (1990b). (iv) It always takes time and effort to expand our competence sets. An expansion would not help much if it takes too much time to complete. Thus the time needed in the expansion is also an important issue. MCDM techniques may be employed to modify the proposed competence set analysis.

ACKNOWLEDGEMENT

We are very grateful to an anonymous referee for helpful comments on our preliminary drafts.

REFERENCES

Busacker, R. G. and T. L. Saaty, *Finite Graphs and Networks*, McGraw-Hill Book Company, 1965.

Hillier, F. S. and G. L. Lieberman, *Introduction to Operations Research*, 3rd edition, Holden-Day, Inc., San Francisco, 1980.

Yu, P. L. *Forming Winning Strategies - An Integrated Theory of Habitual Domains*, Springer-Verlag, Heidelberg, 1990.

Yu, P. L. and D. Zhang, "Competence Set Analysis for Effective Decision Making," *Control: Theory and Advanced Technology,* Vol.5, No.4, pp.523-547, 1989.

Yu, P. L. and D. Zhang, "Multicriteria optimization," in: G. Salvendy, eds., *Handbook of Industrial Engineering*, 2nd edition, Wiley, New York, to appear 1990a.

Yu, P. L. and D. Zhang, "A foundation for Competence Set Analysis," *Mathematical Social Sciences*, Vol.20, No.3, pp.251-299, 1990b.

Yu, P. L. and D. Zhang, "Optimal expansion process of competence sets and decision support," *Information Systems and Operational Research*, to appear 1990c.